Advances in
CHEMICAL ENGINEERING
CHARACTERIZATION OF FLOW, PARTICLES AND INTERFACES

VOLUME **37**

ADVANCES IN
CHEMICAL ENGINEERING

Editor-in-Chief

GUY B. MARIN
Department of Chemical Engineering
Ghent University
Ghent, Belgium

Editorial Board

DAVID H. WEST
Research and Development
The Dow Chemical Company
Freeport, Texas, U.S.A.

PRATIM BISWAS
Department of Chemical and Civil Engineering
Washington University
St. Louis, Missouri, U.S.A.

JINGHAI LI
Institute of Process Engineering
Chinese Academy of Sciences
Beijing, P.R. China

SHANKAR NARASIMHAN
Department of Chemical Engineering
Indian Institute of Technology
Chennai, India

Advances in
CHEMICAL ENGINEERING
CHARACTERIZATION OF FLOW, PARTICLES AND INTERFACES

VOLUME **37**

Edited by

JINGHAI LI

Institute of Process Engineering
Chinese Academy of Sciences
Beijing, P.R. China

Amsterdam • Boston • Heidelberg • London • New York • Oxford
Paris • San Diego • San Francisco • Singapore • Sydney • Tokyo
Academic Press is an imprint of Elsevier

Academic Press is an imprint of Elsevier
Radarweg 29, PO Box 211, 1000 AE Amsterdam, The Netherlands
32 Jamestown Road, London NW1 7BY, UK
30 Corporate Drive, Suite 400, Burlington, MA 01803, USA
525 B Street, Suite 1900, San Diego, CA 92101-4495, USA

First edition 2009

Copyright © 2009 Elsevier Inc. All rights reserved

No part of this publication may be reproduced, stored in a retrieval system
or transmitted in any form or by any means electronic, mechanical, photocopying,
recording or otherwise without the prior written permission of the publisher

Permissions may be sought directly from Elsevier's Science & Technology Rights
Department in Oxford, UK: phone (+44) (0) 1865 843830; fax (+44) (0) 1865 853333;
email: permissions@elsevier.com. Alternatively you can submit your request online by
visiting the Elsevier web site at http://www.elsevier.com/locate/permissions, and selecting
Obtaining permission to use Elsevier material

Notice
No responsibility is assumed by the publisher for any injury and/or damage to persons
or property as a matter of products liability, negligence or otherwise, or from any use
or operation of any methods, products, instructions or ideas contained in the material
herein. Because of rapid advances in the medical sciences, in particular, independent
verification of diagnoses and drug dosages should be made

Library of Congress Cataloging-in-Publication Data
A catalog record for this book is available from the Library of Congress

British Library Cataloguing in Publication Data
A catalogue record for this book is available from the British Library

ISBN: 978-0-12-374738-9
ISSN: 0065-2377

For information on all Academic Press publications
visit our website at elsevierdirect.com

Printed and bound in USA

09 10 11 12 13   10 9 8 7 6 5 4 3 2 1

CONTENTS

Contributors   vii
Preface   ix

1. Ultrasound-Based Gas–Liquid Interface Detection in Gas–Liquid Two-Phase Flows   1

S. Roberto Gonzalez A, Yuichi Murai and Yasushi Takeda

1. Introduction   2
2. Gas–Liquid Interface Inferred from Liquid Velocity   4
3. Gas–Liquid Interface Inferred from the Peak Ultrasound Echo Intensity   11
4. Summary   24
List of Symbols   25
Acknowledgments   26
References   27

2. Micromanipulation in Mechanical Characterisation of Single Particles   29

Z. Zhang, J.D. Stenson and C.R. Thomas

1. Introduction   30
2. Description of the Techniques   31
3. Status and Applications   51
4. Other Applications   68
5. Perspectives on Future Development   77
Nomenclature   78
Acknowledgements   81
References   81

3. Particle Image Velocimetry Techniques and its Applications in Multiphase Systems   87

Feng-Chen Li and Koichi Hishida

1. Introduction   88
2. Fundamentals of Particle Image Velocimetry   90
3. Various Types of Particle Image Velocimetry   103

4.	Measurement of Multiphase Flow Using Particle Image Velocimetry	118
5.	Summary and Outlook	140
	Notation	141
	References	142

4. Positron Emission Imaging in Chemical Engineering — 149

J.P.K. Seville, A. Ingram, X. Fan and D.J. Parker

1.	Introduction	150
2.	Positron Emission Techniques and their Recent Development	151
3.	Applications	156
4.	Portable PEPT	168
5.	Summary and Future Plans	174
	Symbols and Abbreviations	176
	Acknowledgements	177
	References	177

5. Electrical Capacitance, Electrical Resistance, and Positron Emission Tomography Techniques and their Applications in Multi-Phase Flow Systems — 179

Fei Wang, Qussai Marashdeh, Liang-Shih Fan and Richard A. Williams

1.	Introduction	180
2.	Electrical Capacitance Tomography	182
3.	Electrical Resistance Tomography	196
4.	Positron Emission Tomography	209
5.	Concluding Remarks	216
	Nomenclature	216
	References	217

6. Time-Resolved Laser-Induced Incandescence — 223

Alfred Leipertz and Roland Sommer

1.	Introduction	224
2.	Measurement Principle	225
3.	Flame Investigations	236
4.	Technical Applications	237
5.	Conclusions	265
	Nomenclature	267
	Acknowledgements	267
	References	268

Subject Index — 271
Contents of Volumes in this Serial — 279
See Color Plate Section at the End of this Book

CONTRIBUTORS

Liang-Shih Fan, *Department of Chemical and Biomolecular Engineering, The Ohio State University, Columbus, OH 43210, USA*

X. Fan, *School of Physics and Astronomy, University of Birmingham, Birmingham B15 2TT, UK*

S. Roberto Gonzalez A., *Division of Energy and Environmental Systems, School of Engineering, Hokkaido University, N13W8, Sapporo 060-8628, Japan*

Koichi Hishida, *Department of System Design Engineering, Keio University, Yokohama 223-8522, Japan*

A. Ingram, *School of Chemical Engineering, University of Birmingham, Birmingham B15 2TT, UK*

Alfred Leipertz, *Department of Engineering Thermodynamics and Erlangen Graduate School in Advanced Optical Technologies, University of Erlangen-Nuremberg, Am Weichselgarten 8, 91058 Erlangen, Germany*

Feng-Chen Li, *School of Energy Science and Engineering, Harbin Institute of Technology, Harbin 150001, China*

Qussai Marashdeh, *Department of Chemical and Biomolecular Engineering, The Ohio State University, Columbus, OH 43210, USA*

Yuichi Murai, *Division of Energy and Environmental Systems, School of Engineering, Hokkaido University, N13W8, Sapporo 060-8628, Japan*

D.J. Parker, *School of Physics and Astronomy, University of Birmingham, Birmingham B15 2TT, UK*

J.P.K. Seville, *School of Engineering, University of Warwick, Coventry CV4 7AL, UK*

Roland Sommer, *Department of Engineering Thermodynamics and Erlangen Graduate School in Advanced Optical Technologies, University of Erlangen-Nuremberg, Am Weichselgarten 8, 91058 Erlangen, Germany*

J.D. Stenson, *School of Chemical Engineering, University of Birmingham, Edgbaston, Birmingham B15 2TT, UK*

Yasushi Takeda, *Division of Energy and Environmental Systems, School of Engineering, Hokkaido University, N13W8, Sapporo 060-8628, Japan*

C.R. Thomas, *School of Chemical Engineering, University of Birmingham, Edgbaston, Birmingham B15 2TT, UK*

Fei Wang, *Department of Chemical and Biomolecular Engineering, The Ohio State University, Columbus, OH 43210, USA*

Richard A. Williams, *Institute of Particle Science and Engineering, School of Process, Environmental and Materials Engineering, University of Leeds, Leeds LS2 9JT, UK*

Z. Zhang, *School of Chemical Engineering, University of Birmingham, Edgbaston, Birmingham B15 2TT, UK*

PREFACE

Chemical engineering science has played an important role in modern civilization in the provision of energy, the making of materials and the protection of environment. However, its current knowledge base is still insufficient to support future development of economy and society. Although it is still difficult for us to quantitatively design and optimize conventional chemical processes, we have come to face new challenges of depletion of fossil and non-renewable energy resources and difficulties in coping with chemical processes related to nano-, bio- and other emerging technologies. We need to think how to better serve society by adapting ourselves to these new developments in science and technology. We need to ask what the bottleneck is in upgrading our knowledge base, and what breakthrough is needed.

Gradually we have come to recognize that understanding of the spatio-temporal multi-scale structures in various chemical processes is essential for upgrading our ability. In fact, as much as we know about macro-scale chemical processing, e.g., radial and axial profiles, and what happens at micro-scales, e.g., chemical reaction and even molecular structures, we really know little about what happens at the meso-scale in between. That is, meso-scale is the key to understanding multi-scale structures. This is the bottleneck!

Stimulated by this recognition, multi-scale analysis and simulation have received unprecedented attention in recent years, as shown by the dramatic increase in related publications. However, measurement technology focused on multi-scale structures, particularly, on meso-scale phenomena, has not been sufficiently tackled. Without breakthroughs in this aspect, theories and simulations could not be verified and validated, and upgrading the knowledge base for chemical engineering would be futile.

Recognizing this problem, *Advances in Chemical Engineering* decided to organize the present thematic issue on *"Characterization of Flow, Particles and Interfaces"* to alert the chemical engineering community to this challenging issue. We selected the following six meso-scale measurement technologies.

1. *Ultrasound-based Gas-liquid Interface Detection in Gas–liquid Two Phase Flows* (by Prof. Yasushi Takeda et al.) introduces two ultrasonic-based detection methods for gas-liquid interface of gas-liquid two-phase flows in horizontal pipes, based on ultrasonic velocity profiler (UVP) measurements. One approach using ultrasonic peak echo intensity information to predict gas-liquid interface has wider application range and has been validated. Another approach based only on liquid velocity information is a relatively new technique and is still at intermediate stage of an ongoing development.
2. *Micromanipulation in Mechanical Characterization of Single Particles* (by Prof. Zhibing Zhang et al.) reviews a number of micromanipulation-based techniques and their applications for measuring mechanical strength of single particles, particle-particle interaction as well as particle-surface interactions. Among the techniques introduced, AFM, optical trapping and diametrical compression are more attractive, and the description of models for extracting intrinsic mechanical property information from compression testing data is useful.
3. *Particle Image Velocimetry Techniques and Its Applications in Multiphase Systems* (by Prof. Koichi Hishida et al.) reviews particle image velocimetry (PIV) technique and describes the progress in applying this technique from single phase flow system to two-phase flow system, such as liquid-liquid two-fluid flow, gas-liquid two-phase flow and particle-fluid flow systems, providing whole-field velocity profiles based on the captured images of seeded particles.
4. *Positron Emission Imaging in Chemical Engineering* (by Prof. J. P. K. Seville et al.) reviews the state-of-the-art positron emission particle tracking (PEPT) technique and its extensive applications in chemical engineering, demonstrating its versatile features, that is, the capability to track tracer particles down to about 60 µm in size, moving at up to 10 m/s, yielding locations to within ± 1 mm at frequencies better than 100 Hz, and also revealing its application prospects such as development of multi-particle PEPT, applications in liquid-phase-continuous systems, in validation of computational codes, as well as in situ large-scale use of mobile PEPT at industrial sites.
5. *Electrical Capacitance, Electrical Resistance, and Positron Emission Tomography Techniques and Their Applications in Multi-Phase Flow Systems* (by Prof. L.S. Fan and Prof. Richard Williams et al.) reports three tomography measurement techniques of capacitance (ECT), resistance (ERT) and positron emission (PET), with ECT and ERT being electrical modalities and PET radioactive modality. Their applicability in imaging multiphase flow dynamics

of industrial processes are exemplified by widespread applications, including fluidized beds, pneumatic solid conveying, slurry bubble column, hydrocyclone, oscillatory baffled reactor, stirred tank reactors.

6. ***Time-Resolved Laser-Induced Incandescence*** (by Prof. Alfred Leipertz *et al.*) introduces an online characterization technique (time-resolved laser-induced incandescence, TIRE-LII) for nano-scaled particles, including measurements of particle size and size distribution, particle mass concentration and specific surface area, with emphasis on carbonaceous particles. Measurements are based on the time-resolved thermal radiation signals from nanoparticles after they have been heated by high-energetic laser pulse up to incandescence or sublimation. The technique has been applied in *in situ* monitoring soot formation and oxidation in combustion, diesel raw exhaust, carbon black formation, and in metal and metal oxide process control.

We hope the publication of this issue would stimulate the application of these measurement techniques in chemical engineering and further our understanding of multi-scale structures. I thank Prof. Guy B. Marin, the editor-in-chief of this book series, for his advice in selecting the present topic and, Prof. Zhuyou Cao and Dr. Jiayuan Zhang, for their contributions made in editing this issue. Thanks are also extended to reviewers and authors for their efforts and time.

Jinghai Li
Institute of Process Engineering,
Chinese Academy of Sciences,
Beijing 100190,
P. R. China

CHAPTER 1

Ultrasound-Based Gas–Liquid Interface Detection in Gas–Liquid Two-Phase Flows

S. Roberto Gonzalez A*, **Yuichi Murai** and **Yasushi Takeda**

Contents

1.	Introduction	2
2.	Gas–Liquid Interface Inferred from Liquid Velocity	4
	2.1 Experimental setup	4
	2.2 Experimental approach	6
	2.3 Validation of the proposed experimental method	10
3.	Gas–Liquid Interface Inferred from the Peak Ultrasound Echo Intensity	11
	3.1 Experimental setup	12
	3.2 Experimental method	13
	3.3 Validation of the experimental approach	18
4.	Summary	24
	List of Symbols	25
	Acknowledgments	26
	References	27

Abstract   Ultrasonic velocity profiler (UVP) measurements are performed for gas–liquid two-phase flow. The UVP measurements are conducted in a rectangular channel and in a pipe, both horizontally oriented. The liquid velocity measurements and ultrasound echo intensity

Division of Energy and Environmental Systems, School of Engineering, Hokkaido University, N13W8, Sapporo 060-8628, Japan

*Corresponding author: Currently at Nanyang Technological University, Division of Physics and Applied Physics, School of Physical and Mathematical Sciences, 21 Nanyang Link 05–13, Singapore 637371
E-mail address: sroberto@ntu.edu.sg

are used to locate the position of the gas–liquid interface information. Liquid velocity data are analyzed to obtain the bubble interface for the study of the mechanism of drag reduction. The peak ultrasound echo intensity is used to locate the position of the gas–liquid interface in a pipe for the estimation of the liquid flow rate. No gas-flow rate calculations are performed. The experiments in the pipe are performed using three UVP–DUO systems simultaneously to obtain a more accurate shape for the gas–liquid interface. The tests conducted belong to three different flow regimes: stratified, elongated bubble, and slug flow. The results show very good agreement with the actual liquid flow rates.

1. INTRODUCTION

The use of sound in science and technology can be traced back to as early as the 1950s. At that time, devices that used ultrasonic waves to detect flaws in metals were commercially available. The use of sound has spread to different fields such as medicine (Satomura, 1957; Wild and Reid, 1953), meteorology (Lhermitte, 1973), and oceanography (Pinkel, 1979). Takeda (1986) began to apply ultrasound to several problems of interest in fluid mechanics and heat transfer. The ability of an ultrasound-based technique that can provide velocity profile measurements in opaque flows gives it a crucial advantage over optical methods such as laser Doppler anemometry, particle tracking velocimetry, and particle image velocimetry (PIV). This was recognized by researchers who later applied ultrasound to other fluid-flow-related problems. Another field where the ultrasonic velocity profiler (UVP) has been well received is that of two-phase flows. Although originally developed to be used in single-phase flows, it can be applied to study multiphase flows, particularly gas–liquid flows. Zhou et al. (in Murakawa et al., 2003), Suzuki et al. (2002), and Murakawa et al. (2003, 2005) suggested a statistical approach to separate the velocity information of the gas and liquid phases. Gas–liquid two-phase flows are common in engineering and scientific applications; gas–liquid interface detection has been investigated using techniques such as hot-wire anemometry, laser Doppler velocimetry (Kitagawa et al., 2005), PIV (Suzuki et al., 2002), and imaging tomography (Murai et al., 2005, 2007). Ultrasound also can be applied in the detection of the gas–liquid interface. Wada et al. (2006) demonstrated the use of echo intensity for pattern recognition of two-phase flow. In this chapter, the detection of the gas–liquid interface is used to approach two different engineering problems that nowadays are the focus of extensive research. The first problem is the determination of the mechanism of bubble-based drag reduction. The second problem is the detection of the gas–liquid interface for estimating the liquid flow rate of a gas–liquid two-phase flow. Ideally, we want to estimate both the gas- and liquid-phase flow rates, but current research has focused on the estimation of the liquid flow rate. Future research should lead to the accurate estimation of the gas-phase flow rate as well.

Recently, there have been papers published that report new ultrasound-based techniques for visualizing fluid flow. One technique is a high-frame-rate technique referred to as speckle correlation velocimetry. It was originally developed for imaging the propagation of low frequency shear waves in soft tissues (Sandrin et al., 1999). It was later used to obtain echographic images of a stretched vortex and a jet flowing through a hole 3 mm in diameter (Sandrin et al., 2001). To obtain local measurements of the flow, Manneville et al. (2001) proposed a technique based on the combination of classical ultrasonic Doppler velocimetry and two-dimensional echographic techniques. This appears to be an interesting method for extending the capability of ultrasonic Doppler velocimetry to two-dimensional local measurements and increasing the frame rates of normal scanners ($\sim$50–5,000 frames per second). Using two perpendicular transducer arrays, they mapped the flow of a stretched vortex.

This technique appears particularly attractive because the high frame rate allows the dynamics of fast changing liquid flows to be studied and the spatial resolution is significantly reduced using a high ultrasound frequency (Manneville et al., 2005). This technique can be adapted to small-scale, high-speed gas–liquid two-phase flows that are not presently subject to ultrasound-based techniques.

Section 2 describes the methodology for locating a gas–liquid interface. In this approach, only the liquid velocity is used to detect the interface. As previously mentioned, the motivation for developing this method is application to the investigation of the mechanism of bubble-based frictional drag reduction, which is the focus of a large research effort aimed at reducing the cost of transportation by ships. Using the idea suggested by Wada et al. (2006), Section 3 presents a method for detecting the gas–liquid interface by means of the peak ultrasound echo intensity. This method is used to estimate the liquid flow rate of gas–liquid two-phase flow in a horizontal pipe. Although the flow rate in a vertical configuration is more commonly measured (Oddie and Pearson, 2004), the flow rate of a two-phase mixture in a horizontal orientation is measured in the transport of oil–gas mixtures from wells to reservoirs (Cook and Behnia, 2000; Oddie and Pearson, 2004; Shemer, 2003) and in chemical and power generation processes. Therefore, a noninvasive device capable of accurately measuring the flow rate of the component phases in real time may be well received in these important engineering activities.

2. GAS–LIQUID INTERFACE INFERRED FROM LIQUID VELOCITY

The UVP system emits ultrasonic pulses along its measuring line. It provides velocity and echo intensity information for every profile captured. The accuracy of the measurements obtained using the UVP method are considered to be $\sim 3.5\%$ for velocity and $\sim 1.1\%$ for position under optimized conditions (Takeda, 1986). The overall experimental error expected in estimating the liquid flow rate is $\pm 5.1\%$; this value does not take into account the fact that the liquid flow is assumed to be one-dimensional in the calculation of the liquid flow rate. In liquid slugs and the volume of liquid below the bubbles, this appears to be a good approximation; however, in the regions near the leading edge and trailing edge of a bubble, the liquid flow is not one-dimensional. The principle of the UVP was described by Takeda (1986).

In this section, the UVP is applied to obtain the location of the gas–liquid interface of two-phase flow in a horizontal square channel. In this method, only the velocity data are used to obtain the position of the gas–liquid interface.

We propose signal processing of the UVP output. The gas–liquid interface is detected without using the echo intensity or any optical information. The method presented in this section represents an intermediate and necessary step in the development of an ultrasound-based sensor for reducing frictional drag. In the near future, a complete monitoring system for a bubbly two-phase boundary layer is to be developed. The system is to be applied to ships and pipelines.

2.1 Experimental setup

Figure 1 is a schematic diagram of the experimental setup. The test section is a horizontal rectangular channel 40 mm in height (H), 160 mm in width (W), and 6,000 mm in length (L). The rectangular channel is completely constructed of transparent acrylic resin, as shown in Figure 2. Tap water and air are used as the gas and liquid phases, respectively. Water is circulated by a 2.2 kW pump fed by a water reservoir 4.2 m away. Air bubbles are injected into the horizontal channel from the upper inner surface of the channel. An array of capillary needles produces bubbles 10–100 mm in length. Before the air and water are mixed, their volumetric flow rates are measured. After leaving the horizontal channel, the gas–liquid mixture is dumped into a tank that acts as a bubble remover; when the liquid phase is recirculated it is free of bubbles. At the end of the horizontal channel tracer particles are added to the water to act as ultrasound reflectors. The mean particle diameter is 200 µm and the particle density is 1020 kg/m^3. These tracer particles are assumed to

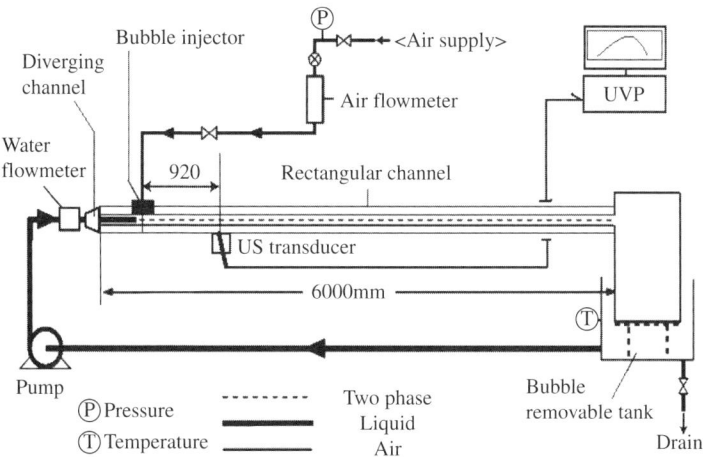

Figure 1 Schematic diagram of the experimental setup.

Figure 2 Horizontal channel for the gas–liquid two-phase flow experiments.

follow the liquid flow faithfully; this ability has been assessed by Melling (1997). The superficial velocity of the liquid varies from $V_l = 0.3\,\text{m/s}$ to $1.5\,\text{m/s}$; Reynolds number ranges from $\text{Re}_H = 1.2 \times 10^4$ to $\text{Re}_H = 6 \times 10^4$ on the basis of $\text{Re}_H = V_l H / \nu$, where H is the channel height, V_l the superficial velocity of the liquid, and ν the kinematic viscosity of the liquid. The ultrasound transducer is 920 mm ($x/H = 23$) downstream of the injection point.

2.2 Experimental approach

Figure 3 shows three images taken from the top of the channel at $Re_H = 2.8 \times 10^4$. The size of the images is 160 mm × 160 mm. The flow direction is from left to right. α is defined as $\alpha = Q_g/(Q_a + Q_g)$, where Q_g is the gas-phase flow rate and Q_a the actual liquid-phase flow rate. In addition, the relative beam thickness is shown in the images. The maximum Weber number, We, is 10^3; $We = \rho V_l^2 d/\sigma$, where ρ is the liquid density, d the bubble diameter, and σ the surface tension of the liquid. Owing to the high value of We, individual bubbles are considerably deformed by the inertia of the fluid. In addition, images taken in the spanwise direction confirm (a) the translational velocity of the bubbles is lower than the mean velocity of the liquid and (b) all bubbles are confined to the upper part of the channel. Therefore, there is significant friction between the bubbles and the upper wall of the channel.

The gas–liquid interface is detected by taking advantage of the ultrasound behavior in the vicinity of the interface. At the gas–liquid interface, there is a large difference in acoustic impedance (density multiplied by the sound speed). Therefore, an ultrasound beam is mostly reflected at the gas–liquid interface. In addition, the ultrasound phase is reversed owing to the fixed pressure. The incoming ultrasound waves and those reflected at the interface create standing waves. The standing waves occur within a length of the pulsed ultrasound from the gas–liquid interface. Figure 4 is a schematic representation of the ultrasound beam reflected at the interface. The horizontal axis is the time evolution. Considering a four-cycle pulse as used in this study, four layers of weak pressure are created as a result of the standing waves. There are tracer particles in the regions where the standing waves occur; however, the echo scattered from the particles in these regions is rather weak. The particles that cross the standing waves produce two echoes; one is produced by the emitted wave and one is produced by the receding wave. The net result is that no net Doppler shift is produced although component shifts exist. Consequently, the output of the UVP system

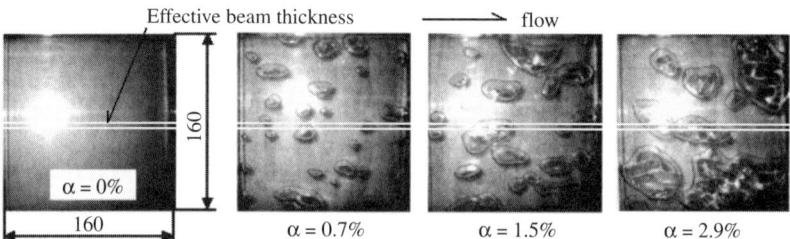

Figure 3 Bubble size for different α values compared with the effective thickness of the ultrasound beam.

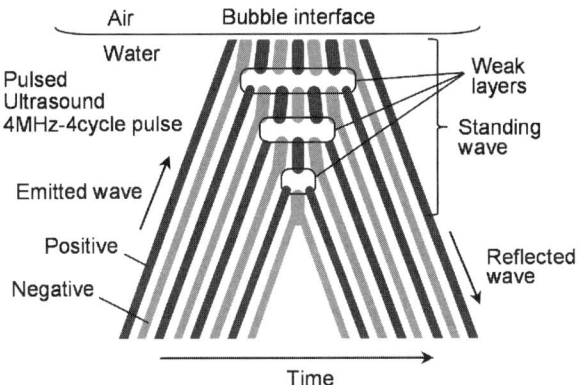

Figure 4 Reflected four-pulse ultrasound wave at the gas–liquid interface.

(UVP–DUO, Met-Flow Inc.) is near zero velocity in this region. This phenomenon can be exploited to detect the position of the gas–liquid interface. The thickness of the standing wave perpendicular to the interface can be estimated by

$$\varphi = \frac{1}{2} \lambda N_{UC} \cos\theta \qquad (1)$$

where λ is the ultrasound wavelength, φ the thickness of the standing wave, N_{UC} the number of cycles per ultrasonic pulse, and θ the inclination angle with respect to the flow. The estimated standing wave thickness is 0.74 mm for a 4 MHz four-cycle pulse ultrasound system.

As previously mentioned, the UVP–DUO system outputs near zero velocity in the vicinity of the gas–liquid interface. Therefore, large velocity gradients exist in this region. In image processing, several methods have been proposed to detect changes in pixel intensity and perform pattern recognition. Such an approach is used here. A linear filter is applied to the velocity values to detect rapid changes in the velocity near the gas–liquid interface. The filters applied are 3×3 square regions centered at every point (i,j) as shown in Figure 5 (Yang, 1994). Figure 6a shows the raw velocity data obtained by the UVP. In the upper part of the image, the gas phase is present but not visible. The ultrasound transducer is placed at the bottom of the acrylic channel in a water jacket. Multiple reflections occur at $y/H \sim 0.5$ and a horizontal line (the original liquid-phase velocity values) is observed in Figure 5a. This line is magnified in Figures 6b and c, after applying the Laplacian and first-order spatial differentiation filters, respectively; however, this line is outside the region of interest of the gas–liquid interface. Figure 6b shows the resulting image after the Laplacian filter is applied to the velocity values; this filter enhances intensity variations. In the upper part of the

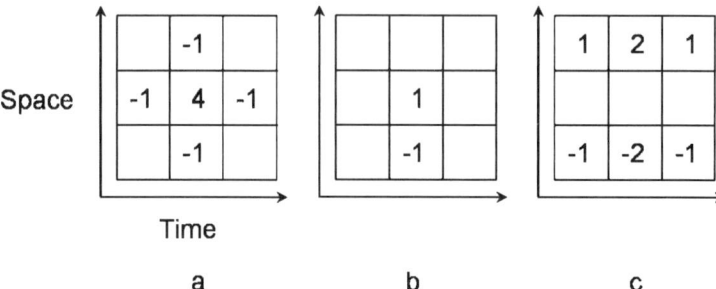

Figure 5 Filters applied to the UVP velocity values: (a) Laplacian, (b) forward differentiation, and (c) Sobel.

image, the regions of light color indicate possible locations where the gas–liquid interface may be located; however, this region does not provide a good contrast for extracting the bubbles present in the image. Figure 6c shows the results for the first-order spatial differentiation filter. Unfortunately, this filter is not able to accurately determine the location of the interface. The upper part of the pipe, which is occupied by the gas phase, appears to be occupied by a continuous medium; this is not so, the gas phase consists of discrete bubbles separated by short liquid slugs. Consequently, it is not possible to apply a threshold to the image because (a) the contour of the gas phase cannot be accurately obtained and (b) too much liquid-phase information is discarded. Figure 6d shows the results of applying a Sobel filter and threshold level. This filter appears to work best in detecting spatial changes. The gas phase can be seen in the upper part of the pipe. After the filter is applied, the image is binarized. Each column of pixels is examined; pixels on and above the gas–liquid interface are given the value 1; pixels below the gas–liquid interface are given the value zero. Figure 6e shows the resulting Sobel image after binarization. This image displays more clearly the location and number of bubbles present in the liquid flow. Finally, Figure 6f shows the two-phase flow image with the gas phase highlighted in the upper part of the pipe. In this image, only the liquid velocity and gas–liquid interface are available; the translational velocity of the bubbles present in the flow is not obtained. In summary, the Sobel filter provides the most accurate representation of the distribution of the bubbles in the flow.

It is worth mentioning that there are two necessary conditions for this method to be applied successfully. One is that there must be enough particles to act as ultrasound reflectors along the measurement line, or else the near zero velocities near the interface are not detected. The second is that the wavelength of the gas–liquid interface shape must be larger than $\sim 10\lambda$. Therefore, the leading and trailing edges of individual bubbles cannot be detected owing to their strong curvature. Small solid

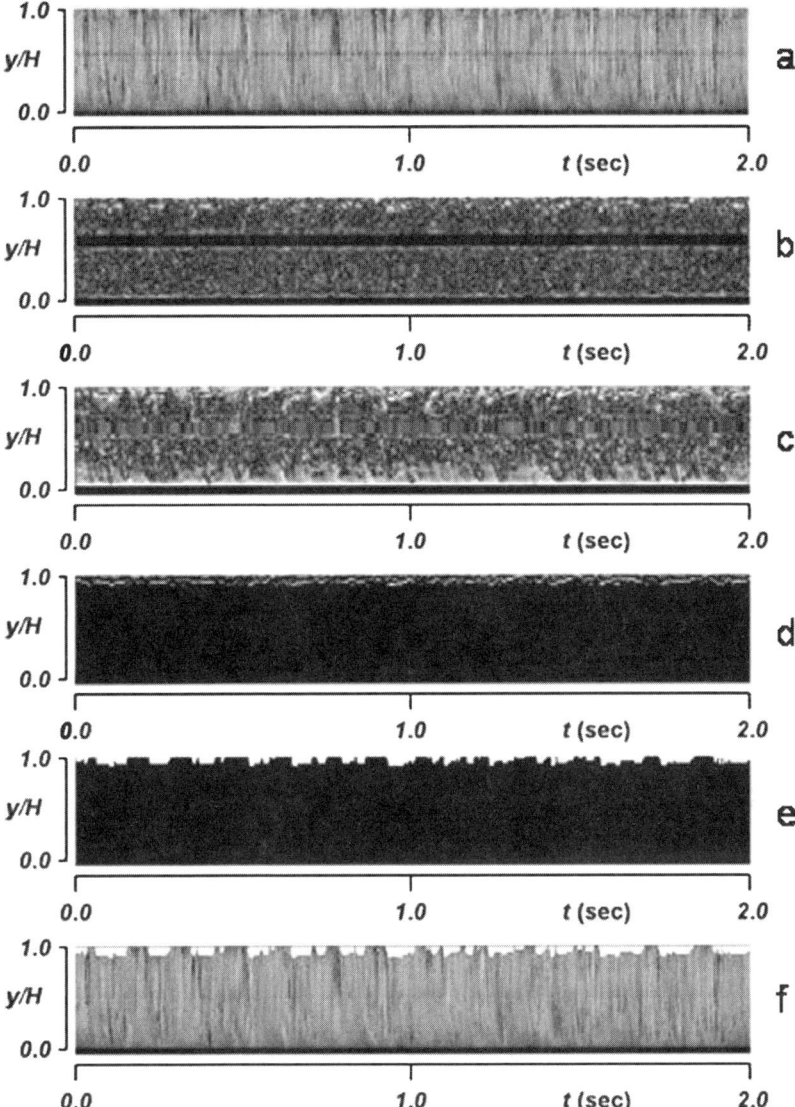

Figure 6 UVP data: (a) raw velocity data, (b) data using a Laplacian filter, (c) data using a forward differentiation filter, (d) data using a Sobel filter, (e) binarized Sobel data, and (f) gas-phase detected after the application of the Sobel filter.

particles can act as ultrasound scatterers since they reflect enough of the ultrasound energy back to the transducer; however, the leading and trailing edges of large bubbles (much larger than the solid particles) do not reflect the ultrasound beam strongly enough to the transducer.

Consequently, it is more difficult to detect the gas–liquid interface accurately in the vicinity of the leading and trailing edge of the bubbles.

2.3 Validation of the proposed experimental method

Figure 7a shows an example of the applicability of the method described; the horizontal axis only shows the first 500 ms of the UVP data acquired so that detail of the shapes of bubbles (black) at the top of the channel can be seen. Figure 7b is a qualitative image of the liquid velocity distribution. Low velocities are indicated by blue and high velocities by green. The velocities in this image are subtracted from the mean liquid velocity of each velocity profile. Figure 7c shows the velocity gradients along the height of the channel, du/dy. As in Figure 7b, blue indicates low du/dy values and green high du/dy values. Once the gas–liquid interface is detected, various studies can be performed. The relevance of this method is that it allows the detection of the gas–liquid interface without the need to have ultrasound echo intensity information. However, this method is best suited for flows where the void fraction is low (i.e., up to

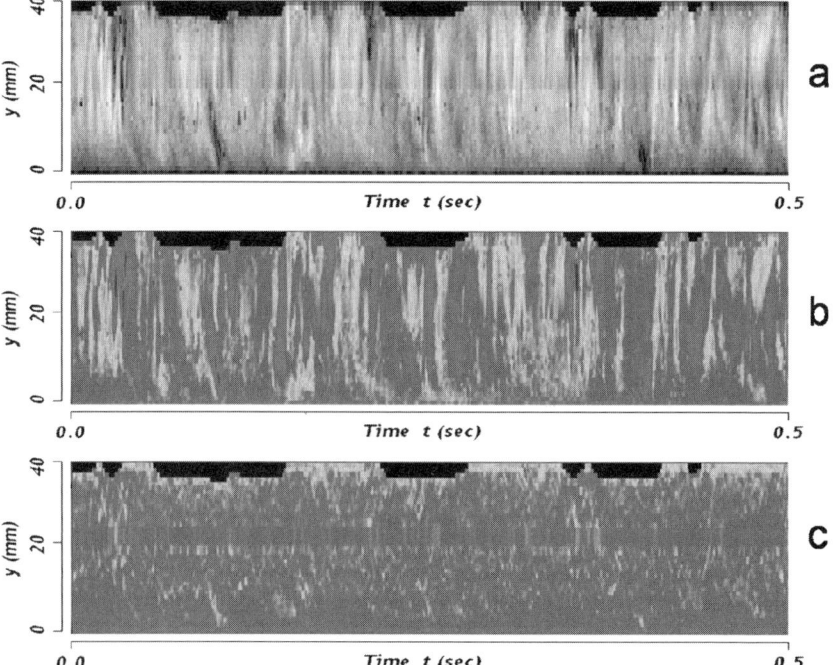

Figure 7 UVP data: (a) gas–liquid interface detected, (b) u distribution (subtracted from mean liquid velocity of each velocity profile), and (c) du/dy profile.

10%). In many engineering applications, gas–liquid flows of larger gas void fractions are encountered. In the next section, a method is described that makes use of the peak ultrasound echo intensity to detect the location of the gas–liquid interface in such flows.

3. GAS–LIQUID INTERFACE INFERRED FROM THE PEAK ULTRASOUND ECHO INTENSITY

In this section, the peak ultrasound echo intensity is used to locate the position of the gas–liquid interface flowing in a horizontal pipe. The accurate detection of the gas–liquid interface in pipe flow has important implications in engineering.

(a) The gas–liquid interface is an important parameter for predicting the behavior of these types of two-phase flows. Suggested models for predicting the unsteady behavior of two-phase flows (Duckler and Hubbard, 1975; Fossa et al., 2003; Taitel and Barnea, 1990) require the average length of the slugs to characterize the flow. Accurate gas–liquid interface detection may lead to better models for predicting two-phase flows. Moreover, two-phase flow process equipment can be improved if the statistical expected values of the liquid slug length are available to design engineers. Not only more reliable but also more efficient equipment can be developed.

(b) The position of the gas–liquid interface can be used to determine the portion of the cross-sectional area of the pipe occupied by the liquid phase; the liquid flow rate can be estimated from the liquid phase velocity distribution and area occupied. This may lead to the development of a flow metering device that can be applied in a gas–liquid two-phase flow.

In gas–liquid two-phase flow, several flow patterns exist such as bubbly, slug, plug, and annular flow depending on the pipe configuration, geometry, and flow conditions. Of these types of flows, slug flow is one of the most complex, owing to its intermittent and transient nature. Despite its complexity, industrial processes often require an online, accurate, and noninvasive estimation of such flow. If this is accomplished, industrial processes can be kept within acceptable quality limits and additionally, financial losses may be reduced.

With regard to multiphase flow measurement, two of the most commonly used flow meter devices in industry are the Coriolis flow meter and electromagnetic flow meter. The Coriolis flow meter is widely used owing to its high accuracy (Tavoularis, 2005) but it assumes (a) the phases do not slip with respect to each other when oscillated and (b) the phases are not compressible (Oddie and Pearson, 2004). Therefore, the device is suitable in solid–liquid and liquid–liquid flow applications, but it may not be acceptable in solid–gas or gas–liquid application.

On the other hand, electromagnetic flow meters process signals that depend on the electromagnetic conductivity of the phase of the flow. Cha et al. (2002) reported good results using an electromagnetic flow meter designed by their research team. However, they noticed a decrease in accuracy between experimental and theoretical values as the void fraction increased. This two-phase flow metering method appears attractive. It is nonintrusive and has no moving parts and is therefore maintenance free, but it may require a separate measurement of the density for the mass flow rate measurements (Oddie and Pearson, 2004). Approaching the gas–liquid flow metering problem by means of the use of ultrasound is advantageous. It is also a nonintrusive method and has no moving parts. In addition, the mass flow rate can be measured by adding a temperature sensor to the system and correlating the temperature to a density value to obtain the corresponding mass flow rate.

3.1 Experimental setup

The experimental setup can be seen in Figure 8. It mainly consists of the gas–liquid two-phase flow loop and three UVP–DUO systems. The pipe loop is made of plexiglass and has an inner diameter D of 40 mm. It consists of a water reservoir, a water pump controlled by a frequency inverter, liquid- and gas-phase flow meters, a pneumatic valve, and the

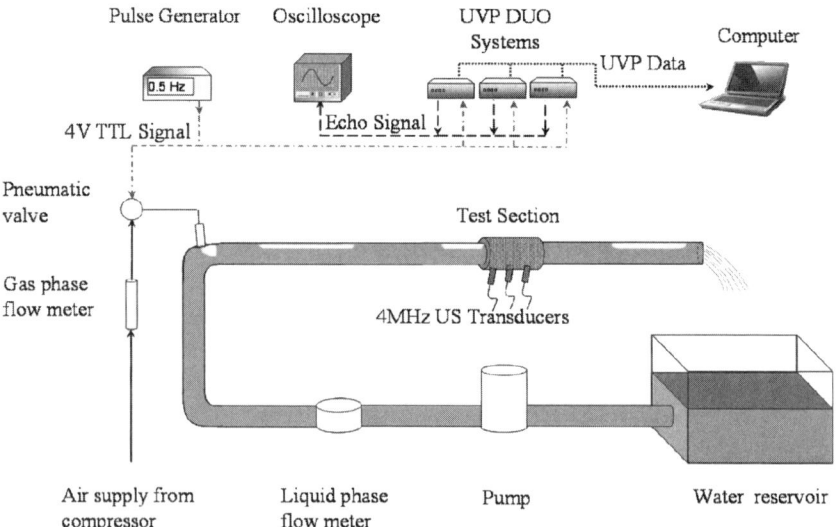

Figure 8 Experimental setup.

test section for the UVP measurements. For the UVP measurements, the three UVP–DUO systems, a pulse generator, an oscilloscope, and a personal computer are used. Water and air are used as the liquid and gas phases, respectively. The water is fed into the pump from the water reservoir. The liquid-phase flow meter is installed 1.4 m downstream of the pump. The water flows through the pipe loop and returns in a horizontal path. The gas phase is added at the beginning of the horizontal path. Before the gas is added to the liquid phase, its volumetric flow rate is measured. In this way, the desired void fraction for each test is set. The test section housing the three ultrasonic transducers is located $60D$ from the entrance of the gas phase. It is a cylinder with an inner diameter of 40 mm and length of 90 mm made of a material that absorbs ultrasonic sound. The absorbent material of the test section is selected to avoid interference among the UVP transducers due to their proximity to each other. A schematic diagram of the test section can be seen in Figures 2a and b.

The transmitting frequency f of the UVP–DUO systems is 4 MHz in all tests. The ultrasound wavelength λ is 370 µm and the sound velocity in water c is 1,480 m/s. 100 mm ion exchange (Diaion) particles are added to the flow as flow tracers; their ability to follow the liquid flow has been assessed using Basset's analysis (Melling, 1997). Owing to theoretical considerations, the size of the flow tracers must be larger than one quarter of the emitted ultrasonic burst (Met-Flow, 2002).

Each test begins when a 4 V transistor–transistor logic signal from the pulse generator triggers the three UVP–DUO systems and pneumatic valve; the pneumatic valve allows the entrance of the gas phase into the pipe loop and this creates the bubbles that are added to the liquid phase. The frequency of the signal from the pulse generator is 0.5 Hz in all tests. When the tests are conducted, both the gas flow rate and the liquid flow rate are set to the desired values. The actual liquid flow rate Q_a is measured by collecting water in a bucket for a certain period. The UVP measurement is performed and the estimated liquid flow rate Q_e is compared against the actual liquid flow rate.

Re_D based on the superficial velocity of the liquid V_l and pipe diameter D ranged from 5.3×10^3 to 6.4×10^4. V_l ranged from 0.13 m/s to 1.6 m/s. Re_D is defined as $Re_D = V_l D/v$, where v is the kinematic viscosity of the liquid. The superficial gas velocity V_g ranged from 0.01 m/s to 0.7 m/s.

3.2 Experimental method

For each UVP file created, one gas–liquid interface height time series $h_n(N)$ is obtained; $n = 1,2,3$ (one time series for each transducer) and N is the velocity or ultrasound echo intensity profile number. The gas–liquid interface height is obtained as follows. Each UVP file contains velocity and echo intensity information. From each echo intensity profile, the position of the maximum absolute echo intensity is located and labeled as the location where the gas–liquid interface is found; that is

$$h_n(N) = Max|E_I(k)|, \quad 0 \leq k \leq k_{max} \qquad (2)$$

where $h_n(N)$ is the gas–liquid interface location, $E_I(k)$ the raw ultrasound echo intensity, k the channel number, and k_{max} the closest channel to the upper wall of the pipe. The ultrasound echo intensity values are then binarized; above the channel of maximum ultrasound echo intensity, the gas–liquid interface, the intensity is set to zero; at the channel of maximum ultrasound echo intensity and below the intensity is set to 1. This process is done for each of the three UVP files created in each flow condition; therefore, three gas–liquid interface height time series are obtained. Although their data acquisitions begin simultaneously, each ultrasonic transducer detects the gas–liquid interface at a different time; this is due to their positions in the test section (Figure 9a). The gas–liquid interface is first detected by transducer 2 (connected to UVP–DUO system 2), then transducer 1 and finally transducer 3.

It is desirable to consider the interface heights of the three transducers with transducers 1 and 3 being to the left and to the right of transducer 2, respectively, as shown in Figure 9b. When only one transducer is used, the gas–liquid interface is assumed to be flat as in Figure 10a (line p–p'); then the area occupied by the liquid phase near the wall of the pipe and above the gas–liquid interface is ignored. With three transducers, however, a more accurate gas–liquid free surface shape may be obtained

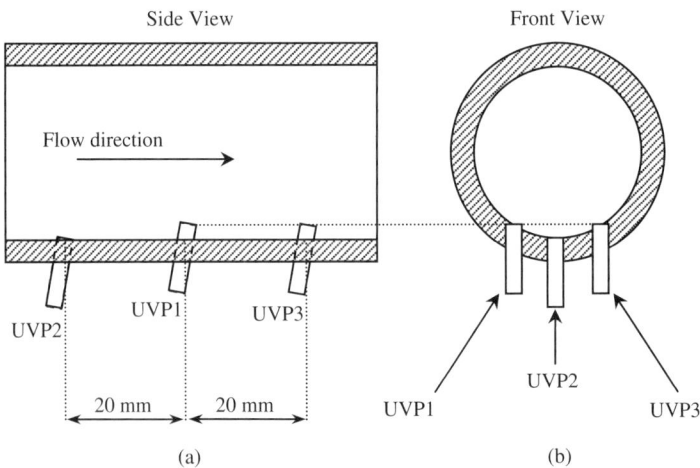

Figure 9 Ultrasound transducer arrangement in the test section: (a) side view and (b) front view.

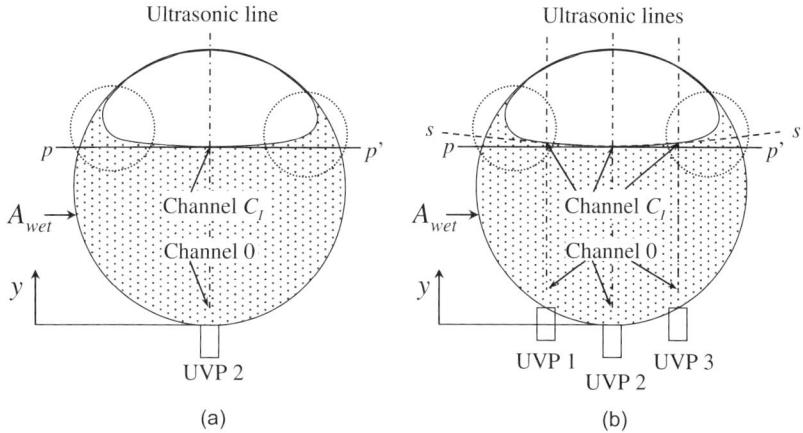

Figure 10 (a) One transducer and (b) three transducers.

(line s–s' in Figure 10b. With three points of the gas–liquid interface, the area ignored in the calculation of A_{wet} is reduced.

3.2.1 Correlation analysis

A correlation analysis is performed to align the positions of the gas–liquid interface detected by the UVP transducers as if they are located on the same x–y plane. As mentioned earlier, the gas–liquid interface height series from the UVP transducers 1, 2, and 3 are $h_1(N)$, $h_2(N)$, and $h_3(N)$, respectively. Each series consists of 4,096 elements (N_{max}), i.e., the number of velocity/ultrasound echo intensity profiles captured, $N = 0, 1, \ldots, 4{,}095$.

The correlation coefficient is defined as

$$C_c(i) = \frac{\sigma_{j,k}}{\sigma_j \sigma_k} \tag{3}$$

where $\sigma_{j,k}$ is the covariance of the gas–liquid interface height for series j and k, σ_j the standard deviation of the gas–liquid interface height for series j (transducer 2) where series j consists of $N-i$ elements and $j = 2$, and σ_k the standard deviation of the gas–liquid interface height for series k (transducers 1 and 3) where series k consists of $N-i$ elements and $k = 1$, 3 and $0 \leq i \leq 10$.

As mentioned in Section 3.2, owing to the position of the UVP transducers along the pipe, the gas–liquid interface height is determined first by transducer 2, then transducer 1 and finally transducer 3. The positions of the gas–liquid interface heights determined by the transducers are similar but differ slightly. The value of i in Equation (3) that maximizes the correlation coefficient between series 2 and 1 is denoted i_{2-1} and the value that maximizes the correlation coefficient between series 2 and 3 is denoted i_{2-3}. Once i_{2-1} and i_{2-3} are found, the series can be aligned. The maximum of the two values determines the final length of the series. Once the interface height series are aligned and resized, they consist of the following elements:

$$h_1(N) \quad i_{2-1} \leq N \leq N_{max} - (i_{2-3} - i_{2-1}) \quad (4a)$$

$$h_2(N) \quad 0 \leq N \leq N_{max} - i_{2-3} \quad (4b)$$

$$h_3(N) \quad i_{2-3} \leq N \leq N_{max} \quad (4c)$$

Not only do the series have the same number of elements but also the gas–liquid interface heights of the series appear as if the three transducers are located in the x–y plane where transducer 2 is originally located. The next step determines the curvature of the interface from these three points.

3.2.2 Calculation of the portion of the pipe occupied by the liquid phase A_{wet}

We consider a square area as shown in Figure 11. This area is divided into 40,000 square elements. When summing the areas of the square elements in the circle in Figure 11a, the value obtained differs from the circular area by 0.1%. The locations of transducers 1, 2, and 3 and the locations of the gas–liquid interface heights are x_1, x_2, and x_3 and $s_1(x_1,y_1)$, $s_2(x_2,y_2)$, and $s_3(x_3,y_3)$, respectively. In the range $x_1 \leq x \leq x_3$, the free surface is calculated by a cubic polynomial function (spline interpolation), as shown by Kreyszig (1999). In the ranges $0 \leq x \leq x_1$ and $x_3 \leq x \leq x_f$, the gas–liquid interface is calculated by a linear extrapolation. The slope of

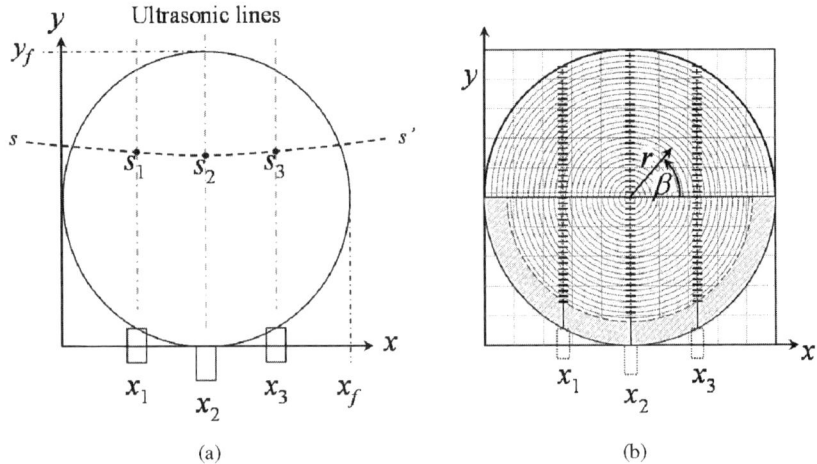

Figure 11 (a) Gas–liquid interface height position and (b) channel distribution.

the height in these ranges is the slope of the curve at x_1 and x_3, respectively.

Next, the values inside the circular area are binarized. The values above the gas–liquid interface are set to zero. The values below and those of the gas–liquid interface are set to 1. Summing all areas of the square elements of value 1, the portion of the pipe occupied by the liquid phase A_{wet} is obtained.

The velocity values obtained from the three UVP transducers are distributed along the vertical lines of $x = x_1$, $x = x_2$, and $x = x_3$, as shown in Figure 11b. Next, the velocity values are distributed radially. The cross-sectional area of the pipe is considered to be composed of three areas: the upper part of the pipe (above the middle of the pipe), the lower part of the pipe (from the first measurement to the channel closest the middle of the pipe), and the bottom region of the pipe, the region below the location of the first measurement (the gray shaded region). If the gas–liquid interface is above the middle of the pipe then the velocity values (from transducers 1 and 3) above the middle of the pipe and the same number of channels below the middle of the pipe are averaged. If the gas–liquid interface is below the middle of the pipe then the velocity values from transducers 1 and 3 are averaged. With regard to the transducer 2 values, if the gas–liquid interface is above the middle of the pipe then the values above the middle of the pipe and the same number of channels below the middle of the pipe are averaged. If the gas–liquid interface height is below the middle of the pipe then the transducer 2 values are simply distributed radially. It can be seen from Figure 11b that some overlap takes place; however, the number of channels from transducers 1 and 3 that overlap with those of transducer 2 is no more than 25%.

At the bottom of the pipe, the gray shaded region, the following liquid velocity values apply:

$$\text{For } 180 \leq \theta < 250, \quad V(r,\theta) = 0.7\overline{V}_{UVP1} \tag{5a}$$

$$\text{For } 250 \leq \theta < 290, \quad V(r,\theta) = 0.7\overline{V}_{UVP2} \tag{5b}$$

$$\text{For } 290 \leq \theta \leq 360, \quad V(r,\theta) = 0.7\overline{V}_{UVP3} \tag{5c}$$

$\overline{V}_{UVP1}$, $\overline{V}_{UVP2}$, and $\overline{V}_{UVP3}$ are the average liquid velocities for transducers 1, 2, and 3, respectively, from channel 0 to the channel where the gas–liquid interface is located. The constant 0.7 is obtained in the region $0 \leq y < 3.6$ from the power law equation (Munson et al., 1990). This equation is used in single-phase turbulent flow; the assumption made here is that the gas phase is located in the upper part of the pipe and the liquid velocity, not disturbed by the gas phase, develops in the lower part of the pipe as it does in single-phase turbulent flow.

The positions of the velocity values, so far distributed in cylindrical coordinates, are converted to Cartesian coordinates. The liquid flow rate is estimated by multiplying each velocity value $V(x,y)$ with its corresponding binary element $B(x,y)$.

$$Q_e = \frac{\sum_{t=t_i}^{t=t_f} \sum_{x=0}^{x=x_f} \sum_{y=0}^{y=y_f} V(x,y) \cdot B(x,y) \cdot t}{\sum_{t=t_i}^{t=t_f} t} \tag{6}$$

In the liquid flow rate calculation, one-dimensional flow is assumed; however, it is acknowledged this is not true in the vicinity of the bubbles (especially at their leading and trailing edges) where the relative velocity between the bubbles and the liquid phase creates a complex liquid motion that is far from one dimensional.

3.3 Validation of the experimental approach

The experimental approach described earlier is validated by performing experiments at different liquid flow rates and void fractions. The liquid flow rates tested are 0.6, 1.8, 2.8, 3.4, 4.5, 5.2, 6.2, and 7.5 m³/h. The void fraction α in the experiments is 0% (liquid phase only), 10%, 20%, 30%, 40%, and 50%. Here again the void fraction α is defined as

$$\alpha = \frac{Q_g}{Q_a + Q_g} \tag{7}$$

where Q_g is the gas-phase flow rate and Q_a is the actual liquid-phase flow rate. The difference between the actual liquid flow rate Q_a and the estimated liquid flow rate Q_e is expressed by

$$\delta_e = \frac{Q_e - Q_a}{Q_a} \tag{8}$$

The sample rate of the experiments ranged between 8 and 12 ms. The flow rates of the gas phase and liquid phase are set to desired values before each test is performed and then the flow rate is measured by pouring the mixture into a bucket for a certain period. The volumetric flow rate of the liquid phase is then calculated and recorded; this is the actual flow rate Q_a. Next, the UVP measurements are performed.

Figure 12 shows the estimated liquid flow rate versus the actual liquid flow rate. The tests for $Q_a = 0.6\,\text{m}^3/\text{h}$ are of the stratified flow type. The tests of $1.8 \leq Q_a \leq 7.5\,\text{m}^3/\text{h}$ are of the elongated bubble and slug flow type (Brennen, 2005; Govier and Aziz, 1972). In this study, the flow conditions are compared with the flow regime charts of Govier and Omer (1962 in Govier and Aziz, 1972) and Mendhane (1974 in Brennen, 2005).

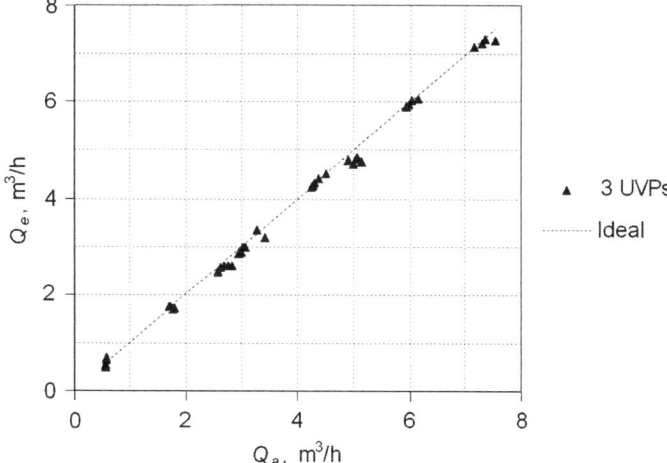

Figure 12 Q_a vs Q_e.

The chart of Mendhane contains flow regime boundaries for several pipe diameters, i.e., 25, 30, and 50 mm (horizontally oriented). The inner diameter of the pipe used in this study is 40 mm; therefore, the chart can be used as a reference to compare against. However, the chart does not indicate a boundary between the elongated bubble and slug flow regimes. The chart by Govier and Omer, although it is for air–water mixtures flowing in a 26 mm inner diameter horizontal tube does indicate a border between the elongated and slug flow regimes. It is acknowledged that the boundaries of the flow regimes are empirical transition lines and may vary slightly for a 40 mm inner diameter pipe, but as previously mentioned, the cited charts are used as references to describe the type of flow visualized by the UVP.

Figure 12 shows a good comparison between the estimated and actual liquid flow rates in the range tested although in some tests, i.e., those of $Q_a = 5\,\text{m}^3/\text{h}$ and $Q_a = 7.5\,\text{m}^3/\text{h}$, there is a slight underestimation. However, this graph does not show clearly how the different void fractions tested affect the ability of the method to estimate the liquid flow rate. Figure 13 is a surface plot of δ_e under each flow condition (Q_a and α). Although the graph shows a wide δ_e range, −10 to 20%, this is due to the tests of $Q_a = 0.6\,\text{m}^3/\text{h}$ where some values for δ_e are larger than in other tests. The experiments conducted in the range $1.8 \leq Q_a \leq 7.5\,\text{m}^3/\text{h}$ have a narrower range; in this range, the largest value for δ_e of 1.9% is at $Q_a = 1.8\,\text{m}^3/\text{h}$ and $\alpha = 50\%$ and the lowest value for δ_e of −8.2% is at $Q_a = 2.8\,\text{m}^3/\text{h}$ and $\alpha = 0\%$. From normal distribution relations, the 95% confidence interval for δ_e is

$$R_e = \mu_{\delta_e} \pm 2\sigma_{\delta_e} \tag{9}$$

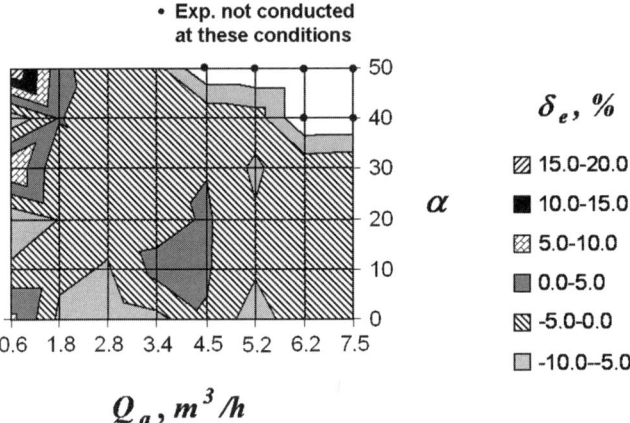

Figure 13 δ_e at different α values.

where R_e is the 95% confidence interval of the δ_e values, μ_{δ_e} the average of the δ_e values, and σ_{δ_e} the standard deviation of the δ_e values.

From Equation (9), $R_e = -1.9 \pm 2*5.1\%$ in the range $0.6 \leq Q_a \leq 7.5\,m^3/h$. This may be not suitable for an industrial or scientific application. However, in the range $1.8 \leq Q_a \leq 7.5\,m^3/h$, the 95% confidence interval is $R_e = -2.6 \pm 2*2.6\%$. Although the absolute average of the tests performed increases, more importantly, the standard deviation of these tests decreases by 42%.

Figures 14–16 show results of selected tests. These tests are selected because they are for three different flow regimes (Govier and Aziz, 1972). Figure 14 shows the test for $Q_a = 0.6\,m^3/h$ and $\alpha = 0\%$; the superficial liquid velocity is $V_l = 0.13\,m/s$ and the intensity graphs, Figures 14a and b, show the first 1,000 velocity profiles (or 24%) of the 4,096 profiles captured. If all velocity profiles are displayed, important details of the flow may be lost. The color scale is the liquid velocity, the x-axis is the velocity profile number, and the y-axis is the pipe transverse distance. This test is of the stratified flow regime, and thus it is known that the liquid phase does not fill the entire cross-sectional area of the pipe; however, this is not clear from Figure 14a. After the ultrasound echo intensity is processed and the maximum ultrasound echo intensity of each velocity profile is detected, the gas–liquid interface is easily observed. The gas–liquid interface is located at $y = 25\,mm$, as shown in Figure 14b. Both Figures 14a and b show the same velocity information, but Figure 14b shows the velocities after the gas–liquid interface has been detected; the interface height shown in this figure corresponds to that obtained by transducer 2, the transducer at the bottom of the pipe; the flat, quasi uniform interface clearly seen in the velocity profiles along the horizontal axis is shown in Figure 14b. There are a few velocity profiles for which the gas–liquid interface was not

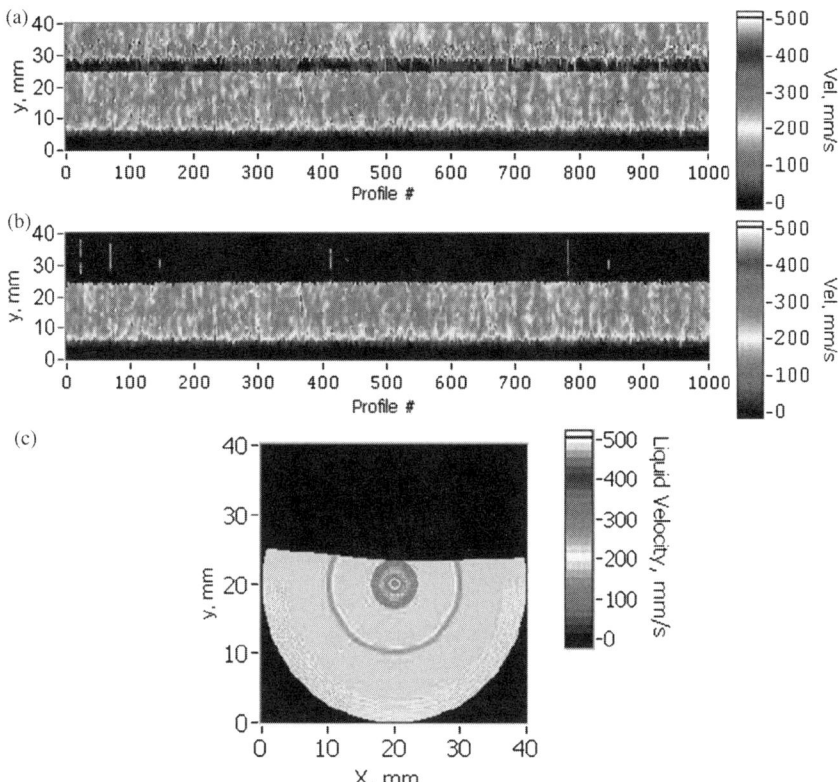

Figure 14 $Q_a = 0.6\,\text{m}^3/\text{h}$, $\alpha = 0\%$: (a) raw velocity values, (b) liquid velocity values after gas–liquid interface is detected, and (c) flow map, velocity profile # 50 (see Plate 1 in Color Plate Section at the end of this book).

accurately detected, but they form a low percentage ($< 1\%$) of the velocity profiles displayed. There are no liquid velocity values for $0 \leq y \leq 3.6\,\text{mm}$; the first UVP measurement is taken 3.6 mm from the bottom of the pipe. Figure 14c shows a flow map (velocity profile #50) of the cross-sectional area of the pipe. The gas–liquid interface heights detected by the three transducers are very similar; the interface is not flat as expected but instead shows a very small curvature. The velocity of the liquid is higher near the center of the pipe. The lighter (yellow) lines in the lower part of the pipe and near the wall correspond to values of the liquid velocity obtained from the power law equation (Equations 5a–c).

Figure 15a shows the raw data of the test of $Q_a = 1.8\,\text{m}^3/\text{h}$ and $\alpha = 10\%$; the superficial liquid and gas velocities are 0.4 and 0.04 m/s, respectively. This flow condition is in the elongated flow regime (Brennen, 2005; Govier and Aziz, 1972). Even though the flow is turbulent, the turbulence in the flow is not sufficiently large to overcome the rising drift

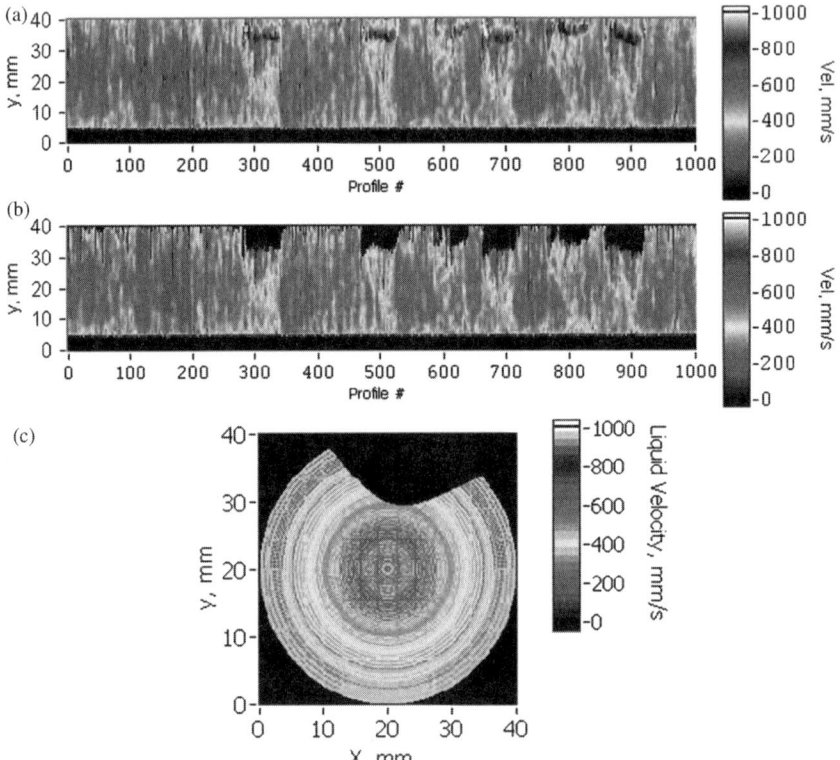

Figure 15 $Q_a = 1.8 \, m^3/h$, $\alpha = 10\%$: (a) raw velocity values, (b) liquid velocity values after gas–liquid interface is detected, and (c) flow map, velocity profile # 495 (see Plate 2 in Color Plate Section at the end of this book).

of the bubbles due to the high-density ratio. Therefore, long bubbles located at the upper wall of the pipe are expected. Figure 16b shows the bubbles present in the flow (black regions in the upper part of the pipe) after the UVP data are processed. The first ∼280 velocity profiles correspond to the liquid phase only, and then bubbles can be seen at periodic intervals. All bubbles are elongated and located in the region $29 \leq y \leq 40$ mm. In addition, it is to be noted that the liquid slugs (lighter colored regions) move faster than the liquid layer below the bubbles. These elongated bubbles are analogous to the Taylor bubbles observed in vertical flow. Figure 15c shows the cross-sectional area of the pipe. Velocity profile #495 is shown. This liquid velocity profile belongs to the liquid below the middle section of the second bubble. It shows that the gas phase is not symmetrically distributed at the top of the pipe. In addition, the liquid velocity is higher at the center of the pipe and tends to decrease radially.

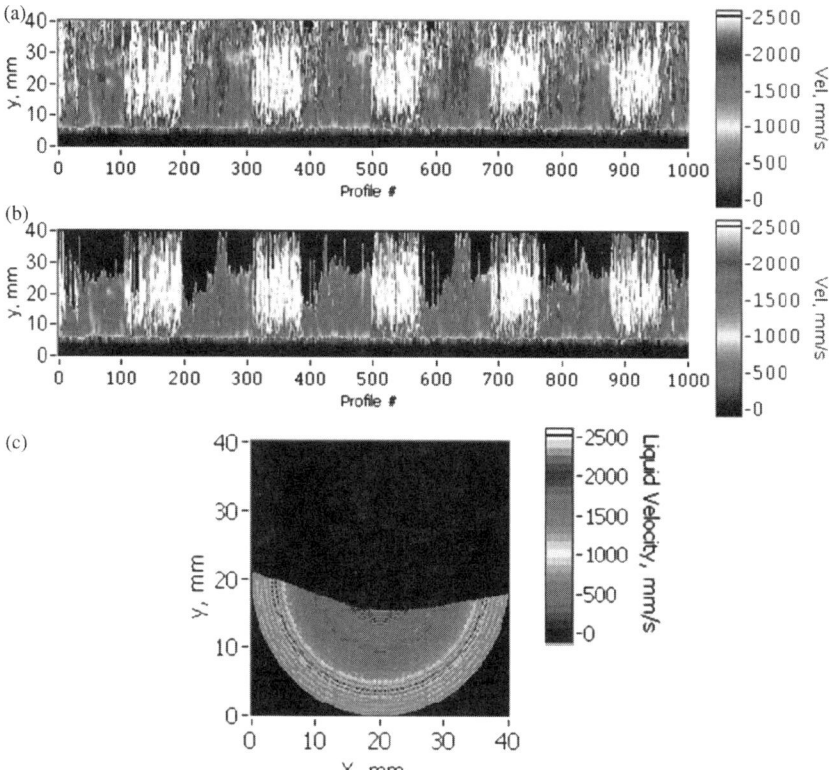

Figure 16 $Q_a = 6.2\,\text{m}^3/\text{h}$, $\alpha = 30\%$: (a) raw velocity values, (b) liquid velocity values after gas–liquid interface is detected, and (c) flow map, velocity profile # 221 (see Plate 3 in Color Plate Section at the end of this book).

Figure 16a shows the raw data of the test of $Q_a = 6.2\,\text{m}^3/\text{h}$ and $\alpha = 30\%$; the superficial liquid and gas velocities are 1.3 and 0.6 m/s, respectively. The flow condition is of the slug type, although it is close to the boundary between elongated bubble and slug flow regimes, according to the flow map developed by Govier and Omer in 1962 (in Govier and Aziz, 1972). In Figure 16a, regions of high liquid velocity (light color) and lower velocity (dark green) can be observed; the gas–liquid interface is difficult to infer from this image. After processing the echo intensity information, the gas–liquid interface is clearly revealed, as shown in Figure 16b; the interface can be found below the middle of the pipe at this void fraction. Even though the flow is of the slug type, the flow displays some features appropriate to the slug flow type. Namely, the flow exhibits large and deformed bubbles; and these bubbles fill a large part of the cross-sectional area of the pipe. The layer of liquid below the bubbles shows a slower velocity than the liquid slug does; however, the liquid slug does not fill the entire cross-section of the pipe and small bubbles can be observed in the liquid slugs. These are small dispersed bubbles trailing the larger bubbles. The small bubbles may not ride at the top of the pipe; however, the lower extent of the distribution of these bubbles appears to be the gas–liquid interface.

Figure 16c is a sample flow map of the liquid velocity profile #221. The gas phase occupies more than 50% of the cross-sectional area of the pipe, and it is not symmetrically distributed above the liquid phase. Figure 16c also shows a higher liquid velocity near the center of the pipe that decreases radially. The lighter (yellow) lines in the lower part of the pipe and near the wall correspond to the liquid velocity values obtained from the power law equation; the darker lines (dark gray) above the power law equation values correspond to the near field effect of the transducer.

4. SUMMARY

The results presented show the gas–liquid interface can be detected by two different methods. First, we can detect changes in the velocity gradient near the interface. In this method, only velocity information of the liquid is needed; however, this method is restricted to low gas void fractions and bubbles larger than 10λ. The main motivation for applying this technique is in the development of a UVP-based sensor to reduce frictional drag. This sensor can be incorporated into ships or pipelines. Our results are part of an intermediate step crucial in the development of such a device.

Second, a peak-intensity ultrasound echo can be used to detect the gas–liquid interface, but in this case the aim is the development of a flow meter capable of estimating the ratio of component phases accurately and in real time. Our results are promising for the estimation of the liquid flow rate of gas–liquid two-phase flow; further research will produce valuable data that will allow the estimation of flow rates for the two phases simultaneously. The results presented here show the liquid flow rate estimated by the peak echo intensity method can provide an accurate estimate of the actual liquid flow rate. This method can be applied to pure liquid as well as to a two-phase flow where the void fraction is as high as 50%. The flows tested are of the stratified, elongated bubble, and slug flow types. Other types of flow such as wave flow and dispersive flow were not tested; the present experimental setup does not provide the gas and liquid flow rates needed to achieve such flows.

The expected average and standard deviations between the actual liquid flow rate and the estimated liquid flow rate are -1.9% and 5.1%, respectively, in the range $0.6 \leq Q_a \leq 7.2 \, \text{m}^3/\text{h}$. The large standard deviation is due to the tests of $Q_a = 0.6 \, \text{m}^3/\text{h}$. Small differences between the estimated liquid flow rate and the actual liquid flow rate represent a large δ_e value.

In the range $1.8 \leq Q_a \leq 7.2\,\text{m}^3/\text{h}$, the expected average and standard deviations between the actual liquid flow rate and the estimated liquid flow rate are -2.6% and 2.6%. Although the average value increases, the standard deviation value decreases by 42%.

This method for locating the position of the gas–liquid interface can be applied in parametric studies; the gas–liquid interface is an important parameter for predicting the behavior of these types of two-phase flows. Accurate gas–liquid interface detection may help improve existing two-phase flow models; this may lead to the development of more reliable and efficient equipment for handling two-phase flow.

LIST OF SYMBOLS

A_wet	portion of the cross-sectional area of the pipe occupied by the liquid phase (m^2)
$B(x,y)$	binary matrix
c	speed of sound (m/s)
$C_c(i)$	correlation coefficient
C_I	channel where the gas–liquid interface is located
d	bubble diameter (m)
D	pipe inner diameter (m)
E_I	raw echo intensity
f	transmitting frequency (Hz)
H	channel height (m)
$h_n(N)$	gas–liquid interface location (mm)
k	channel number
k_max	channel number of the closest channel to the upper wall of the pipe
L	channel length (m)
N	velocity profile number
N_max	number of velocity profiles captured
N_UC	cycles per ultrasonic pulse
Q_a	actual liquid flow rate (m^3/h)
Q_e	estimated liquid flow rate (m^3/h)
Q_g	gas flow rate (m^3/h)
R_e	95% confidence interval range of the δ_e values
Re_D	Reynolds number based on the superficial velocity and pipe diameter of the liquid
Re_H	Reynolds number based on the average liquid velocity and channel height
$s(x,y)$	gas–liquid interface height (mm)
t	time delay between transmitted signals (s)

u	liquid velocity distribution subtracted from mean liquid velocity of each velocity profile (m/s)
$\overline{V}$	average liquid velocity (m/s)
V_g	gas phase superficial velocity (m/s)
V_l	liquid phase superficial velocity (m/s)
$V(r,\theta)$	liquid velocity values, cylindrical coordinates (m/s)
$V(x,y)$	liquid velocity values, Cartesian coordinates (m/s)
W	channel width (m)
We	Weber number
y	vertical distance from the tip of the transducer to the gas–liquid interface (mm)

GREEK SYMBOLS

α	gas phase void fraction
δ_e	difference between the actual liquid flow rate Q_a and the estimated liquid flow rate Q_e
φ	thickness of the standing wave (mm)
λ	ultrasound wavelength (m)
$\mu_{\delta e}$	average of δ_e values
v	kinematic viscosity of the liquid
θ	angle of the ultrasonic transducer with respect to the flow (degrees)
ρ	liquid-phase density (kg/m^3)
σ	surface tension of the liquid (N/m)
$\sigma_{\delta e}$	standard deviation of δ_e values
$\sigma_{j,k}$	covariance of the gas–liquid interface height for series j and k
σ_j	standard deviation of the gas–liquid interface height for series j (transducer 2); series j consists of $N–i$ elements, where $j = 2$
σ_k	standard deviation of the gas–liquid interface height for series k (transducers 1 and 3); series k consists of $N–i$ elements, where $k = 1, 3$

ACKNOWLEDGMENTS

The authors wish to thank the New Energy Development Organization of Japan for their support of this study, project number 05A45002d. The technical support of Mr. T. Sampo and support of the members of the *Laboratory for Flow Control*, Hokkaido University, is also appreciated. Professor Yasunori Watanabe of the Civil Engineering Department, Hokkaido University contributed to the research work concerning the third UVP–DUO system; his assistance in conducting this research is highly appreciated.

REFERENCES

Brennen, C. E., "Fundamentals of Multiphase Flow". Cambridge University Press, New York, NY (2005).
Cha, J. E., Ahn, Y. C., and Kim, M. H. *Flow Meas. Instrum.* **12**, 329–339 (2002).
Cook, M., and Behnia, M. *Chem. Eng. Sci.* **55**, 2009–2018 (2000).
Duckler, A. E., and Hubbard, M. *Ind. Eng. Chem. Fundam.* **14**, 337–347 (1975).
Fossa, M., Guglielmini, G., and Marchitto, A. *Flow Meas. Instrum.* **14**, 161–168 (2003).
Govier, G. H., and Aziz, K., "The flow of complex mixtures in pipes". R. E. Krieger Publishing Co., Malabar, Florida (1972).
Kitagawa, A., Hishida, K., and Kodama, Y. *Exp. Fluids* **38**, 466–475 (2005).
Kreyszig, E., "Advanced Engineering Mathematics". Wiley, New York (1999).
Lhermitte, R. *Science* **182**, 258–262 (1973).
Manneville, S., Becu, L., Grodin, P., and Colin, A. *Colloids Surf. A* **270–271**, 195–204 (2005).
Manneville, S., Sandrin, L., and Fink, M. *Phys. Fluids* **13**, 1683–1690 (2001).
Melling, A. *Meas. Sci. Technol.* **8**, 1406–1416 (1997).
Met-Flow., "UVP Monitor User's Guide". Lausanne, Switzerland (2002).
Munson, B. R., Young, D. F., and Okiishi, T. H., "Fundamentals of Fluid Mechanics". Wiley (1990).
Murai, Y., Inaba, K., Takeda, Y., and Yamamoto, F. *Flow Meas. Instrum.* **18**, 223–229 (2007).
Murai, Y., Oiwa, H., Sasaki, T., Kondou, K., Yoshikawa, S., and Yamamoto, F. *Meas. Sci. Technol.* **16**, 1459–1468 (2005).
Murakawa, H., Kikura, H., and Aritomi, M. *J. Nucl. Sci. Technol.* **40**, 644–654 (2003).
Murakawa, H., Kikura, H., and Aritomi, M. *Exp. Therm. Fluid Sci.* **29**, 843–850 (2004).
Oddie, G., and Pearson, J. R. A. *Annu. Rev. Fluid Mech.* **36**, 149–172 (2004).
Pinkel, R. *J. Phys. Oceanogr.* **9**, 675–686 (1979).
Sandrin, L., Catheline, S., Tanter, M., Hennequin, X., and Fink, M. *Ultrason. Imaging* **21**, 259–272 (1999).
Sandrin, L., Manneville, S., and Fink, M. *Appl. Phys. Lett.* **78**, 1155–1157 (2001).
Satomura, S. *J. Acoust. Soc. Am.* **29**, 1181–1185 (1957).
Shemer, L. *Int. J. Heat Fluid Flow* **24**, 334–344 (2003).
Suzuki, Y., Nakagawa, M., Aritomi, M., Murakawa, H., Kikura, H., and Mori, M. *Exp. Therm. Fluid Sci.* **26**, 221–227 (2002).
Taitel, Y., and Barnea, D. *Chem. Eng. Sci.* **45**, 1199–1206 (1990).
Takeda, Y. *Int. J. Heat Fluid Flow* **7**, 313–318 (1986).
Tavoularis, S., "Measurement in Fluid Mechanics". Cambridge University Press, New York (2005).
Wada, S., Kikura, H., and Aritomi, M. *Flow Meas. Inst.* **17**, 207–224 (2006).
Wild, J. J., and Reid, J. M. *J. Acoust. Soc. Am.* **25**, 270–280 (1953).
Yang W. J., "Computer-Assisted Flow Visualization". CRC Press, Inc., Boca Raton, Florida (1994).

CHAPTER 2

Micromanipulation in Mechanical Characterisation of Single Particles

Z. Zhang*, J.D. Stenson and **C.R. Thomas**

Contents		
	1. Introduction	30
	2. Description of the Techniques	31
	2.1 Pressure probe	31
	2.2 Micropipette aspiration	32
	2.3 Cell poking and atomic force microscopy (AFM)	33
	2.4 Optical trapping (also known as laser tweezers)	35
	2.5 Diametrical compression (also called compression testing by micromanipulation)	37
	3. Status and Applications	51
	3.1 Biological particles	51
	3.2 Biocompatible particles	58
	3.3 Non-biological particles	59
	4. Other Applications	68
	4.1 Particle–particle adhesion	68
	4.2 Particle adhesion to a surface	70
	4.3 Fouling deposits on surfaces	72
	4.4 Nanomanipulation of sub-micron/nanoparticles	75
	5. Perspectives on Future Development	77
	Nomenclature	78
	Acknowledgements	81
	References	81

School of Chemical Engineering, University of Birmingham, Edgbaston, Birmingham B15 2TT, UK

*Corresponding author.
E-mail address: Z.Zhang@bham.ac.uk

Abstract Many functional industrial products are in the form of microscopic particles. Mechanical characterisation of these particles is essential if physical damage to them in manufacturing processes is to be minimised, and their performance is to be optimised. Several experimental techniques can be used to characterise the mechanical properties of single microparticles, including micropipette aspiration, particle poking, atomic force microscopy, optical trapping and diametrical compression. The details of these techniques and their applications are presented in this review. Among them, diametrical compression has many advantages. It is capable of determining the mechanical properties of both biological and non-biological particles as small as 400 nm in diameter, and can be used for measurements at high deformations, including up to rupture. The technique can be enhanced by mathematical modelling to allow intrinsic mechanical properties to be estimated, for example, the particle (or particle wall) elastic modulus, and viscoelastic and plastic parameters. For biological materials, present and potential applications include studying mechanical damage to animal cells in suspension cultures, yeast and bacterial disruption in downstream processing equipment, changes of the morphology of filamentous microorganisms in submerged fermentations, plant cell behaviour in food processing, flocculation processes, cell mechanics, biomaterials and tissue engineering. For non-biological materials, applications include understanding and controlling particle breakage in processing equipment, handling and transport, and end-use applications.

1. INTRODUCTION

Both biological and non-biological microscopic particles (or microparticles) are used to produce functional products over a wide range of industrial sectors including chemical, agrochemical, food and feed, pharmaceutical and medical, human care and household care. These particles should have desirable structures and properties, for example, the appropriate size, porosity, mechanical strength and surface charge. Understanding the mechanical properties of microparticles is essential to predict their behaviour in manufacturing, and their performance in end-use applications.

In principle, the mechanical properties of particles can be determined by measuring their deformation under a mechanical load. This may be realised on single particles or a population. Mechanical characterisation of single particles is relatively difficult, particularly when their sizes are in the micro or nano ranges. The basis of many techniques for mechanical characterisation of single particles is micromanipulation. This can be used to perform a variety of mechanical operations on tiny objects under high magnification through the use of manipulators and a microscope (Madgar et al., 1996). Typical examples are the use of a micropipette to inject a single sperm into an ovum to aid fertilisation, and moving a single cell/particle in suspension from one location to another using optical trapping. Combination of micromanipulation of single particles with force measurement has enabled characterisation of the mechanical properties of micro/nanoparticles.

2. DESCRIPTION OF THE TECHNIQUES

The techniques that have been used to characterise the mechanical properties of microparticles may be classified as indirect and direct. The former includes measurement of breakage in a "shear" device, for example, a stirred vessel (Poncelet and Neufeld, 1989) or bubble column (Lu et al., 1992). However, the results from these indirect techniques are rather difficult to use since the mechanical breakage depends not only on the mechanical properties but also the hydrodynamics of the processing equipment, and the latter are still not well understood. To overcome this problem, a cone and plate viscometer that can apply well-defined shear stresses has been used to study breakage of hybridomas (Born et al., 1992), but this is not a widely applied or applicable technique because the forces are too small to break most cells.

Another possibility is to compress a layer of particles between two glass plates with a given applied force (Ohtsubo et al., 1991). The percentage of particle breakage can be related to the force. This gives some global measure of the strength of a sample of particles.

For biological particles, for example, cells which have a semi-permeable membrane and semi-permeable microcapsules, their mechanical integrity can be characterised by exposuring them to media with different osmotic pressures (Van Raamsdonk and Chang, 2001).

None of these indirect techniques relate the breakage of an individual particle to its mechanical properties. To achieve this, direct techniques are required. These are described later, and their capabilities and limitations discussed. Direct techniques also allow more sophisticated mathematical modelling to be undertaken, which is particularly valuable when the particles are not homogeneous, for example, cells with walls and membranes surrounding cytoplasm, or a liquid-filled microcapsule.

2.1 Pressure probe

The pressure probe is frequently used to study the hydraulic behaviour of plant tissues and individual cells, and their response to water stress.

High turgor (hydrostatic) pressures within cells lead to tensions in the cell wall, which maintain the cellular shape and function. Turgor pressure measurements can provide useful information about the physiology of cells.

In this technique, a capillary probe penetrates a cell, usually in a tissue, and allows the turgor pressure to be measured. It is assumed that the penetration of the cell does not affect the mechanical behaviour of the cell wall. By manipulating the pressure within the cell it is possible to derive the "volumetric elastic modulus" of the cell (Tomos, 2000; Tomos and Leigh, 1999). The primary use of the volumetric elastic modulus is to describe water transport in plant cells under non-steady conditions (Cosgrove, 1988). This parameter is often confused with the bulk elastic modulus, which (in solid mechanics) relates the fractional volumetric change of a solid to some external pressure change. However, in studies of plant–water relations, the mass of the cell is not conserved, and this definition of the bulk elastic modulus is invalid. Owing to this confusion, the volumetric elastic modulus has been described as physically "meaningless" (Wu et al., 1985), which is correct in relation to conventional mechanical analysis, notwithstanding the term still being of value in plant physiology.

Wang et al. (2006b) used the pressure probe to determine the internal turgor pressure and volumetric elastic modulus of single tomato cells from suspension cultures, rather than cells from tissues. However, this single cell technique is limited to cells over ca. 20 µm in diameter, because otherwise it would be challenging to obtain a sufficiently small probe, which would in any case probably block when it penetrates the cell wall. Besides, cell deformations achievable with a pressure probe are not great enough to cause cell failure.

2.2 Micropipette aspiration

The mechanical behaviour of living cells has been studied widely using the micropipette aspiration technique. In the most common application, a cell suspended in solution is partially or wholly aspirated into the mouth of a pipette (Figure 1). The inner diameter of the pipette may range from less than 1 µm to 10 µm (Lim et al., 2006). Video microscopy is used to record the change in shape of the cell as it is drawn into the micropipette, and the edges of the cell can be tracked to an accuracy of less than 25 nm. Measuring the elongation and suction pressure allows the mechanical properties of the cell to be determined, using a mechanical model of the behaviour (Hochmuth, 2000).

Micropipette aspiration has been used widely to measure the mechanical properties of many types of cells including red blood cells (RBCs), leukocytes and chondrocytes (Lim et al., 2006). One important

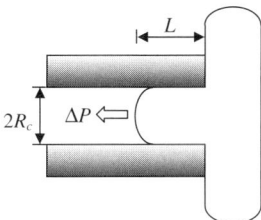

Figure 1 Diagram of the micropipette aspiration technique (modified from Lim et al., 2006). L is the length of extension into the pipette, R_c the inner radius of the pipette and ΔP the suction pressure.

application of the micropipette aspiration technique involves measuring the Poisson ratio of biological cells. Recently, He et al. (2007) used analytical simulations and finite element analysis to determine the Poisson ratio and elastic modulus of a single neutrophil cell wholly aspirated into a tapered cylindrical micropipette. Boudou et al. (2006) have also presented a finite element method for determining the Poisson ratio and Young's modulus of biological films; though this method was only implemented on polyacrylamide gels.

Compared to atomic force microscopy (AFM) or optical tweezers (described later) the micropipette aspiration technique appears simple. However, a major disadvantage of this method is the friction that occurs between the cell membrane and the glass pipette; accounting for this is difficult. In addition, the pipette aspiration technique produces stress concentrations at the edge of the pipette and the cell membrane. Recent advances in the use of finite element analyses have made it possible to allow for these stress concentrations, although direct measurement of the forces is impossible (Boudou et al., 2006). Other disadvantages of micropipette aspiration are that curvature in the cell membrane as it is drawn into the pipette can cause damage to the cell, that it has only been used on cells that are large and easily deformable, and that it does not allow failure criteria to be determined, which is essential for improving our understanding of cell disruption.

2.3 Cell poking and atomic force microscopy (AFM)

The cell poking technique involves the indentation of a cell with a micro tip of a probe and measuring the loading and unloading forces involved. Owing to the localised bending that this causes within the cell wall, analysis of this method would require the use of a very complicated mechanical model (Daily et al., 1984; Duszyk et al., 1989). Cell poking has been used to measure the mechanical response of a number of mammalian cell types including erythrocytes (Daily et al., 1984), and stretching and breaking events of plant cell walls leading to determination of mean cell diameter and cell wall fracture energy (Hiller et al., 1996). Goldman (2000) showed that cell poking, when used in conjunction with AFM, could be used to obtain qualitative information about the mechanical properties of mouse embryonic carcinoma cells. However, cell poking has largely been supplanted by the use of AFM using probes significantly smaller than the sample to determine local mechanical properties.

Figure 2 shows a schematic of a typical AFM instrument that consists of a cantilever-mounted tip, a Piezoelectric scanner, four position-sensitive photo detectors, a laser diode and a control unit. The process of operation of an AFM is relatively simple. The beam from the laser is directed against the back of the cantilever beam onto the quadrants of the photo detector. As the tip is moved across a sample, this causes the cantilever beam to bend or be twisted in manner that is proportional to the interaction force. This bending or twisting of the cantilever causes the position of the laser on the photo detector to be altered. The deflection of the cantilever beam can then be converted into a 3D topographical image of the sample surface (Gaboriaud and Dufrene, 2007; Kuznetsova et al., 2007; Lim et al., 2006).

Initially this technique was intended purely as a high-resolution imaging device, down to the sub-nanometre level (Alessandrini and Facci, 2005). However, it was soon adapted for measuring the interactive forces in a process known as force spectroscopy. In this process, the sample is moved towards the tip and then retracted, with the vertical displacement of the Piezoelectric scanner being recorded. This produces voltage data recorded by the photo detector as a function of the displacement of the Piezoelectric stage. A force curve can be produced from this which provides information about the interactions between the tip of the probe and the sample. This force data can be interpreted to

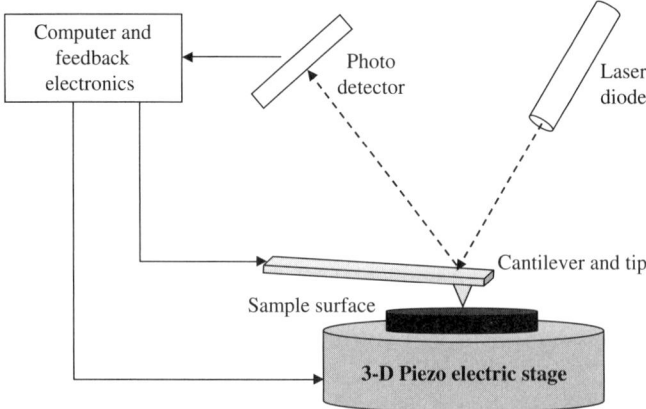

Figure 2 Schematic of an AFM setup.

allow information about the mechanical properties of the sample to be obtained. This method can be used to determine the effect of different cell materials on the mechanical properties of the cell wall. For example, this technique was employed by Touhami et al. (2003) to determine the Young's modulus of different areas of the Brewer's yeast cell wall (*Saccharomyces cerevisiae*). In their work, Touhami et al. (2003) found that a chitin bud scar had a Young's modulus value ten times higher than the value for the surrounding cell wall area.

Furthermore, a rigid glass bead can be glued to a tipless cantilever, and used as a force probe to compress single particles, such as Jurket T lymphomas cells (Lulevich et al., 2006) and polyelectrolyte microcapsules (Lulevich et al., 2003).

There are many related applications of AFM. For example, it has been used to determine the adhesive forces between *S. cerevisiae* cells and surfaces in biofilm formation, as well as many other organisms and non-biological particles (Bowen et al., 2001). An additional method has used AFM to detect local nanomechanical motion of the cell wall of *S. cerevisiae* caused by the active metabolic processes of the cell (Pelling et al., 2004). AFM has also been used to produce information on the conformational changes of single pyranose rings, which are one of the main monomer constituents of the yeast cell wall (Marszalek et al., 1998). In this work, the pyranose ring was identified as the unit controlling the molecular elasticity. This technique has the potential to elucidate the tensile properties of individual polymer chains of the polysaccharides that make up the yeast cell wall.

AFM has become an indispensable tool when investigating the forces associated with cellular and molecular biomechanical events. However, AFM does not currently have the ability to cause cell wall failure since it can only measure relatively small forces, in a range of nano- to micro-Newtons.

2.4 Optical trapping (also known as laser tweezers)

The optical trapping method uses a highly focused laser beam to trap and manipulate particles of interest in a medium (illustrated in Figure 3). The laser is focused on a dielectric particle (e.g., a silica microscopic bead), the refractive index of which is higher than the suspension medium. This produces a light pressure (or gradient force), which moves the particle towards the focal point of the beam, that is, the beam waist (Lim et al., 2006).

Optical trapping is a very sensitive method that is capable of the manipulation of sub-micron particles such as individual viruses and bacteria. Figure 4 is an illustration of the use of the optical trapping method to measure the elastic properties of RBCs (Mills et al., 2004).

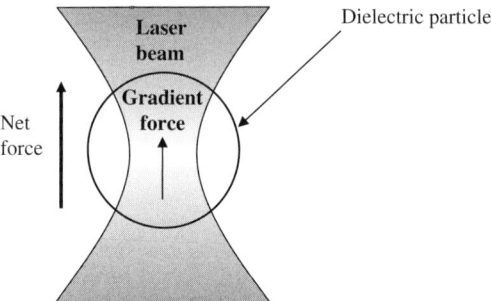

Figure 3 Optical trapping of a dielectric particle (simplified from Lim et al., 2006).

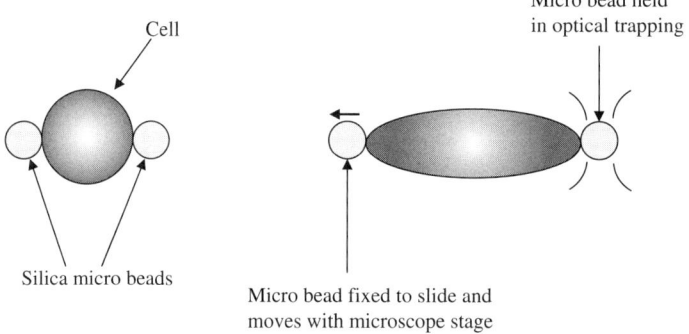

Figure 4 Stretching of a particle (e.g., a RBC) using the optical trapping method (modified from Mills et al., 2004).

Initially silica microbeads were attached to opposite sides of the cell with the bead on the left-hand side being attached to the surface of a glass slide and the right-hand bead held in the optical beam. The trapped bead remained stationary as the glass slide was moved to the left causing the cell to stretch. From this measurements of the stretching force of the cell could be determined.

There are numerous examples of the use of the optical trapping to gain insight into the mechanical properties of cells. As well as measuring the mechanical properties of RBCs, the stretching of DNA molecules has also been investigated (Bryant et al., 2003; Smith et al., 1996). A dual wavelength optical trapping method has also been described in which one beam was used to trap a *S. cerevisiae* (yeast) cell, whereas the other beam was used to generate Raman spectra (Creely et al., 2005). Raman spectroscopy produces an individual spectrum of the macromolecules within the cell and can observe changes occurring during cell replication. The budding of a yeast cell was observed for 150 min and biological changes in the cell were monitored. The use of optical trapping in this case prevented problems of fluorescence that have been previously reported for cells adhering to a glass slide (Xie and Li, 2003). As well as this, the optical trapping method has been used to make the first direct measurements of the Young's modulus of bacterial macrofibers of *Bacillus subtilis* (Mendelson et al., 2000).

One of the advantages of the optical trapping method is that there is limited physical contact with the cell, although it is possible that it could be damaged by the laser (Lim et al, 2006). In addition, it is possible to measure forces in the sub pico-Newton range, which is extremely difficult to do by other methods. However, these small forces mean that it is not appropriate for investigating the response of samples to large deformations, especially if the sample has a cell wall.

2.5 Diametrical compression (also called compression testing by micromanipulation)

2.5.1 Experimental setup

In this technique, an individual particle is compressed between two flat parallel surfaces, usually until it ruptures. By measuring the force being applied to the particle and the displacement of the surfaces, force-deformation data can be found. Figure 5 shows an illustration of a particle during compression where a force (F) is deforming it between two surfaces separated by a known distance ($2\bar{\eta}$).

Initial experiments using this method involved the compression of relatively large sea urchin eggs and visual measurement of the surface displacement to determine the cell deformation to the forces being imposed (Cole, 1932; Hiramoto, 1963; Yoneda, 1964, 1973). Recently, compression testing by micromanipulation was developed, which is capable of determining the mechanical properties of small cells or particles with sizes down to approximately 1 µm, employing forces of 1 µN or greater (Zhang et al., 1992a).

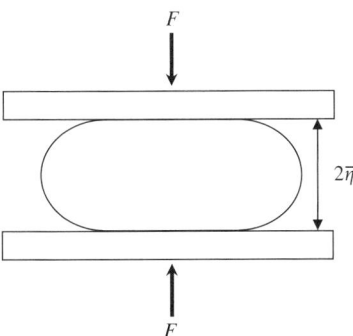

Figure 5 Compression experiment. Particle compressed by force F between surfaces separated by the distance $2\bar{\eta}$.

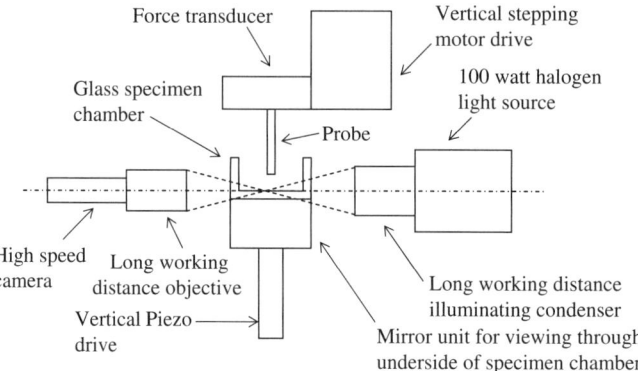

Figure 6 Schematic diagram of the newly built micromanipulation rig (reproduced from Wang et al., 2005). Permitted by Elsevier.

Figure 6 shows a diagrammatic representation of the most recent apparatus. In comparison with previous micromanipulation equipment (Blewett et al., 2000), the key feature of this rig is its capability of generating high-speed compressions using a Piezoelectric stack with a relatively large motion (of the order of 100 μm). In a recent study of the mechanics of tomato fruit cells (Wang et al., 2006a), compression speeds up to 1,500 μm s^{-1} were used, although even higher speeds might have been achievable. The Piezoelectric stack is fixed on the base of the rig, and the glass chamber holding a suspension of the particles is mounted on top of the stack. The particles are actually compressed against the force transducer probe by upward displacement of the stack and chamber. This design also allows images of the particles to be obtained, viewing through the bottom and side. The other major modification from earlier equipment is the connection of a digital high-speed camera to the side view microscope, which in the work of Wang et al. (2005) allowed 510 × 484 pixel resolution images to be collected and recorded at 500 frames per second for later image analysis. The latter can be used to measure the extent of particle deformation and possibly changes in particle volume during compression and subsequent relaxation.

A range of experiments can be carried out using this micromanipulation equipment, for example, compression at different speeds, repeated compression, and loading and unloading. These allow the identification of elastic, viscoelastic and elastic-plastic properties. Figure 7 shows compression of an ion-exchange resin particle to different deformations, and then unloading. The corresponding force versus displacement data for loading and unloading are presented in Figure 8.

Figures 7 and 8 show this ion-exchange resin particle was very elastic up to a deformation of 46%. The contact radius and lateral extension of the particle were also measured as a function of time (data not shown),

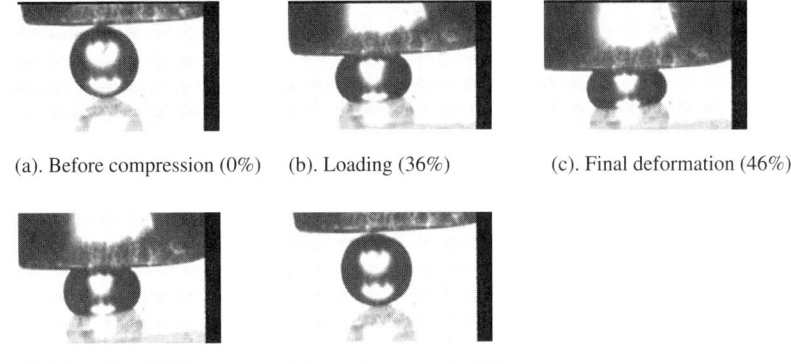

(a). Before compression (0%) (b). Loading (36%) (c). Final deformation (46%)

(d). Unloading (36%) (e). Totally released (0%)

Figure 7 Images of an ion-exchange resin particle (DOWEX 1X8-200, Sigma-Aldrich, UK) compressed to different deformations (values in brackets) and then released. The particle diameter was 163 μm (images provided by T. Liu, University of Birmingham, UK).

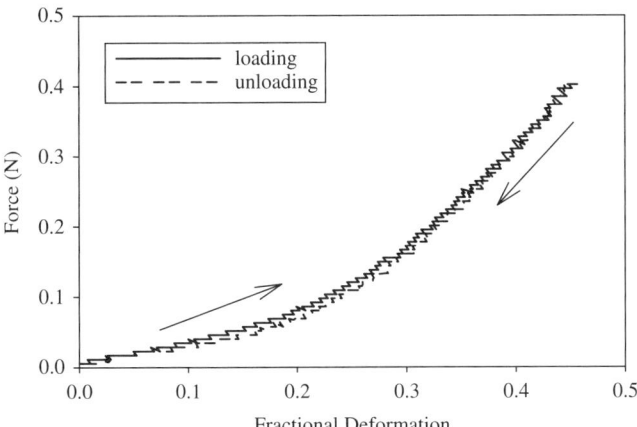

Figure 8 Loading and unloading curves for a single 163 μm diameter ion-exchange resin particle (DOWEX 1X8-200, Sigma-Aldrich, UK) obtained at a speed of 22.4 μm s^{-1} (data provided by T. Liu).

and this information can be used for validation of later theoretical modelling of the deformation (see Section 2.5.2).

However, some particles, for example, single Eudragit microparticles, show significant hysteresis on loading and unloading, as shown in Figure 9 (Yap et al., 2008). The dotted line is the best fit to the loading data using the Hertz equation, which will be described later. Figure 9 also shows that the force dropped to zero before the displacement reduced to zero, which indicates the particle had undergone some plastic deformation.

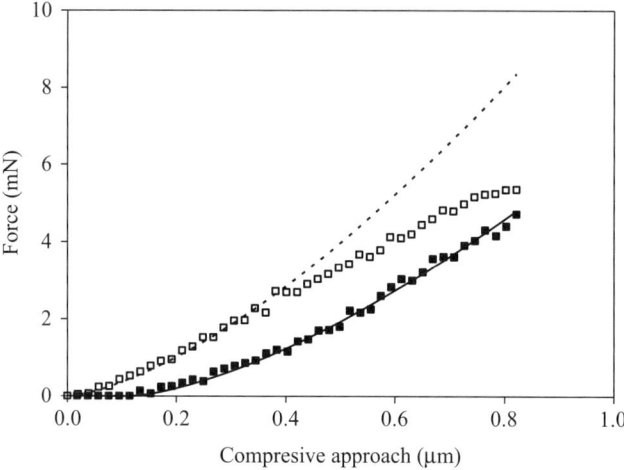

Figure 9 Typical loading (□) and unloading (■) data for a single Eudragit® L100-55 particle (Degussa Rohm, UK) compressed to a final strain of 6.0% (diameter = 27.4 µm) (Yap et al., 2008). Permitted by Elsevier.

Some particles can be viscoelastic or lose liquid under compression, for example, hydrogels and plant cells, and the force versus displacement data may therefore vary with the compression speed.

Force versus displacement data are directly useful in comparing some simple mechanical properties of particles from different samples, for example, the force required to break the particle and the deformation at breakage. However, these properties are not *intrinsic*, that is, they might depend on the particular method of measurement. Determination of intrinsic mechanical properties of the particles requires mathematical models to derive the stress–strain relationships of the material.

2.5.2 Mathematical modelling of the compression of single particles

2.5.2.1 Hertz model. The mechanics of a sphere made of a linear elastic material compressed between two flat rigid surfaces have been modelled for the case of small deformations, normally less then 10% strain (Hertz, 1882). Hertz theory provides a relationship between the force F and displacement h_p as follows:

$$F = \frac{4E\sqrt{R}}{3(1-v^2)}\left(\frac{h_p}{2}\right)^{\frac{3}{2}} = \frac{4}{3}\frac{\sqrt{R}}{2^{1.5}}E^* h_p^{3/2} \qquad (1)$$

where R, E and v are the radius, Young's modulus and Poisson's ratio of the sphere, respectively. E^* is the reduced modulus:

$$E^* = \frac{E}{1-v^2} \qquad (2)$$

which can be estimated by fitting Equation (1) to the experimental data using non-linear regression (see Figure 9). Hertz theory has been applied to describe data corresponding to small deformations of microparticles used for chromatography media (Müller et al., 2005). However, the applicability of Hertz theory is only to small displacements/strains and to linear elasticity (Tatara, 1991). At higher deformations, Tatara theory is more appropriate.

2.5.2.2 Tatara analysis. For larger deformations (up to 60%) of soft solid spheres made from linear and non-linear elastic materials, Tatara (1991, 1993) proposed a model to describe the relationship between the force and displacement. For the case of linear elasticity

$$h_p = \frac{3(1-v^2)F}{2Ea} - \frac{2Ff(a)}{\pi E} \quad (3)$$

where a is the contact radius, given by

$$a = \left[\frac{3(1-v^2)RF}{4E}\right]^{\frac{1}{3}} \quad (4)$$

$f(a)$ is given by

$$f(a) = \frac{2(1+v)R^2}{(a^2+4R^2)^{3/2}} + \frac{1-v^2}{(a^2+4R^2)^{1/2}} \quad (5)$$

Equation (3) reduces to Equation (1) if the second term on the right-hand side is ignored, which corrects the Hertz's displacement since Hertz's assumption of near sphericity is not valid at large deformations. For a given force, the displacement predicted by the Tatara model is always smaller than that from Hertz theory since the value of $f(a)$ is positive.

From Equations (3)–(5), the relationship between the displacement and force can be determined if the values of R, E and v are known. In other words, if experimental data of force versus displacement are available, they can be used to determine the value of E given values of R and v.

The relationship between the force and displacement in the Tatara model (Equation (3)) may be approximated by a semi-empirical equation (Andrei et al., 1996):

$$F = c_1 h_p^{3/2} + c_2 h_p^3 + c_2 h_p^5 \quad (6)$$

where c_1 is equal to the coefficient of the Hertz equation:

$$c_1 = \frac{4E}{3(1-v^2)}\frac{\sqrt{R}}{2^{1.5}} = \frac{4}{3}\frac{\sqrt{R}}{2^{1.5}}E^* \quad (7)$$

However c_2 and c_3 are arbitrary constants. Mathematically, Equation (6) may fit a given set of experimental data well since it has 3 adjustable constants, but the values of c_2 and c_3 are physically meaningless.

2.5.2.3 Viscoelastic model.
Some microspheres may be viscoelastic or contain a large amount of liquid that can flow during slow compression or during force relaxation (during holding) after compression. In either case, they can show time-dependent behaviour, which may be described by a viscoelastic model. The simplest theoretical model of a viscoelastic solid is represented by a combination of a linear spring with a Kelvin–Voigt element (Haddad, 1995):

$$\sigma(t) = E_\infty \varepsilon_\infty \left(1 - e^{-t/\tau}\right) + \sigma_0 e^{-t/\tau} \qquad (8)$$

where $\sigma(t)$ is the time-dependent stress, E_∞ and ε_∞ are, respectively, the Young's modulus and strain when the relaxation re-establishes equilibrium, σ_0 is the stress at time zero and τ the relaxation time. A more general and semi-empirical form of Equation (8), the so-called Maxwell model, which includes several exponential terms, is given by

$$\sigma(t) = \sigma_\infty + \sum_{i=1}^{n} \sigma_i \, e^{-t/\tau_i} \qquad (9)$$

where σ_∞ is the stress at the equilibrium, σ_i are proportionality constants and τ_i relaxation times.

In terms of force, a similar equation to Equation (9) can be written (Mattice et al., 2006) as

$$F(t) = F_\infty + \sum_{i=1}^{n} F_i \, e^{-t/\tau_i} \qquad (10)$$

where $F(t)$ is the time-dependent force, F_∞ the force when the relaxation reaches equilibrium and F_i are proportionality constants.

Mattice et al. (2006) extended Hertz analysis to spherical indentation of a viscoelastic material:

$$F(t) = \frac{4E\sqrt{R}}{3(1-v^2)} h_p^{3/2} = \frac{8}{3}\sqrt{R} h_p^{3/2} [2G(t)] \qquad (11)$$

where the Poisson ratio v equals 0.5 for an incompressible material and $G(t)$ is the shear relaxation modulus (Mattice et al., 2006). It was proposed that Equation (10) can be used to fit the force relaxation data from typical loading and holding experiments using compression testing, and that the shear relaxation modulus also relaxes with the same relaxation times as in Equation (10), so that

$$G(t) = K_0 + \sum_{i=1}^{n} K_i \, e^{-t/\tau_i} \qquad (12)$$

K_0 is related to F_∞ (Mattice et al., 2006) via

$$K_0 = \frac{F_\infty}{h_{p\,\text{max}}^{3/2}(8\sqrt{R}/3)} \tag{13}$$

and

$$K_i = \frac{F_i}{(\text{RCF}_i)\,h_{p\,\text{max}}^{3/2}(8\sqrt{R}/3)} \quad (i = 1, 2, 3, \ldots, n) \tag{14}$$

where RCF_i is called the "ramp correction factor" given by

$$\text{RCF}_i = \frac{\tau_i}{t_R}\left[e^{t_R/\tau_i} - 1\right] \quad (i = 1, 2, 3, \ldots, n) \tag{15}$$

where t_R is the time taken to compress the material. From the values of $K_0, K_1, K_2, \ldots, K_n$, the instantaneous shear modulus (G_0) and long-term shear modulus (G_∞) can be estimated.

$$G_0 = \frac{\sum_{i=0}^{n} K_i}{2} \tag{16}$$

and

$$G_\infty = \frac{K_0}{2} \tag{17}$$

The corresponding Young's moduli are

$$E_0 = 2(1+v)G_0 = 3G_0 \tag{18}$$

$$E_\infty = 2(1+v)G_\infty = 3G_\infty \tag{19}$$

The theoretical force versus time relationship for loading may be given by the generalised Boltzmann integral expression (Mattice et al., 2006)

$$F(t) = \frac{8\sqrt{R}}{3}\int_0^t G(t-u)\left[\frac{d}{du}h_p^{3/2}(u)\right]du \tag{20}$$

The relationship between h and t is linear if the deflection of the compression probe due to its compliance is negligible, given by

$$h_p(t) = V_0 t \tag{21}$$

where V_0 is the moving speed of the probe.

Substituting Equations (10), (12) and (21) into (20) and integrating by parts gives

$$F(t) = 4\sqrt{R}\,V_0^{3/2}\left[\frac{2}{3}K_0 t^{3/2} + \sum_{i=1}^{n} K_i\,e^{-t/\tau_i}\tau_i^{3/2}\left(\sqrt{\frac{t}{\tau_i}}\,e^{t/\tau_i} - \int_0^{\sqrt{t/\tau_i}} e^{x^2}dx\right)\right] \tag{22}$$

where $\int_0^{\sqrt{t/\tau_i}} e^{x^2} dx$ cannot be solved analytically, and has to be determined numerically.

It should be pointed out that Equation (11) should be modified to Equation (1) when compression of a single particle between two surfaces is modelled, that is, spherical indentation is replaced by compression between two surfaces. Applications in which these models were applied to experimental data from compression testing are described later.

2.5.2.4 Models to describe microparticles with a core/shell structure. Diametrical compression has been used to measure the mechanical response of many biological materials. A particular application has been cells, which may be considered to have a core/shell structure. However, until recently testing did not fully integrate experimental results and appropriate numerical models. Initial attempts to extract elastic modulus data from compression testing were based on measuring the contact area between the surface and the cell, the applied force and the principal radii of curvature at the point of contact (Cole, 1932; Hiramoto, 1963). From this it was possible to obtain elastic modulus and surface tension data. The major difficulty with this method was obtaining accurate measurements of the contact area.

Following on from this work two types of mathematical model were developed that do not rely on measuring the contact area. These models are the "liquid-drop" model (Yoneda, 1973) and the elastic membrane model (Cheng, 1987a; Feng and Yang, 1973; Lardner and Pujara, 1980).

The liquid-drop model was used to model the deformation of sea urchin eggs (Yoneda, 1973). This theory assumes that the tensions in the wall during the compression are uniform and isotropic as is stated by Cole (1932) and Yoneda (1964). However, Hiramoto (1963) suggested that the circumferential tensions are actually up to two times greater than the tensions in the meridian direction. This result suggests that the use of the liquid-drop model may not be appropriate to determine material properties of cells.

An analytical elastic membrane model was developed by Feng and Yang (1973) to model the compression of an inflated, non-linear elastic, spherical membrane between two parallel surfaces where the internal contents of the cell were taken to be a gas. This model was extended by Lardner and Pujara (1980) to represent the interior of the cell as an incompressible liquid. This latter assumption obviously makes the model more representative of biological cells. Importantly, this model also does not assume that the cell wall tensions are isotropic. The model is based on a choice of cell wall material constitutive relationships (e.g., linear-elastic, Mooney–Rivlin) and governing equations, which link the constitutive equations to the geometry of the cell during compression.

The constitutive equations relate the stresses and strains that are generated in the cell wall during the compression of a cell. Constitutive equations can be obtained by assuming that the cell wall is hyperelastic, in which case the stress components can be derived from a strain energy function. Strain energy is the energy stored in a body because of its elastic deformation, and is equal to the work required to produce the strain in the body. There are many choices of strain energy function that can be used depending on the cell wall material properties. In the case of Feng and Yang (1973), the constitutive equations were based on a Mooney–Rivlin model previously used to describe rubber material properties. On the other hand, Lardner and Pujara (1980) employed a Skalak–Tozeren–Zarda–Chien (STZC) material relationship. These relationships have been used to represent the material properties of RBCs and sea urchin eggs. Cheng (1987a) gave a generalised Hooke's law strain energy function, which allows various ways of deriving the strain energy function to be considered. The assumption of Hooke's law implies a linear relationship between the stress and the strain (i.e., a constant elastic modulus) although a non-linear force-deformation curve will be observed because of changes in the geometry of the cell as it is deformed. The generalised Hooke's law strain energy function may be written as an explicit function of the strain, but this requires that the definition of strain be considered carefully. In addition to this, the equations of Feng and Yang (1973) and Lardner and Pujara (1980) are expressed in terms of stretch ratios, which are a measure of deformation, and are often used for materials that can support high deformations before failure. Stretch ratios are defined as the length of a section in a given direction following deformation divided by its original length. Each strain measure also has a different relationship to the stretch ratio.

The elastic membrane model can be solved using a finite element approach (Cheng, 1987). The finite element approach has the advantage of being able to take into account irregularities such as non-spherical cells, local and geometric material differences (e.g., bud scars on the surface of yeast cells), assuming that detailed information is available about such irregularities. However, it is advisable to validate any finite element model against a corresponding analytical solution. As well as this, all but the most expert users treat the finite element model as a "black box" without fully understanding how the model results are generated.

The elastic membrane model assumes that the cell is a thin-walled sphere filled with incompressible fluid. Because the wall is thin, it may be treated as a mechanical membrane. It can be presumed that the wall cannot support out-of-plane shear stresses or bending moments. This situation is described as plane stress, as the only non-zero stresses are in the plane of the cell wall. Furthermore, the stresses can be expressed as

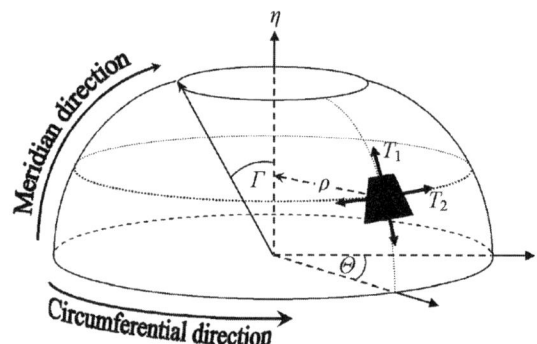
(a). Compound view

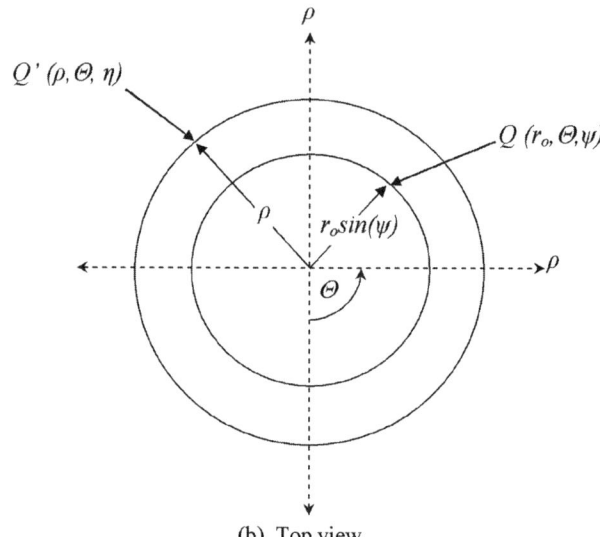
(b). Top view

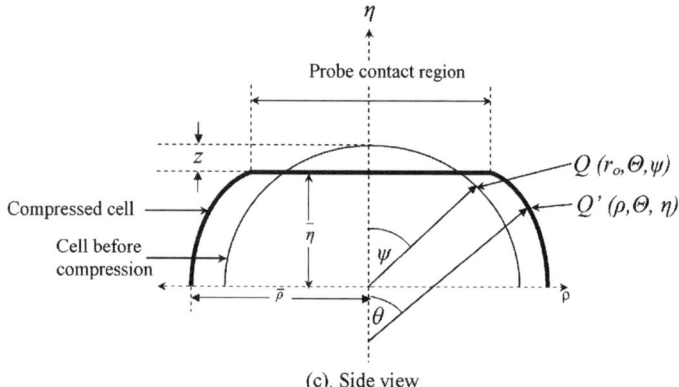
(c). Side view

wall tensions. It is usually assumed that a cell is inflated before compression (i.e., there is an initial pressure difference across its wall), but microcapsules are usually not initially inflated.

It is assumed that the cell is symmetrical across the equatorial plane and axi-symmetrical around the axis of compression, the η axis. This symmetry allows the compression of the cell to be fully represented by a 2D curve in the positive η and ρ axis. To understand the elastic membrane model, the geometry of the spherical cell under compression can be represented by Figure 10.

From Figure 10 there are two directions that need to be considered when deriving governing equations for the elastic membrane model, the meridian and the circumferential directions. In addition, there are separate groups of governing equations for the contact and non-contact regions of the cell during compression. The independent variable in these equations is ψ, which relates the position of any point on the boundary of the cell back to the original position of that point in the inflated but uncompressed cell.

The governing equations corrected from those presented by Lardner and Pujara (1980) are defined as

Contact region:

$$\frac{d\lambda_1}{d\psi} = -\frac{\lambda_1}{\lambda_2 \sin \psi}\left(\frac{f_3}{f_1}\right) - \left(\frac{\lambda_1 - \lambda_2 \cos \psi}{\sin \psi}\right)\left(\frac{f_2}{f_1}\right) \quad (23)$$

$$\frac{d\lambda_2}{d\psi} = \frac{\lambda_1 - \lambda_2 \cos \psi}{\sin \psi} \quad (24)$$

where f_1, f_2 and f_3 are functions of the principal tensions and are defined later.

Figure 10 Geometry of a thin-walled spherical cell under compression as represented by the elastic membrane model. (a) Compound view showing the upper half of the cell corresponding to the positive η axis of a cell under compression. T_1 and T_2 represent the tensions in the meridian and circumferential directions, respectively, Γ represents the angle between the positive η axis and the edge of the contact region, ρ is the horizontal coordinate, and Θ the angle of rotation in the circumferential direction. (b) Top view of a cell in compression where $Q(r_o,\Theta,\psi)$ is a point on the inflated cell, $r_o\sin(\psi)$ the radius at point Q, and $Q'(\rho,\Theta,\eta)$ the corresponding point on the compressed cell. (c) Side view of the cell in compression where $\bar{\eta}$ is the distance between the compression surface and the equatorial plane, $\bar{\rho}$ the distance between the η axis and the edge of the cell on the equatorial plane, z is half the displacement of the probe, also termed compressive approach, ψ the angular position of the point measured from the η axis before compression and θ angle between the normal to the surface and the η axis following compression.

Non-contact region:

$$\frac{d\lambda_1}{d\psi} = \left(\frac{\delta \cos\psi - \omega \sin\psi}{\sin^2\psi}\right)\left(\frac{f_2}{f_1}\right) - \left(\frac{\omega}{\delta}\right)\left(\frac{f_3}{f_1}\right) \qquad (25)$$

where

$$\delta = \lambda_2 \sin\psi \qquad (26)$$

and

$$\omega = \frac{d\delta}{d\psi} \qquad (27)$$

$$\frac{d\lambda_2}{d\psi} = \left(\frac{\omega \sin\psi - \delta \cos\psi}{\sin^2\psi}\right) \qquad (28)$$

The turgor pressure (P) inside the cell can be related to ω using

$$\frac{d\omega}{d\psi} = \frac{d\lambda_1}{d\psi}\frac{\omega}{\lambda_1} + \frac{(\lambda_1^2 - \omega^2)}{\delta}\left(\frac{T_2}{T_1}\right) - \frac{\lambda_1(\lambda_1^2 - \omega^2)^{1/2} Pr_0}{T_1} \qquad (29)$$

where r_0 is the uninflated radius of the cell. Equation (29) is used to link the deformation of the cell to the forces imposed through the internal pressure P.

The functions of the principal tensions in Equations (23) and (25) are defined as

$$f_1 = \frac{\partial T_1}{\partial \lambda_1} \qquad (30a)$$

$$f_2 = \frac{\partial T_1}{\partial \lambda_2} \qquad (30b)$$

and

$$f_3 = T_1 - T_2 \qquad (30c)$$

where T_1 and T_2 are the tensions in the meridian and circumferential directions, respectively. Equations (30a–30c) link the geometry (Equations (23)–(29)) to the tensions, which depend on the constitutive equation of the cell wall material. Since the compression of the cell consists of axi-symmetric deformation, T_1 and T_2 are principal tensions, which are defined with respect to the current deformed shape of the cell.

For the analysis of experimental force-deformation data, it is necessary to use a suitable constitutive equation for the material under test. The constitutive equation relates the stresses and strains that are generated in the wall during compression, and therefore relates the tensions and stretch ratios. For example, Liu et al. (1996) used a Mooney–Rivlin constitutive equation to investigate the compression of polyurethane microcapsules and the functions f_1, f_2 and f_3 are produced in corrected form below.

$$f_1 = \frac{\partial T_1}{\partial \lambda_1} = 2hC_1(1 + \beta\lambda_2^2)\left(\frac{1}{\lambda_2} + \frac{3}{\lambda_1^4 \lambda_2^3}\right) \quad (31a)$$

$$f_2 = \frac{\partial T_1}{\partial \lambda_2} = 2hC_1\left[\left(\frac{3}{\lambda_1^3 \lambda_2^4} - \frac{\lambda_1}{\lambda_2^2}\right)(1 + \beta\lambda_2^2) + 2\beta\lambda_2\left(\frac{\lambda_1}{\lambda_2} - \frac{1}{\lambda_1^3 \lambda_2^3}\right)\right] \quad (31b)$$

$$f_3 = T_1 - T_2 = 2hC_1\left[\frac{\lambda_1}{\lambda_2} - \frac{\lambda_2}{\lambda_1} - \beta\left(\frac{1}{\lambda_1^3 \lambda_2} - \frac{1}{\lambda_1 \lambda_2^3}\right)\right] \quad (31c)$$

where C_1 and C_2 are material constants, $\beta = C_2/C_1$, h is the cell wall thickness, λ_1 and λ_2 are the principal stretch ratios in the meridian and circumferential directions, respectively.

When using the generalised Hooke's law strain energy function there are a number of possible strain definitions that can be used depending on the situation. When material deformation is very small the infinitesimal strain approach is a valid approximation with the strain defined as

$$\varepsilon_i = (\lambda_i - 1) \quad (32)$$

where ε_i is the infinitesimal strain in the ith principal direction. In situations where the deformation of the material is greater then alternative strain definitions need to be considered. When deformation is non-infinitesimal, then the most common strain definition is Green strain defined as

$$E_i = \frac{1}{2}(\lambda_i^2 - 1) \quad (33)$$

where E_i is the Green strain in the ith principal direction. This strain measure is more appropriate for large deformations and large rotations. It is computationally convenient. However, when large deformation is accompanied by large strain then the most appropriate strain measure is the Hencky or true strain, which is defined as

$$H_i = \ln \lambda_i \quad (34)$$

where H_i is the Hencky strain in the ith principal direction. From this it is possible to define the functions f_1, f_2 and f_3 for the Hencky strain as

$$f_1 = \frac{\partial T_1}{\partial \lambda_1} = \frac{2Eh_o}{3\lambda_1^2 \lambda_2}[2 - \ln(\lambda_1^2 \lambda_2)] \quad (35a)$$

$$f_2 = \frac{\partial T_1}{\partial \lambda_2} = \frac{2Eh_o}{3\lambda_1 \lambda_2^2}[1 - \ln(\lambda_1^2 \lambda_2)] \quad (35b)$$

$$f_3 = T_1 - T_2 = \frac{2Eh_o}{3\lambda_1 \lambda_2}\ln\left(\frac{\lambda_1}{\lambda_2}\right) \quad (35c)$$

where E is the elastic modulus, h_o the cell wall thickness and λ_1 and λ_2 are the principal stretch ratios in the meridian and circumferential directions, respectively.

The method for solving these equations was outlined by Wang et al. (2004) and Liu et al. (1996). The governing equations were solved by the Runge–Kutta method using the MATLAB ode45 solver (The MathWorks Inc., Cambridge, UK). The simulations can be solved as a series of static equilibrium problems as the cell wall material is considered to be time independent (i.e., negligible viscoelasticity). In each step of a simulation, the probe is displaced a set amount depending on the time length of each step and the speed of compression. This probe displacement compresses and deforms the cell a certain amount following which the pressure inside the cell is adjusted until a number of boundary conditions are met including a constant cell volume. The fluid contents of the cell are assumed to be incompressible and the cell wall is assumed to be impermeable. At the end of each step the volume of the cell is determined and the cell boundary co-ordinates ρ and η calculated. Once all of the boundary conditions have been satisfied corresponding values for the pressure (P) and the deformation ($\bar{\eta}$) are generated. From this force (F) and deformation (X) data can be obtained using the following equations:

$$F = PA_c \tag{36}$$

where P is the turgor pressure and A_c the contact area between the probe and the cell. The deformation of the cell is defined using

$$X = \frac{z}{r_o \lambda_s} = 1 - \frac{\bar{\eta}}{r_o \lambda_s} \tag{37}$$

where z is half the distance that the cell has been compressed (Figure 10c) and λ_s the initial stretch ratio of the cell.

The expressions in Equations (31a, 31b and 31c) provide the functions required to solve Equations (23)–(29), using the following boundary conditions (Liu et al., 1996):

$$\psi = 0, \; \lambda_1 = \lambda_2 = \lambda_0$$

$$\psi = \Gamma, \; \lambda_1(\text{contact region}) = \lambda_1(\text{non-contact region})$$

$$\psi = \Gamma, \lambda_2(\text{contact region}) = \lambda_2(\text{non-contact region})$$

$$\psi = \Gamma, \; \eta = \bar{\eta}$$

$$\psi = \Gamma, \eta' = 0 \text{ or } \omega = \lambda_1$$

$$\psi = \frac{\pi}{2}, \; \omega = 0$$

where Γ identifies the angle of the points on the edge of the contact region between the compression surface and the cell (Figure 10a), η the horizontal coordinates and $\bar{\eta}$ the distance between the compression surface and the equatorial plane (Figure 10c).

Numerical simulations produce force-deformation data whose shape and magnitude is dependent on the initial parameters defined within the model, including the elastic modulus (E), the uninflated cell radius (r_o) and the initial stretch ratio (λ_s). Experimental data are fitted to these numerical simulations allowing intrinsic material properties to be derived.

Applications in which these models were applied to experimental data from compression testing are described later.

3. STATUS AND APPLICATIONS

3.1 Biological particles

3.1.1 Animal cells in suspension culture

Animal cells lack a cell wall and so are considered to be relatively weak in suspension cultures. There have been inconsistent reports concerning the mechanisms by which animal cells might be damaged in bioreactors and about the protective effects of medium additives (Michaels et al., 1991; Murhammer and Goochee, 1990), but it is now well-established that the predominant damaging mechanism in suspension cultures is bubble disengagement (Papoutsakis, 1991). To predict the breakage of animal cells in bioreactors or other processing equipment, it is necessary to know the intrinsic mechanical properties of the cells and the hydrodynamics of the process of interest. Intrinsic mechanical properties are independent of the method of measurement, and can be found by compression testing combined with appropriate mathematical modelling. Assuming the hydrodynamics are known, which is not always the case, and given some model of how the cells might distort and break in the flow, it should be possible to predict cell breakage in fluid flow. This has been demonstrated for disruption of animal cells in both laminar and turbulent flows (Born et al., 1992; Thomas et al., 1994), but it is yet to be shown that this approach would be successful in predicting cell damage caused by bubble disengagement.

Until recently there was only limited knowledge on the mechanical properties of animal cells. Micromanipulation has allowed some progress in this area. Figure 11 shows typical force-sampling time data for a hybridoma, with a bursting force of a few micro-Newtons. From such data the intrinsic mechanical properties of cell diameter, membrane tension at bursting and elastic area compressibility modulus can be

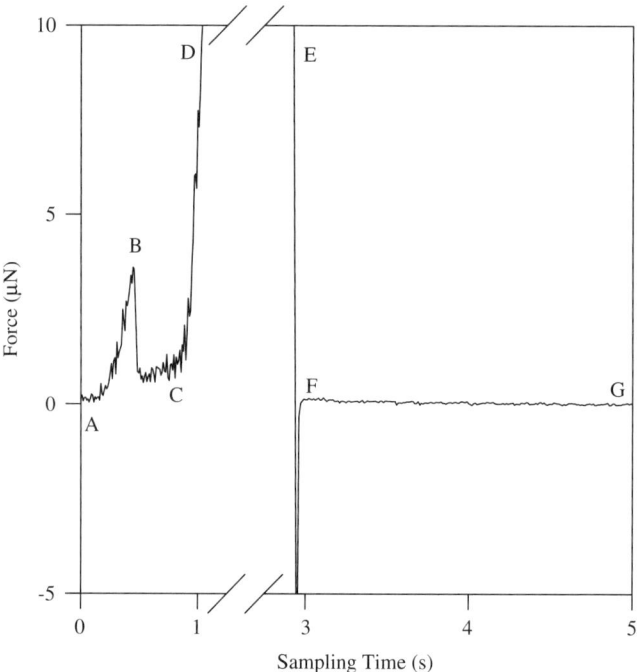

Figure 11 Typical force-sampling time curve for the compression of a TB/C3 murine hybridoma. (AB) cell compression, (B) cell bursting, (C) cell debris compression, (DE) probes touching each other and (FG) two probes separated (reproduced from Thomas et al., 2000). Permitted by Springer.

obtained (Zhang et al., 1992a). Micromanipulation has also been used to show how the culture additive Pluronic F68 can strengthen animal cell membranes (Zhang et al., 1992b), and to investigate the role of the plasma membrane, F-actin and microtubules in determining the mechanical properties of hybridomas (Welsh, 1998).

The mechanical properties of animal cells, including RBCs and leukocytes, have also been investigated using micropipette aspiration. Using this technique leukocytes are assumed to act as a liquid-drop leading to a surface tension of approximately $30\,\text{pN}\,\mu\text{m}^{-1}$ (Hochmuth, 2000). The mechanical properties of malaria-infected RBCs have been measured using the optical trap method (Lim et al., 2006). It was discovered that RBCs could be stiffened and its shear modulus increased by an order of magnitude after they were invaded by malaria parasite *Plasmodium falciparum*. This is interesting since there may be a relation between biomechanical states and the onset and progression of diseases such as cancer and malaria, and the cell mechanical properties may be used as an indicator of cellular diseases. A wide range of techniques has been used to investigate the mechanical properties of animal cells as only relatively small forces are required to deform these cell types. When cells that possess cell walls are of interest the number of applicable methods is not so wide.

3.1.2 Chondrocytes

Chondrocytes respond biologically to mechanical stimuli, and this is called mechanotransduction. Understanding cellular responses to mechanical stimuli and how they can be related to tissue level characteristics has many applications in biology, tissue engineering and medical science. Information about the mechanical properties of chondrocytes is fundamental to such research. Micropipette aspiration has been applied to investigate the deformation of single chondrocytes (Jones et al., 1999). It was found that human chondrocytes behave like viscoelastic solids. The Young's modulus of normal chondrocytes appeared to be very variable, but was of the order of 0.65 ± 0.63 kPa. However, the viscoelastic behaviour was not quantified.

Leipzig and Athanasiou (2005) designed a creep cytoindentation (i.e., cell poking) apparatus and applied a constant stress to adherent cells, while tracking the resulting cellular deformation using a cantilever. Middle/deep zone chondrocytes were isolated from articular cartilage, which was harvested from the distal metatarsal joint of one- to two-year-old heifers. The shape of the chondrocytes was approximated to be disc-like so that analytical data analyses could be made. Experimental data were fitted using three models. The linear elastic solid model gave a Young's modulus of 2.6 ± 0.8 kPa. The viscoelastic model (adapted from the Kelvin model described in Equation (8)) yielded an instantaneous modulus of 2.5 ± 0.8 kPa, a relaxed modulus of 1.5 ± 0.4 kPa and an apparent viscosity of 1.9 ± 1.8 kPa. Leipzig and Athanasiou (2005) also used a linear biphasic model to analyse the data, which was developed by Mow et al. (1980) to describe interstitial fluid flow-dependent viscoelastic responses of hydrated soft tissue, and obtained an aggregate modulus of 2.6 ± 0.9 kPa, a permeability of $2.6 \pm 3.1 \times 10^{-12}\,\mathrm{m^4 N^{-1} s^{-1}}$ and a Poisson's ratio of 0.07 ± 0.02. However, direct measurement of the volume change of single cells gave a Poisson ratio of 0.26 ± 0.08 (Shieh and Athanasiou, 2006). Nevertheless, the three models seemed to give similar values of the Young's modulus. It was found that chondrocytes taken from the superficial zone were significant stiffer than those from middle/deep zone, which might be due to the different mechanical environments of the cells *in vivo*.

Chondrocytes surrounded by a layer of pericellular matrix (PCM) are called chondrons. Understanding the mechanical properties of the PCM is also important in investigating the biological responses of chondrocytes to mechanical stimulation since PCM determines the mechanical environment of the chondrocytes, and may serve as a transducer to amplify strains, while protecting the chondrocytes from damage. However, little data are available (Choi et al., 2007) and research is required in this area.

3.1.3 Yeast and bacterial cells

Yeast cells are generally considered less susceptible to mechanical damage due to the presence of a cell wall, which maintains the cell shape and provides protection from physical stresses (Klis et al., 2006). Baker's yeast cells (*S. cerevisiae*) have been widely used to produce many biological products. If these products cannot be engineered to be secreted from the cells, disruption is necessary to release the intracellular contents. A common unit operation for this purpose at industrial scale is high-pressure homogenisation (Middelberg, 1995). Modelling and optimisation of this process has been restrained by a lack of understanding of the mechanisms of cell breakage, which has arisen because of a lack of cell mechanical property information. However, Mashmoushy et al. (1998) demonstrated that such information could be obtained by compression testing by micromanipulation. Initially it was only possible to measure the bursting force of cells where a mean value of 101 ± 2 N was measured from stationary phase Baker's yeast. The yeast cells were significantly stronger than hybridomas, presumably because of the presence of a cell wall.

A mathematical model was developed by Smith et al. (1998) to extract data on the elastic Young's modulus and some criteria of failure of stationary phase Baker's yeast from compression testing data. A mean Young's modulus of 150 ± 15 MPa with a corresponding mean von Mises strain at failure of 0.75 ± 0.08 and a mean von Mises stress at failure of 70 ± 4 MPa (Smith et al., 2000b) were obtained.

It was also shown by Smith et al. (2000a) that the bursting force was compression rate independent for speeds of 1.03–$7.68\,\mu m\,s^{-1}$, which suggests little cell wall viscoelasticity or water loss from the cells during compression. However, separate squeeze-hold experiments using longer compression times have shown some relaxation does occur (Mashmoushy et al., 1998). Direct video measurements of cell volume might elucidate the reasons for this relaxation, although clear imaging of small yeast cells is yet to be achieved.

In homogenisation studies (Kleinig, 1997), it was discovered that larger cells were more susceptible to breakage, which suggested that the forces on these cells in the homogeniser were greater or that their ability to resist disruption was lower. This result supported the pressure gradient theory of cell disruption proposed by Kleinig (1997).

Besides compression testing, AFM has been used to derive intrinsic material properties of yeast cell walls by nano-indentation. This technique was employed by Touhami et al. (2003) to determine the

Young's modulus of different areas of the Brewer's yeast cell wall (*S. cerevisiae*). In this work, it was found that chitin bud scars had a Young's modulus of 6.1 ± 2.4 MPa, which is an order of magnitude higher than the value for the surrounding cell wall area of 0.6 ± 0.4 MPa. It is not entirely clear why these moduli are much lower than the global values found by Smith et al. (2000b), although it may be that nano-indentation by AFM gives surface-specific material properties.

The ability to derive the intrinsic mechanical properties of yeast cells should now allow the mechanisms of high-pressure homogenisation to be determined unambiguously, and *a priori* predictions of the extent of cell disruption to be made for given homogeniser conditions. This should allow better process optimisation.

Bacterial cells are generally much smaller than yeast cells, and are often not even approximately spherical. Figure 12 shows an *Escherichia coli* cell positioned beneath the end of a probe ready for compression. A typical force-deformation trace is given elsewhere, with a bursting force of approximately 4 µN (Shiu et al., 1999). Modelling compression data from such cells remains an unsolved challenge, and the poor quality of the images will probably prevent accurate, direct, volume estimations. Nevertheless, it may eventually be possible to produce intrinsic material properties from compression testing data and predict their breakage in high-pressure homogenisation.

In addition to using compression testing, the Young's modulus of bacterial macrofibers of *Bacillus subtilis* has been determined to be 50 MPa using the optical trap method (Mendelson et al., 2000).

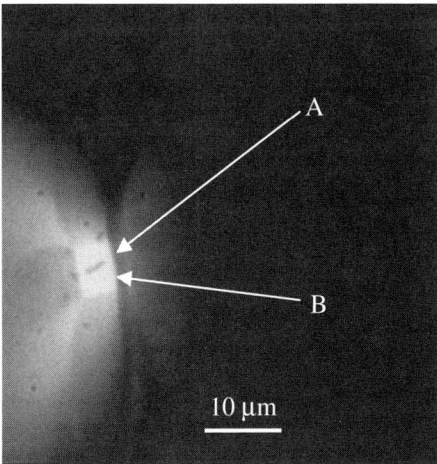

Figure 12 An *E. coli* cell positioned beneath the squared end of a probe ready for compression: (A) cell and (B) probe. Photograph by courtesy of C. Shiu. Permitted by Springer.

3.1.4 Filamentous microorganisms

By using glues, it is possible to attach a filamentous microorganism to two probes to allow it to be stretched by micromanipulation to breakage. Such a method was developed by a student of the authors (A. D. Roberts) and was used to determine directly the tensile strength of single fungal hyphae, specifically of *Fusarium graminearum*. Samples, which were taken from early and late stages of growth and from the stationary phase of a shake flask culture were tested both live, and after heat shock treatment (73°C for 10 min). Live hyphae were weakest (approximately 40 μN) during the late stages of growth, and strongest (approximately 70 μN) during the stationary phase. These forces correspond to Young's moduli for the wall materials of the order of 50–100 MPa, assuming linear elastic behaviour and assuming reasonable values for hyphal diameter and cell wall thickness. Heat-treated, single hyphae were significantly stronger (by approximately 8 μN) than those tested live. If this technique could be applied to other fungi, it might be possible to investigate the mechanisms of agitation damage to fungi in submerged cultures, for example, with antibiotic fermentations. However, many fungi are heavily aggregated in such cultures, and this might prevent unambiguous strength and strain measurements.

A similar technique has been used to investigate the elasticity and mechanical strength of the commercially important erythromycin-producing filamentous bacterium *Saccharopolyspora erythraea* where the mean Young's modulus was estimated to be 100 ± 20 MPa (Stocks and Thomas, 2001).

In addition to using the tensile test, the Young's moduli of filamentous microoganisms have also been determined using AFM to compress the wall, followed by mathematical modelling of the deformation using finite element analysis. For example, the Young's modulus of the cell wall of *Aspergillus nidulans* has been found from AFM testing to be 110–200 MPa (Zhao et al., 2005).

3.1.5 Plant cells

Compression testing by micromanipulation has been used to investigate the mechanical properties of tomato cells, both from suspension cultures and isolated single fruit cells.

In the case of suspension cultures, the mechanical properties of 2-week-old root callus cells of tomato (*Lycopersicon esculentum*) were investigated. These cells are approximately 70 μm in diameter, which allowed a clear side view of the compression to be recorded. Bursting forces were approximately 5 mN, at a compression rate of $23\, \mu m\, s^{-1}$ (Blewett et al., 2000). However, squeeze-hold experiments show a relaxation of the holding force, with a time constant of the order of seconds (Figure 13). This is consistent with water loss from the cells,

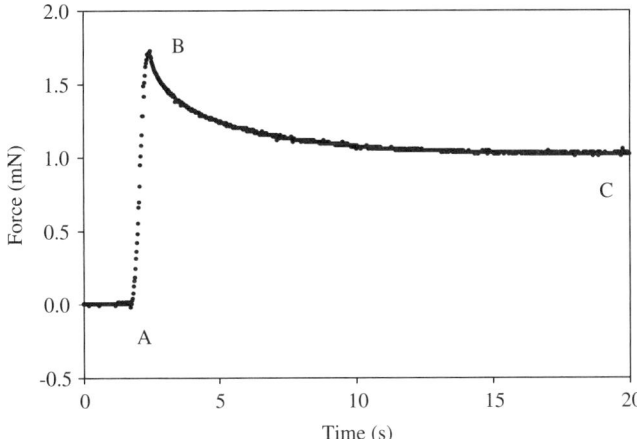

Figure 13 Force versus sampling time for compressing (curve AB) and holding (BC) a single tomato plant cell. The plant cell was 70 μm in diameter and was compressed at 23 μm s^{-1} (reproduced from Thomas et al., 2000). Permitted by Springer.

as expected from pressure probe experiments on other plant cells (Tomos, 2000). Image analysis has shown that water loss does occur (for the relaxation shown in Figure 13, approximately 2% of the volume at the beginning of the holding phase). Wang et al. (2006b) also used the pressure probe to determine the internal turgor pressure of suspension tomato cells. This technique was combined with compression testing to obtain independent turgor (hydrostatic) pressure measurements of cells, to show the pressure within a cell can indeed be found from Equation (29).

Data from compression testing of two-week-old suspension-cultured tomato cells at low strains has been modelled using the analytical elastic membrane model described earlier (Equations (35a)–(35c)). A mean elastic modulus of 2.3±0.2 GPa was found (Wang et al., 2004). The cell wall was assumed *a priori* to be linear elastic in nature and the rate of compression was assumed to be sufficient for water loss from the cell during compression to be considered negligible. This result is significantly different to that found for single fruit tomato cells where Young's moduli of 30–80 MPa were obtained when applying the same assumptions (Wang et al., 2006a). It is not surprising that this difference exists between growing cells from culture and mature fruit cells.

As a single cell technique capable of achieving large deformations, the method complements the pressure probe method. It should be valuable in studies of how plant cell mechanical properties are affected by the molecular composition and structure of the wall, and how these are affected by food processing operations.

3.2 Biocompatible particles

Biocompatible microparticles (spheres or capsules) have been studied as carriers for entrapment of food (Gibbs et al., 1999), drugs (Aslani and Kennedy, 1996; Gonzalez-Rodriguez et al., 2002; Ribeiro et al., 1999), macromolecules (DeGroot and Neufeld, 2001; Kikuchi et al., 1997; Vandenberg et al., 2001) and cells (Read et al., 1999; Strand et al., 2000; Torre et al., 2000) or artificial organs (Chang, 1999). Microparticles for such applications should have not only the desired permeability but also adequate mechanical strength because they may be exposed to mechanical forces generated at the sites where they need to function, such as compression in cartilage and bone, tension in muscle and tendon, and shear forces in blood vessels. For example, microcapsules containing encapsulated islets, acting as an artificial pancreas, should have sufficient strength to prevent premature breakage in a patient's body after implantation. Besides this, the release of active ingredients from microcapsules may also be triggered by mechanical signals (Kuen et al., 2001).

Alginate and chitosan are natural polysaccharides with excellent biocompatibility and biodegradability, and can form hydrogel or polyelectrolyte complexes under very mild conditions. In particular, water-soluble sodium alginate droplets can react with calcium chloride to form insoluble calcium alginate microspheres, which can be coated with a layer of chitosan. The mechanical properties of single calcium alginate microspheres can be characterised by diametrical compression. Microspheres bigger than 500 µm may be tested using commercial equipment, such as a texture analyser (Rehor et al., 2001). However, measurement of smaller microspheres can only be carried out using the micromanipulation-based compression testing described earlier.

Single calcium alginate microparticles with diameters of 20–60 µm were compressed at various speeds to a given deformation and held, or compressed to rupture. The forces corresponding to the deformations were measured by a force transducer (Zhao and Zhang, 2004). The force imposed on these particles increased when they were compressed, but relaxed significantly when they were held. For alginate microspheres, the faster the compression speed, the greater the imposed force for a given deformation. Calcium alginate microspheres coated with a shell of chitosan (called alginate–chitosan microcapsules with a shell thickness up to 11 µm) showed less force relaxation when they were held, compared with alginate microspheres. The thicker the shell, the less significant was the force relaxation exhibited by the microcapsules. The mean rupture force of alginate microspheres increased with compression speed, but in general this effect became less for alginate–chitosan microcapsules, and depended on the shell thickness. This is believed to result from the microspheres being the most permeable, with the permeability of the alginate–chitosan microcapsules decreasing with increasing shell thickness. However, the deformation at rupture was independent of the compression speed, which indicates it is not an elastic or viscoelastic parameter and may only be related to plastic failure. On average, the alginate–chitosan microcapsules were bigger than the alginate microspheres and had a greater rupture force.

To minimise speed-dependent behaviour, calcium alginate microspheres were tested at high speeds up to $1,000\,\mu m\,s^{-1}$, speeds that were achieved using a newly developed micromanipulation rig (Wang et al., 2005). High-speed video cameras also allowed measurement of changes in microsphere volume, caused by loss of liquid under compression. It was found that the reduction in microsphere volume was less than 10% for compression speeds of $800\,\mu m\,s^{-1}$ or higher. At lower speeds, or during relaxation on holding after compression, time-dependent phenomena became important. These phenomena have been modelled recently using finite element analysis based on the software package ABACUS (Nguyen et al., 2009). The force relaxation data were fitted by Equation (10) with $i = 2$, and it was found that the two relaxation times were $\tau_1 = 0.012\,s$ and $\tau_2 = 0.105\,s$. The values were used in subsequent modelling using ABACUS. Figure 14 shows a comparison of experimental force versus displacement/time data for compression and holding of single calcium–alginate microsphere with fitting by finite element analysis. As can be seen, there was a very good agreement, with regression coefficients all greater than 0.99. The model gave an instantaneous elastic modulus of E_0 of 450 kPa.

Modelling of the time-dependent phenomena may also be achieved using the viscoelastic model described in Equation (20). Figure 15 shows a typical force versus time curve for compression and holding of a calcium–shellac microsphere, compared with fitting by the model (Xue, 2008). The microsphere was produced by reaction of ammonium shellac solution with calcium chloride. The agreement can be seen to be very good. The instantaneous Young's modulus and long-time modulus were 1.56 and 0.54 MPa, respectively.

3.3 Non-biological particles

3.3.1 Microspheres

Non-biological microspheres are widely used in chemical, agrochemical, food, pharmaceutical and household products. Here two examples are described: microspheres used as chromatographic resins and pharmaceutical excipients used for tableting.

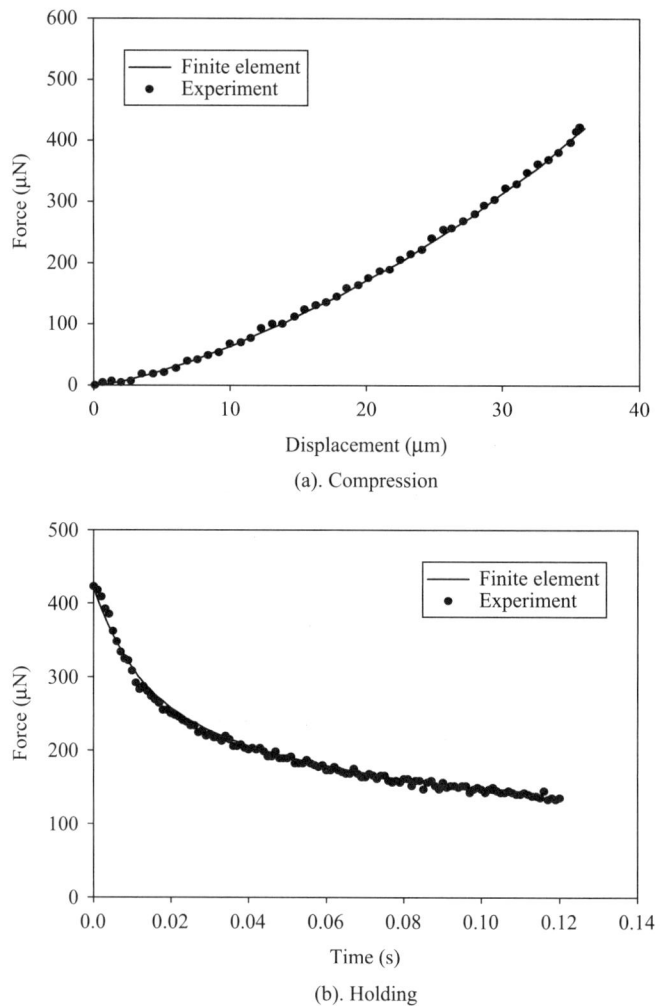

Figure 14 Comparison of experimental force versus displacement/time data with fitting by ABACUS. The microsphere diameter was 125 μm and compression speed 800 μm s^{-1} (Nguyen et al., 2009). Permitted by Elsevier.

3.3.1.1 Chromatographic resins.

Particulate chromatographic resins are additionally functionalised with ionic-, hydrophobic or affinity ligands, and can be used in downstream processing of biologics for purification of proteins and polynucleotides (Müller et al., 2005). These resins are mostly of polymeric nature and should have adequate caustic stability. The life-time of a chromatographic column, which partly depends on the mechanical stability of the resins, determines the overall processing costs to a great extent. Normally mechanically stable particles can be packed

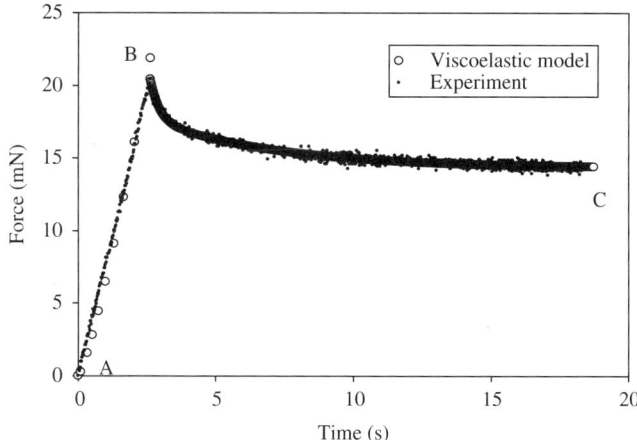

Figure 15 Comparison of force versus time data for compression of a calcium–shellac particle of 40 μm in diameter. The compression speed was $2\,\mu m\,s^{-1}$. Curve AB corresponds to compression and BC holding. The force relaxation data were fitted by Equation (10) with $i = 2$, $F_\infty = 14.3\,\mu N$, $F_1 = 8352\,\mu N$, $\tau_1 = 0.33\,s$, $F_2 = 6.0\,\mu N$ and $\tau_2 = 4.5\,s$ (Xue, 2008).

into a column easily and do not shrink nor swell during chromatographic cycles. Understanding the mechanical properties of chromatographic particles is essential to successful design and operation of such processing equipment. The mechanical properties of single chromatographic particles may be related to their chemical composition and any surface modifications.

Using compression testing, ion exchange resins were shown to be elastic for deformations up to 46% (see Section 2.5.1). To determine the mechanical properties, Tatara's model (Tatara, 1991, 1993) was used to fit the compression data, assuming a material of non-linear properties. The fit is shown in Figure 16. As can be seen, there is a good agreement between the experimental data and fitting by the Tatara model.

Micromanipulation has also been used to determine the mechanical properties of methacrylate-based resins with different surface functionalities (Müller et al., 2005). The particle sizes ranged from 30 to 90 μm, depending on their chemical composition. It was found that the stiffness of the resins depended on the specific chemical modifications, and the difference was more pronounced in the dry state than in the wet state.

3.3.1.2 Pharmaceutical excipients. Pharmaceutical excipients in powders mixed with an active ingredient are often compacted into tablets as the final dosage form (Jivraj et al., 2000; Kotte and Rudnic, 1995; Rubinstein, 2000), and these tablets should have adequate mechanical

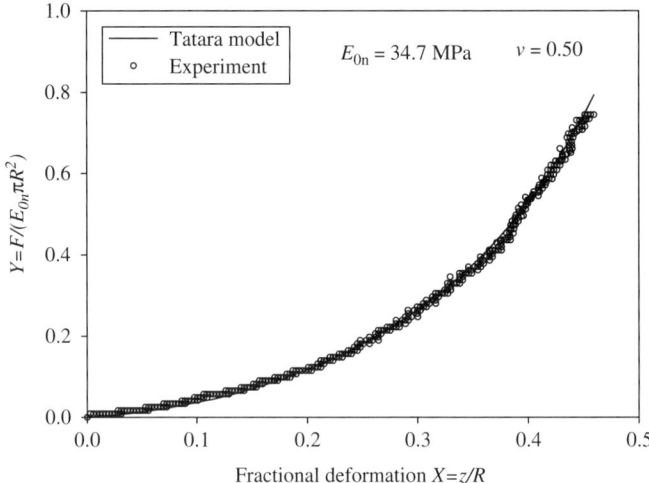

Figure 16 Comparison of the dimensionless force Y and fractional deformation of a single 163 μm diameter ion-exchange resin particle (DOWEX 1X8-200, Sigma-Aldrich, UK) obtained by diametrical compression and by numerical simulation using the Tatara non-linear elastic model. E_{0n} represents the initial Young's modulus at zero strain (data provided by Dr T. Liu).

strength to withstand the various handling operations in the logistic chain from producer to patient. Furthermore, a tablet should disintegrate and release the active ingredient to stomach or intestines at the right time in a reproducible manner (Kotte and Rudnic, 1995). To make rigid tablets, high-compaction pressures are often required. However, some of the active ingredients can be susceptible to high pressure and can be damaged by compaction. Therefore, it is important to understand the relationship between the mechanical properties of primary particles (<100 μm) and their compaction behaviour, to produce tablets with adequate strength at the lowest compression pressure.

During compaction, primary particles are packed, re-arranged and can undergo deformation and possibly breakage. These events can occur sequentially or in parallel. The mechanical strength of a tablet may strongly depend on the mechanical properties of the primary particles and the particle–particle interactions within it. It is essential that the particles deform plastically or rupture since the stored elastic strains can weaken the tablet on release (Roberts and Rowe, 1987).

The mechanical strength of single microparticles relevant to pharmaceutical applications has been measured using compression testing by micromanipulation (Yap et al., 2006). Pharmaceutical excipients, comprising three enteric polymer particles Eudragit® L100-55, Eudragit® L100

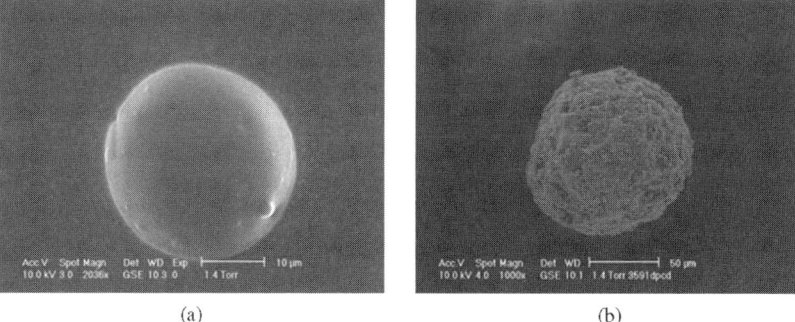

Figure 17 ESEM image of (a) Eudragit® L100-55 and (b) Advantose™ 100 (Yap et al., 2006). Permitted by Elsevier.

and Eudragit® S100 (Degussa Rohm, UK), and three different powders in the form of agglomerates, Advantose™ 100 (SPI Pharma, UK), Barcroft™ CS90 calcium carbonate (SPI Pharma, UK) and Starlac™ (Roquette, UK) were studied. Their diameters ranged from 20 to 90 µm. The enteric polymer particles and the agglomerates had distinct morphologies (Figure 17). The Eudragit® L100-55 particle exhibited a single rupture mode under compression by splitting into two roughly equal halves, as observed by a TV camera. A typical force versus compressive approach curve is presented in Figure 18a. However, particles of Advantose™ 100 showed a different mode of rupture. At the peak force, they split into many fragments. As shown in Figure 18b, there were a number of subsequent ruptures after the first major rupture.

Corresponding to small deformations of the particles, that is, within the elastic limit, the force versus displacement data were fitted by the Hertz model (Equation (1)), and the Young's moduli were determined to be 1.6 ± 0.2 GPa for Eudragit® L100-55 and 0.9 ± 0.4 GPa for Advantose™ 100 (Yap et al., 2006). Moreover, the hardness of a particle, H, could be obtained from the slope of the linear plastic region of the force-displacement curve at which it deviated from Hertzian behaviour, using the following relationship (Johnson, 1985):

$$F = 2\pi HRz \qquad (38)$$

where F, R and z are the imposed force, particle diameter and compressive approach, respectively. The hardness was found to be 95 ± 14 MPa for Eudragit® L100-55 and 191 ± 15 MPa for Advantose™ 100 (Yap et al., 2008).

The same particles were compacted into a tablet using pressures up to 60 MPa (Yap et al., 2006), and pressure versus particle bed displacement data were obtained. It was found that the nominal rupture stress

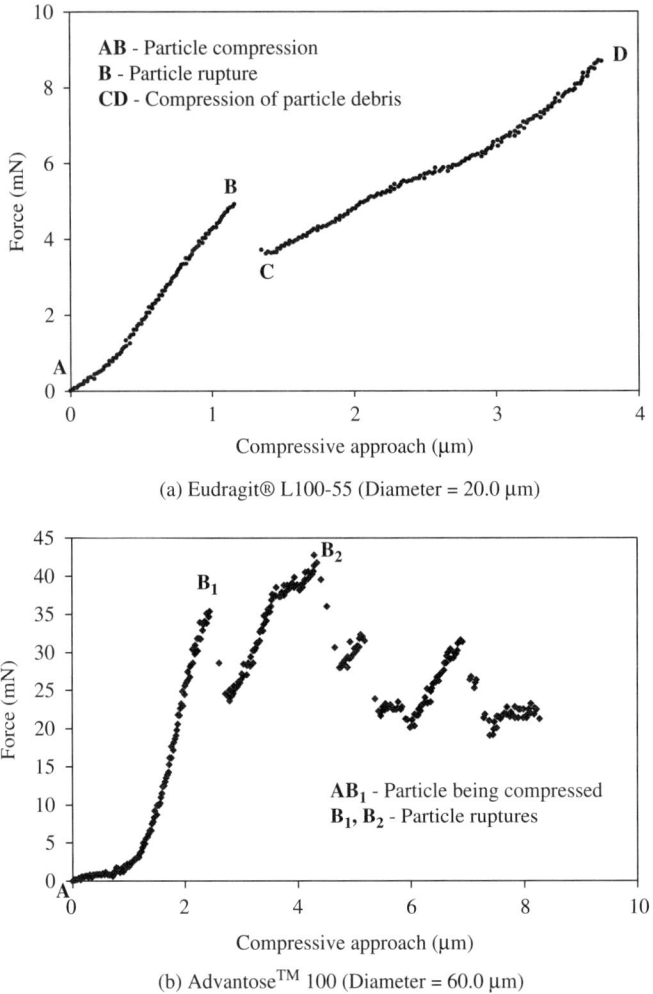

Figure 18 Typical force-compressive approach relationship for the compression of a single particle to rupture (reproduced from Yap et al., 2006). Permitted by Elsevier.

(the ratio of the rupture force to the initial cross-sectional area) of the primary microparticles measured by compression testing was correlated with the compaction behaviour of the particles under confined uniaxial compaction. Single tablets were crushed using a Brazilian test, and scanning electron microscopy images of a crushed tablet were obtained (see Figure 19). As can be seen, the breakage occurred at the interface between feed particles, which implies that the crushing strength of the tablets was determined by the interfacial bonding rather than the mechanical properties of primary particles.

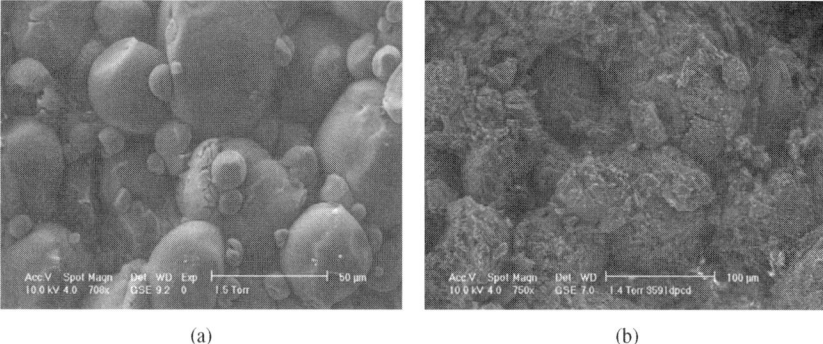

(a)   (b)

Figure 19 Environmental SEM image of (a) Eudragit® L100-55 and (b) Advantose™ 100 tablet in the fracture region after diametrical compression (reproduced from Yap et al., 2006). Permitted by Elsevier.

3.3.2 Microcapsules

Capsules are basically particles with a core-shell structure, although the category also includes other structures, for example, spheres with an active ingredient embedded, a core with multiple shells, or a matrix with multiple embedded cores. Capsules for typical industrial applications can vary from several microns to millimetres in diameter, and these are usually called microcapsules. However, there is an increasing interest in preparing nanocapsules, of sizes less than one micron. Capsules containing active ingredients are widely used or have potential applications for producing household products (e.g., soaps, detergent and brighteners), personal care (health and beauty) products, printing and imaging products (carbonless copy paper, inks, toners, colorants), adhesives, sealants, paints, catalysts, antifouling agents and agrochemicals (herbicides, insecticides, fertilisers, repellents), in enhancing functions of food and feed (aromas and flavours, preservatives, nutrients, antioxidants), and in pharmaceutical and biomedical products to realise sustained (long lasting) drug delivery, vaccine delivery, gene therapy and active drug targeting (Arshady, 1999).

Active ingredients may be encapsulated by various processes, including *in situ* polymerisation, interfacial polymerisation, coacervation, spray drying, extrusion and fluidised bed coating (Madene et al., 2006). There are many different purposes of encapsulation, including stabilisation of an active ingredient, control of its release rate and conversion of a liquid formulation to solid. Capsules should have appropriate physical, mechanical and structural properties, possibly including particle size, size distribution, morphology, surface charge, wall thickness, mechanical strength, glass transition temperature and degree of crystallinity, flowability and permeability (Le Meste et al., 2002). Information about the mechanical properties of capsules is very important to understand their behaviour in different environments, including manufacturing processes and end-use applications. For example, capsules for most industrial applications should be strong enough to withstand the various mechanical forces generated in manufacturing processes, such as mixing, pumping and extrusion but may be required to be weak enough to release the encapsulated active ingredients by mechanical forces at end-use, such as the release of flavour by chewing. The mechanical strength of microcapsules and the release of their active ingredients, are related to microcapsule size, morphology, wall thickness, chemical composition and structure.

Direct methods of determining the mechanical properties of microcapsules include compression of a layer of them between two glass plates. Microcapsule mechanical strength has been characterised by the number of them being broken under a given weight applied to the top plate, but it was observed that the force was applied onto the largest microcapsules first, causing them to break, followed by the smaller ones (Ohtsubo et al., 1991). Although this method is practically useful, it conceals any difference in mechanical strength between microcapsules within a sample. Despite being more time consuming, results attained from assessment of single microcapsules is more accurate and reliable, and can give useful information about sample inhomogeneity (Schuldt and Hunkeler, 2000). Direct methods on single microcapsules include the use of a micropipette aspiration technique or an atomic force microscope probe to measure the elastic properties of single microcapsules. Unfortunately, the former technique cannot be used to determine the force required to rupture the microcapsules (Grigorescu et al., 2002), whereas the latter relies on compression of single microcapsules between a rigid spherical bead and a flat surface (Lulevich et al., 2006), which is difficult to implement. An uniaxial compression of single microcapsules has also been attempted by means of a texture analyser consisting of a penetrometer with a stress gauge (Edwards-Levy and Levy, 1999). This provided a measure of a particle's resistance to compressive force. However, assessment of single microcapsules was often impossible when the size was in the micron range (Martinsen et al., 1989). This limitation may be overcome by using compression testing by micromanipulation because this offers the capability to obtain force versus deformation data for single microcapsules as small as 1 μm, including to rupture.

The micromanipulation technique has been used to measure the mechanical strength of microcapsules of different size, shell thickness and shell composition (Sun and Zhang, 2001, 2002; Xue and Zhang, 2008; Zhao and Zhang, 2004; Zhang et al., 1999). For example, the mechanical properties of single melamine formaldehyde (MF) microcapsules with diameters of 1–12 μm were determined, including their viscoelastic and elastic-plastic properties (Sun and Zhang, 2001). It was found that the microcapsules were mainly elastic up to a deformation of $19\pm1\%$. Beyond this point, the microcapsules underwent plastic deformation and were ruptured at a deformation of $70\pm1\%$. However, the corresponding deformations at the yield point and at the rupture of urea–formaldehyde microcapsules were $17\pm1\%$ and $35\pm1\%$, respectively, which implies that urea–formaldehyde microcapsules were more brittle than those made of MF (Sun and Zhang, 2002). Besides the shell composition, the rupture strength of these microcapsules depended on their size and shell thickness. From compression measurements, the rupture force of different microcapsules can be compared by extracting information from a force-displacement curve up to rupture. The curves can also be fitted with equations derived from a theoretical model, for example, Lardner and Pujara's model (Lardner and Pujara, 1980; Liu et al., 1996). This approach has been used to determine intrinsic mechanical property parameters of microcapsule walls, for example, the Young's modulus, and in further mathematical modelling to determine viscous-elastic and plastic parameters where appropriate.

Liu et al. (1996) used Lardner and Pujara's model to estimate the Young's modulus of polyurethane microcapsules in a size range of 50–100 μm. Both neo-Hookean and non-linear elastic constitutive equations based on rubber-like Mooney–Rivlin materials were used to represent the stress–strain relationships, and it was found that the extensional rigidity of the microcapsule shell Eh (product of the Young's modules E and the shell thickness h) was approximately $540\,\mathrm{N\,m^{-1}}$ and the Young's modulus itself was 2.7 MPa, if the shell thickness was assumed to be 2 μm. However, the model could not be used to describe any plastic behaviour observed at large deformations. A similar approach has been taken to characterise the mechanical properties of urea formaldehyde microcapsules containing a self-healing liquid monomer dicyclopentadiene (DCPD) (Keller and Sottos, 2006). It was found that the mean microcapsule shell modulus was 3.7 ± 0.2 GPa, and this did not depend on whether the microcapsules were dry or immersed in the monomer. The microcapsules clearly showed plastic behaviour at large deformations, which was not modelled.

The mechanical properties of single hydrated dextran microcapsules (< 10 μm in diameter) with an embedded model protein drug have also been measured by the micromanipulation technique, and the information obtained (such as the Young's modulus) was used to derive their average pore size based on a statistical rubber elasticity theory (Ward and Hadley, 1993) and furthermore to predict the protein release rate (Stenekes et al., 2000).

4. OTHER APPLICATIONS

4.1 Particle–particle adhesion

In industry, aggregates are often produced from aqueous solutions to aid processing or to form solid products. Examples of such operations include fermentation, flocculation, precipitation, suspension crystallisation and preparation of stable dispersions, for example, in the production of cement, china clay, paints and pigments (Shamlou and Titchener-Hooker, 1993). The mechanical stability of the aggregates, which is related to their mechanical strength, is a key factor in governing the success of such operations. Conceptually, the aggregate strength (e.g., the compressive or tensile strength) of an aggregate depends on the size and shape of the primary particles, the packing factor, inter-particle bonding strength, and the size, shape and porosity of the aggregate (Jiang and Logan, 1991). The porosity of an aggregate may be related to the packing factor, the shape factor and the fractal dimension of the aggregate, which may be quantified by image analysis, Coulter counting and laser scattering (Logan and Kilps, 1995; Tang et al., 2000; Zhang and Buffle, 1996). Consequently, information regarding the inter-particle bonding strength is crucial to understand the relationship between the properties of primary particles and those of the aggregates. Experimental techniques to characterise particle–particle adhesion include laser trapping (Sugimoto et al., 1997), micropipette aspiration (Fairbrother and Simmons, 1998; Fan et al., 2003; Yeung and Pelton, 1996) and micromanipulation (Fan et al., 2003; Willett et al., 1997).

The interactive force between two polystyrene particles of 2.13 μm in aqueous solution was measured using laser trapping by Sugimoto et al. (1997), who found that the force was a function of the separation distance between the two particles, in a good agreement with predictions by conventional DLVO theory (Castelain et al., 2008). The measured forces were of the order of several pico-Newtons.

The tensile strength of flocs formed by aggregating aqueous precipitated calcium carbonate sols with two water-soluble polymers has been measured by a micromechanical technique based on pulling apart single floc particles using two glass micropipettes (Yeung and Pelton, 1996), one of which worked as a cantilever. The stiffness of the cantilever had been pre-calibrated. The flocs had sizes ranging from 6 to 40 μm in equivalent diameter. Their primary particles were cigar-like in shape and had an average size of 1.34 μm. The tensile strengths ranged from 20 to 200 nN and showed no correlation with floc size. The tensile strengths were related to the fractal dimension and rupture behaviour of the flocs in processing equipment. A similar technique was adopted to measure liquid bridge forces between spherical agglomerates, and a model was developed to predict bridge rupture energies (Fairbrother and Simmons, 1998).

The interparticle forces induced by a liquid film layer between a pair of glass particles of near spherical shape has been measured by a micromanipulation technique (Willett et al., 1997). The two particles of 1.13 mm in diameter were glued to a force transducer probe and glass slide, respectively. The particles were coated with thin layers of silicone oils. The particles were first brought together until the particle surfaces just touched, followed by separation at a fixed velocity. A liquid bridge was formed between the particles. Complete traces of the force–separation relationship were achieved, and the interparticle forces were in very good agreement with a simple theoretical model based on the work of Adams and Perchard (1985).

Measurement of the adhesion between a pair of particles at temperatures from ambient to 580°C was carried out to understand particle processing in high-efficiency coal combustion systems, the integrated coal gasification combined cycle and pressured fluidized bed combustor systems (Masuda et al., 2004). In this work, the adhesive force between fly ash particles from a municipal solid waste incinerator was measured directly using micromanipulation, with the particles being heated on a hot stage under the microscope. The sizes of the coagulating particles ranged from 100 to several hundreds of microns. The results indicated that the adhesion increased with temperature. Moreover, the adhesive forces between particles taken from the incinerator were found to be significantly larger than those from biomass and coal combustors. Thermo-analysis and microscopic evidence suggested that the mechanism of the increasing adhesive force with temperature might be due to formation of liquid bridges at contact points between particles.

Interactions between micrometre-sized ice particles are interesting to researchers who study the preservation of frozen foods (Fan et al., 2003). Work in this area includes understanding the physical mechanisms of ice adhesion, measuring the adhesive strength of ice on surfaces and investigating new materials on which ice adhesion can be reduced. A micromanipulation apparatus similar to the one used for ash particles was used to measure the adhesive force between ice particles in air and sucrose solution. Ice particles were generated in the chamber of a heating/cooling stage under a microscope. The sizes of the ice particles ranged from 10 to several hundreds of microns. The adhesion forces between ice particles increased with their size and contact time, and the adhesion between ice particles in air was significantly stronger than that in sucrose solution. The findings were interpreted in terms of a simple model based on JKR theory (Johnson, 1985).

For completeness, it should be mentioned that laser trapping has been used to quantify the unbinding force between an actin filament and a single motor molecule of muscle (Nishizaka et al., 1995), the isometric forces generated by single kinesin molecules (Kuo and Sheetz, 1993) and the binding forces between *E. coli* bacterial adhesion and galabiose-functionalized beads (Fallman et al., 2003).

4.2 Particle adhesion to a surface

Understanding particle adhesion to a surface has applications in tissue engineering and particle processing. Experimental techniques for charactering particle adhesion to surfaces include laser trapping, AFM and microscopy with force measurement.

The interaction and attachment of human bone cells (human gingival fibroblasts and bone-forming human osteoblast cells) to different types of medical implant materials including glass, titanium and hydroxyapatite have been investigated to facilitate cell attachment and promote migration of progenitor cells to decrease healing times (Andersson et al., 2007). The cells were in contact with each surface for 30–120 s before they were pulled away using the optical trap. Twenty cells were tested on each surface type. It was found that most cells had an adhesion greater than 40–50 pN, exceeding the maximum force that could be measured. Some cells did not adhere, and no adhesive force could be detected. However, some cells had a relatively weak and measurable adhesive force in the range of 5–10 pN. Unfortunately, there was no significant difference in the force between different surfaces, or between the cell types. It seems the data were very scattered, and more cells should have been tested to give statistically representative results. A similar technique has been used to measure forces between a colloidal particle and a phospholipid bilayer (Sharp et al., 2006). Equilibrium and viscous force–distance profiles of silica microspheres of 1–5 μm in diameter on bilayers of dipalmitoyl phosphatidyl choline (DPPC) and on bare mica and DPPC monolayers under same electrolyte conditions were obtained, and they varied significantly with the bilayer.

AFM has been used to measure particle–surface interactions of lactose (Sindel and Zimmermann, 2001), which is a typical excipient for drugs formulated as solids. Information about the interaction forces between particles of crystalline lactose can help to understand and improve their flow properties. Owing to technical difficulties in handling a pair of particles, the interactions between a single particle of crystalline and a tablet of lactose were investigated, rather than studying a pair of particles. It was found that the average adhesion was in the order of 5 nN, which depended on the surface roughness and porosity of the particles. In addition, AFM allowed detailed scans of the substrate surface and could provide quantitative data about the surface roughness (Sindel and Zimmermann, 2001). AFM has also been used to measure the pull-off forces between glass microspheres and flat glass or silica surfaces, used as model systems for studying the behaviour of cohesive powders as a function of relative humidity (Jones et al., 2002). The glass and silicon substrates were treated with either a hydrophobic or a hydrophilic coating. It was found that for the hydrophilic surfaces the pull-off forces increased with relative humidity in the range of 5–90%, and depended on the roughness and asperity contacts at the interface. However, the pulling force did not change significantly with the relative humidity for hydrophobic glass. The results were explained by the theories of capillary bridge formation (Israelachvili, 1992), and could be used to interpret the behaviour of cohesive powders with different coatings or those which show a large humidity dependence (e.g., zeolites).

Recently, the adhesion of MF microparticles on a cellulose film in air as well as in liquid media was characterised using AFM. The cellulose film was made by dissolving cotton powder in N-methylmorpholine-N-oxide (NMMO) solution, followed by spinning on a silicon wafer. Spectroscopic ellipsometry was employed to measure the film thickness, and AFM was also utilised for characterising the film roughness and material distribution (Figure 20). The cotton cellulose film was also

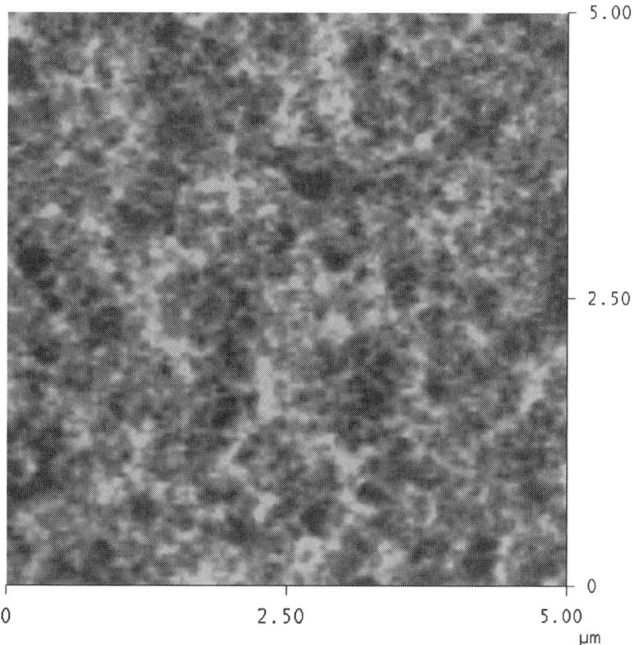

Figure 20 AFM image (tapping mode, height image) of a dry cellulose film. The cellulose concentration was 0.5%w/w. The root-mean-square surface roughness was 5.0 nm (courtesy of M. Liu, University of Birmingham, UK).

analysed with X-ray photoelectron spectroscopy (XPS) to make sure the film was free from residual solvent and confirm the presence of cotton material. It was found that the adhesion in air was in a range of 100–1,000 nN, whereas in water the adhesion was reduced to less than 4 nN.

Liu et al. (2002a, 2002b) developed a theoretical model to estimate adhesion energy between a microcapsule and a flat glass substrate, used as a model system to study cell–substrate interactions. It was found that the adhesion energy was related to the Young's modulus of the particle and the contact area. The Young's modulus of the microcapsule was determined independently by diametrical compression and analytical modelling based on a linear elastic model. The contact area of a urea formaldehyde microcapsule on a hydrophilic-fused silica substrate was measured using high-resolution reflection interference contrast microscopy and it was found that it increased with the osmotic pressure of the suspension liquid. This was due to a reduction in the microcapsule volume. The contact area did not change significantly with temperature in a range of 22–50°C when there was no phase transition of the capsule wall. The adhesion energy was in a range of 0.01–$1.0 \, \text{mJ m}^{-2}$ depending on the osmotic pressure.

4.3 Fouling deposits on surfaces

4.3.1 Biomass and biofilms

Fouling of surfaces by biomass and biofilm formation is often detrimental, although in some cases it is encouraged, for example, for cell immobilisation in bioprocessing. Biofilm formation in industrial water systems can cause significant energy losses due to the increase in frictional resistance through the system, or cause microbial contamination in food processing. The biofilm usually consists of bacterial cells embedded in a network with tangled fibres of exopolysaccharides (EPS). Attachment of a fouling deposit to a substrate is termed adhesion, and molecular interactions within the deposit are defined as cohesion. Understanding the adhesive and cohesive properties of biofilms is crucial for effective control of biofilm growth or removal in industrial water systems, or in ensuring proper hygiene in food processing equipment.

To measure the adhesion strength of bacteria, it is necessary to remove them from the surface. Weiss (1961) measured bacterial adhesion by allowing cells to settle onto a glass surface of a sealed chamber, and then counting them with the aid of a microscope. After a period of incubation the chamber was turned upside down, the unattached cells fell from the surface and the remaining attached cells were recounted. This adhesion number method is purely observational, as it does not measure adhesion directly. Weiss also described a disc-shearing device, which employed a static disc with cells attached and a second disc spinning above the attached cells. The resulting shear stress was dependent on the rotation rate of the disc, the separation distance, the fluid velocity and the radial position. Christie et al. (1970) used a water jet impinged vertically onto the test surface at a fixed velocity to achieve similar ends. Since then there has been a number of modifications to such shearing techniques (Bryers, 1987). Recently, bacterial adhesion measurement has been improved by sophisticated techniques such as micromanipulation and AFM (Boyd et al., 2002; Chen et al., 1998). Micromanipulation can be used to characterise adhesion and cohesion of a layer of fouling deposit on substrate, whereas AFM is very powerful in measuring single cell-substrate adhesion.

Using micromanipulation, a technique has been developed to measure the mechanical properties, including the adhesive strength, of *Pseudomonas fluorescens* biofilms grown in pipe flows (Chen et al., 1998). A T-shaped probe was specially designed to pull the biofilms away from the inner surface of a pipe to which they were attached. The adhesive strength between a biofilm and the substratum was defined as the work required per unit area to remove the biofilm. It was seen that the biofilms exhibited viscoelastic behaviour. The adhesive strength was found to depend on the conditions under which the biofilm was grown. Increases in the fluid velocity, the concentration of suspended cells and the roughness of the attached surfaces resulted in greater biofilm adhesion, whereas pH did not appear to have any significant effect (Chen et al., 2005).

This technique was also used to assess the adhesive integrity of immobilised bacterial populations (biomass) of *P. fluorescens*, harvested at different growth times and then placed on a stainless steel substrate (Garrett et al., 2008). Each sample was tested in a flow chamber. After the biomass was exposed to the flow for a specific period, the sample was removed from the chamber and the apparent adhesion of the remaining biomass was measured using micromanipulation. The surface area of the substrate covered by the biomass was monitored by a digital camera and quantified by image analysis (Garrett et al., 2008). The results indicated a strong correlation between micromanipulation measurements of the adhesive strength and these flow chamber experiments. Both showed that the apparent adhesive strength of the biomass harvested from a given volume of cell suspension increased with growth time, which may be attributed to the increase in viable cell number. Moreover, the cohesive strength of the biomass was also characterised by measuring the force required to break the biomass layer by layer (Garrett et al., 2008). It was found that the apparent adhesive strength was greater than the cohesive strength. This implies that in cleaning by fluid flow or similar processes, the biomass would be removed from the top layer downwards. This has been validated by observations in the flow chamber. In general, the biomass adhesion onto a substrate was found to be much weaker than biofilm adhesion, suggesting a reduced presence of EPS in the biomass. Using these techniques, specific mechanisms of biomass detachment from a surface and optimised cleaning strategies may be tested.

The data produced using micromanipulation can be compared with data acquired using alternative techniques such as AFM and the flow techniques. These techniques complement each other by broadening the measurement range of the forces required to remove biomass/biofilm of different scales, that is, from single bacteria to a layer of fouling deposit. For example, micromanipulation can measure forces within the micro-Newton range, which is relevant to removal of a layer of deposit from substrate, whereas AFM can measure forces within the nano-Newton range, relevant to single cells. Flow cell techniques can be used to observe the mechanical and cohesive properties of biofilms, but cannot be used to measure directly the forces required to remove biomass/biofilms from a surface. As micromanipulation allows the measurement of adhesive and cohesive properties of biofilms directly, it is hoped that this technique can be used to fill the gap left by the limitations of AFM and flow cell devices.

4.3.2 Food fouling deposits

Food process plant rapidly becomes fouled with deposits on the interior surfaces of process equipment (Wilson et al., 2002). This may be caused by crystallisation of insoluble components, such as calcium phosphate in ultra-high-temperature milk fouling, reaction, for example, the polymerisation of β-lactoglobulin in milk or gelation of starch, and the growth of biofilms (Watkinson et al., 2004). Fouling deposits lower process efficiency through increased pressure drop and decreased heat transfer, and endanger both food quality and safety through potential product contamination. The problem of fouling is widespread. Consequently, cleaning-in-place (CIP) is a ubiquitous operation within the food industry, in which cleaning chemicals (e.g. up to 2% NaOH) followed by rinsing water are circulated, at temperature up to 70°C and high flow rate. This is expensive in time and has a tremendous environmental impact. Moreover, cleaning protocols are empirical and rarely optimised.

The micromanipulation technique of Chen et al. (1998) for measuring the adhesive strength of biofilms has been modified by Liu et al. (2002c) and used to measure the force required to remove food deposits from surfaces of stainless steel. Four model food systems were used, that is, tomato and dough representing starchy materials, and whey and egg proteins (Liu et al., 2007). Using micromanipulation, the adhesive strength of deposits that have been immersed in a cleaning chemical or water can be monitored. The force required to disrupt and remove the deposit (usually in the range of 2–10 Jm^{-2}) can be measured directly.

The effect of process variables such as temperature, time and chemical concentration has been determined in such studies, and the forces and the balance between cohesion and adhesion can be directly related to the ways in which surfaces are cleaned. For example, tomato deposits appeared to be largely cohesive in that the force to break the deposit in half exceeded that for total removal. It would seem that cleaning in this case occurs by removal of deposits by shear stresses overcoming the forces between the deposit and the surface (Liu et al., 2002c). However, milk protein deposits appeared largely adhesive and the force required to remove the deposits increased with thickness. Here, cohesive forces between elements of deposit were weaker than adhesive force between the deposit and surface, and cleaning occurred by breakdown of deposit cohesion (Liu et al., 2006a). These results are in agreement with those from liquid jets (Chew et al., 2004).

The adhesive strength of a food fouling deposit may be related to the surface free energy of the substrate. Zhao et al. (2004) have developed a theory that gives the minimum adhesion energy between a deposit and a surface:

$$\sqrt{\gamma_{\text{surface}}^{\text{LW}}} = \left(\frac{1}{2}\right)\left(\sqrt{\gamma_{\text{foulant}}^{\text{LW}}} + \sqrt{\gamma_{\text{fluid}}^{\text{LW}}}\right) \qquad (39)$$

where $\gamma_{\text{surface}}^{\text{LW}}$, $\gamma_{\text{foulant}}^{\text{LW}}$ and $\gamma_{\text{fluid}}^{\text{LW}}$ are the Lifshitz–van der Waals surface free energies of the surface, foulant and fluid (e.g. water), respectively. Liu et al. (2006b) demonstrated that such a minimum adhesive strength (measured by micromanipulation) did exist for tomato paste deposits, at around $25\,\text{mN}\,\text{m}^{-1}$ predicted from Equation (39). However, the greater the thickness of the deposit, the greater the apparent adhesive strength and the lesser the apparent effect of the surface. The results indicate that cleaning depends on both adhesion between a deposit and surface and cohesion between its elements. This work may lead to increased food safety by improving and ensuring cleaning efficiency, minimising environmental impact by reducing the amount and concentration of effluents, and increased manufacturing efficiency through extensions of run length and minimisation of cleaning times.

4.4 Nanomanipulation of sub-micron/nanoparticles

Nanoparticles have numerous applications in the chemical, food, pharmaceutical, biomedical and semiconductor industries. For example, nanoparticles as drug carriers can increase drug efficacy, and can reduce toxicity and side effect after parenteral administration (Feng et al., 2002). Nanoparticles used for industrial applications should have desirable physical properties, including appropriate size, surface charge, surface area, porosity and mechanical strength. The functionality of nanoparticles, such as those used for drug delivery, may strongly depend on their mechanical properties. This has been demonstrated for hydrated dextran microspheres loaded with the model compounds myoglobin, ovalbumin, BSA and IgG. The (pseudo-)Young's modulus was correlated directly to the pore size of the particles and the release rate of the model drugs (Stenekes et al., 2000). In addition, it has been reported that controlled drug delivery from polymer carriers can be triggered by mechanical signals (Lee et al., 2001). Understanding the mechanical properties of single nanoparticles is essential to such applications.

To characterise the mechanical properties of a single nanoparticle requires the simultaneous determination of the force exerted on it and its deformation. This allows stress–strain relationships to be determined by modelling. Electron microscopy can be used to examine the surface morphology and structure of nanoparticles. Over the past few years, various tools have been developed for three-dimensional handling, assembly, characterisation and testing of fundamental building blocks like nanotubes and nanowires in scanning electron microscopes (SEMs), which include nanopositioning, nanomanipulation and microgripping devices (Eichhorn et al., 2007; Fahlbusch et al., 2005; Nakabayashi and Silva, 2007), but these devices did not have the capability to do force measurements. There have also been attempts to construct a manipulation device with a force sensor inside the chamber of an SEM (Yu et al., 1999) and an environmental scanning electron microscope (ESEM). Pioneering work on measuring the mechanical properties of materials in an ESEM includes *in situ* mechanical testing of fully hydrated carrots (Thiel and Donald, 1998), dry and hydrated breadcrumbs (Stokes and Donald, 2000), and the deformation of elementary flax fibres (Bos and Donald, 1999). However, in all these cases the samples were still relatively large, sometimes of the order of mm. A similar approach has been used for measurement of the fracture resistance of other brittle materials (Sorensen et al., 2001), but again the specimens were large.

Recently, an ESEM-based technique has been developed for characterising the mechanical properties of single nanoparticles by constructing and assembling a nanomanipulation device with a force probe in the chamber of an ESEM. This was used to demonstrate the feasibility of measuring the force required to compress single particles to a given deformation to infer their mechanical property parameters (Liu et al., 2005; Ren et al., 2007). This new nanomanipulation technique was first validated by comparison with a well established micromanipulation technique (Sun and Zhang, 2001) using both techniques to measure the force required to cause different deformations of single Eudragit microparticles. The validated nanomanipulation technique was then used to determine the mechanical properties of single polymethylmethacrylate (PMMA) nanoparticles. The technique has been

Figure 21 Images of a MF microcapsule before and after it was compressed to rupture under HV of ESEM: 5 kV, spot size 4. The microcapsule diameter was 16.5 μm (Ren et al., 2007). Permitted by Maney.

applied to determine the rupture mode of single melamine formaldehyde microcapsules under compression (Figure 21).

Very recently, *in situ* measurements of mechanical properties of individual W303 wild-type yeast cells have been made using an nanomanipulation system with an AFM cantilever probe (spring constant $0.02\,\text{N}\,\text{m}^{-1}$) integrated into the chamber of an ESEM (Ahmad et al., 2007). Single cells were penetrated under two different conditions, that is, ESEM (600 Pa) and high vacuum (HV, approximately 3 MPa) modes. Data show that the penetration forces under HV conditions were approximately 25 times those under the ESEM mode, that is, 4 μN rather than approximately 150 nN for a cell size of 5 μm, with a penetration distance of 1 μm. The corresponding Young's moduli were estimated to be 2.2 ± 1.2 MPa under the ESEM mode and 21 ± 2 MPa under the HV mode, both based on the Hertz model. This work demonstrates that the apparent mechanical properties of the cells strongly depended on the environmental conditions within the microscope.

Although SEMs or ESEMs are very powerful in imaging nanoscale materials or particles, caution should be taken to avoid electron beam damage to specimen. This is particularly important when a nanomanipulation system with a force measurement device is to be used to characterise the mechanical properties of particles. Ren et al. (2007, 2008) identified that such damage depended on the electron dose and exposure time, as well as the type of materials under test, and it is extremely important to find a time window in which the damage is negligible to obtain reliable mechanical property data.

5. PERSPECTIVES ON FUTURE DEVELOPMENT

A number of micromanipulation-based techniques have been developed and used to measure the mechanical strength of single particles, particle–particle interactions, particle–surface or film–surface interactions, which have been biological or non-biological in nature. Although it is powerful, micromanipulation is very technically demanding, time consuming and not easy to use. Semi- or full automation of micromanipulation equipment should enhance its popularity. The data obtained directly by micromanipulation are very useful in terms of understanding how the properties vary with morphology, structure and chemical composition of the particles or the surface involved. Analytical models or finite element analyses are available to extract intrinsic elastic property parameters of particles with relatively simple structures (e.g., spherical microspheres, capsules or cells). However, there has been no adequate modelling of the plastic deformation of such particles. For particles with more complex structures such as multiple shells, or several particles embedded in a single carrier, analytical modelling may well be too difficult or impossible, and this is where finite element analysis can and will play an increasingly important role.

Both biological and non-biological particles are often exposed to mechanical forces in processing equipment. Understanding and predicting deformation and breakage of particles in such equipment is important to improve process design and operation. However, although understanding the mechanical properties of the particles is essential, modelling of the interactions between the particles and fluid or other particles is also required. So far, there has only been limited work on such modelling.

One of the challenges to chemical engineers in the 21st century is to produce functional products to meet ever increasing demands for consumer care and healthcare, which requires formulation of particulate products with multiphases and complex structures with scales from micro to nanometres. Mechanical characterisation of particles at sub-micro scale is likely to become more and more important, and this will require miniaturisation of micro/nanomanipulation devices, force sensors and the development of affordable electron microscopes. For healthcare, research on biomaterials, tissue engineering and stem cells and engineering–life interface will also require mechanical characterisation, in these cases of cells or biomaterials in their native state. This is difficult to achieve using existing electron microscopes because of beam damage (Ren et al., 2007, 2008). Ingenious solutions are required here, if the direct measurement of the mechanical properties of any and all micro to nanometre-sized particles is going to become a routine tool in research and process development.

NOMENCLATURE

a	contact radius (Equation (3)) (m)
c_1	equal to the coefficient of the Hertz equation (Equation (1))

c_2, c_3	arbitrary constants
f_1, f_2, f_3	functions of the principal tensions defined in Equations (30a)–(30c) (N m^{-1})
h	cell wall thickness during compression (m)
h_0	initial (uninflated) cell wall thickness (m)
h_p	displacement (m)
h_{pmax}	maximum displacement (m)
$h_p(t)$	time-dependent displacement (m)
r_0	uninflated cell radius (m)
t	time (s)
t_R	time taken to compress a particle (Equation (15)) (s)
u	arbitrary time variable (Equation (20)) (s)
x	arbitrary variable (Equation (22))
z	half the displacement of the probe, also termed compressive approach (m)
A_c	contact area between the probe and the cell (Equation (32)) (m^2)
C_1, C_2	material constants in the Mooney–Rivlin (1940) strain energy function (Pa)
E	Young's or elastic modulus (Pa)
E_0	instantaneous Young's modulus corresponding to G_0 (Equation (18)) (Pa)
E_{0n}	initial Young's modulus for non-linear elastic material (Pa)
E_i	principal component of Green strain in direction i
E^*	reduced modulus (Pa)
E_∞	long-term Young's modulus corresponding to G_∞ (Equations (8) and (19)) (Pa)
F	force on particle (N)
$F(t)$	time-dependent force (N)
F_i	proportionality constants (N)
F_∞	force when the relaxation re-establishes equilibrium (N)
$G(t)$	shear relaxation modulus (Pa)
G_0	instantaneous shear modulus (Pa)
G_∞	long-term shear modulus (Pa)
H	Hardness of particle (Pa)
H_i	principal component of Hencky strain in direction i
K_i (where $i = 1, 2, 3, \ldots, n$)	proportionality constants (Equation (12)) (Pa)
L	length of cell extension into micropipette
M_i ($i = 0, 1, 2, 3, \ldots, n$)	proportionality constants (Equation (14)) (Pa)

P	turgor pressure (Pa)
ΔP	micropipette aspiration suction pressure (Pa)
R_c	inner radius of micropipette (m)
R	particle radius (m)
RCF_i	ramp correlation factor
T_1, T_2	tension in the meridian and circumferential directions, respectively (N m^{-1})
V_0	moving speed of the probe (ms^{-1})
X	fractional deformation

GREEK LETTERS

β	C_2/C_1 Mooney–Rivlin material constants
$\gamma_{\text{surface}}^{\text{LW}}$, $\gamma_{\text{foulant}}^{\text{LW}}$, $\gamma_{\text{fluid}}^{\text{LW}}$	Lifshitz–van der Waals surface free energies of the surface, foulant and fluid (e.g. water), respectively (N m^{-1})
δ	$\delta = \lambda_2 \sin\psi$ (Equation (26))
ε_i	principal infinitesimal strain component in direction i
ε_∞	strain when the relaxation re-establishes equilibrium (Equation (8))
η	vertical coordinate of cell wall (m)
$\bar{\eta}$	distance between the compression surface and the equatorial plane (m)
η'	derivative of η with respect to ψ
λ_i ($i = 1, 2$)	principal stretch ratio in direction i
λ_0	stretch ratio at the centre of contact area during compression
λ_s	initial stretch ratio
ν	Poisson's ratio
ρ	horizontal coordinate of cell wall (m)
$\bar{\rho}$	distance between the η axis and the edge of the cell on the equatorial plane (m)
$\sigma(t)$	time-dependent stress (Pa)
σ_0	stress at time zero (Pa)
σ_∞	stress when the relaxation re-establishes equilibrium (Pa)
σ_i	proportionality constant (Pa)
τ	relaxation time (Equation (8)) (s)
τ_i ($i = 1, 2, \ldots, n$)	relaxation times (Equation (9)) (s)
ψ	the angular position of a point on the cell wall from the vertical axis of symmetry before compression (rad)
ω	derivative of δ with respect to ψ

Γ the angle of the point on the edge of the contact region between the compression surface and the cell following compression (rad)
Θ the angle of rotation in the circumferential direction (rad)

ACKNOWLEDGEMENTS

The authors would like to thank Dr T. Liu, Ms M. Liu and Miss J. Xue for allowing using some of their images and data.

REFERENCES

Adams, M. J., and Perchard, V. *Int. Chem. Eng. Symp. Ser.* **92**, 147 (1985).
Ahmad, M. R., Nakajima, M., Kojima, S., Homma, M., and Fukuda, T., IEEE/RSJ International Conference on Intelligent Robots and Systems 1–9, 602 (2007).
Alessandrini, A., and Facci, P. *Meas. Sci. Technol.* **16**, 65 (2005).
Andersson, M., Madgavkar, A., Stjerndahl, M., Wu, Y. R., Tan, W. H., Duran, R., Niehren, S., Mustafa, K., Arvidson, K., and Wennerberg, A. *Rev. Sci. Instrum.* **7**, 78 (2007).
Andrei, D. C., Briscoe, B. J., Luckham, P. F., and Williams, D. R. *J. Chim. Phys. PCB* **93**, 960 (1996).
Arshady, R. *Colloids Surf. A* **153**, 325 (1999).
Aslani, P., and Kennedy, R. A. *J. Microencapsul.* **13**, 601 (1996).
Blewett, J. M., Burrows, K., and Thomas, C. R. *Biotechnol. Lett.* **22**, 1877 (2000).
Born, C., Zhang, Z., Al-Rubeai, M., and Thomas, C. R. *Biotechnol. Bioeng.* **40**, 1004 (1992).
Bos, H. L., and Donald, A. M. *J. Mater. Sci.* **34**, 3029 (1999).
Boudou, T., Ohayon, J., Arntz, Y., Finet, G., Picart, C., and Tracqui, P. *J. Biomech.* **39**, 1677 (2006).
Bowen, W. R., Lovitt, R. W., and Wright, C. J. *J. Colloid Interface Sci.* **237**, 54 (2001).
Boyd, R. D., Verran, J., Jones, M. V., and Bhakoo, M. *Langmuir* **18**, 2343 (2002).
Bryant, Z., Stone, M. D., Gore, J., Smith, S. B., Cozzarelli, N. R., and Bustamante, C. *Nature* **424**, 338 (2003).
Bryers, J. D. *Biotechnol. Prog.* **3**, 57 (1987).
Castelain, M., Pignon, F., Piau, J.-M., and Magnin, A. *J. Chem. Phys.* **128**, 135101 (2008).
Chang, T. M. S. *Ann. N. Y. Acad. Sci.* **875**, 71 (1999).
Chen, M. J., Zhang, Z., and Bott, T. R. *Biotechol. Tech.* **12**, 875 (1998).
Chen, M. J., Zhang, Z., and Bott, T. R. *Colloids Surf. B Biointerfaces* **43**, 59 (2005).
Cheng, L. Y. *J. Biomed. Eng.* **109**, 10 (1987).
Chew, J. Y. M., Cardoso, S. S. S., Paterson, W. R., and Wilson, D. I. *Chem. Eng. Sci.* **59**, 3381 (2004).
Choi, J. B., Inchan, Y., Cao, L., Leddy, H. A., Gilchrist, C. L., Setton, L. A., and Guilak, F. *J. Biomech.* **40**, 2596 (2007).
Christie, A. O., Evans, L. V., and Shaw, M. *Ann. Bot. - Lond.* **34**, 476 (1970).
Cole, K. S. *J. Cell. Comp. Physiol.* **1**, 1 (1932).
Cosgrove, D. J. *Plant. Cell. Environ.* **11**, 67 (1988).
Creely, C. M., Singh, G. P., and Petrov, D. *Opt. Commun.* **245**, 465 (2005).
Daily, B., Elson, E. L., and Zahalak, G. I. *Biophys. J.* **45**, 671 (1984).
DeGroot, A. R., and Neufeld, R. J. *Enzyme Microb. Technol.* **29**, 321 (2001).
Duszyk, M., Schwab, B., Zahalak, G. I., Qian, H., and Elson, E. L. *Biophys. J.* **55**, 683 (1989).
Edwards-Levy, F., and Levy, M. C. *Biomaterials* **20**, 2069 (1999).

Eichhorn, V., Carlson, K., Andersen, K. N., Fatikow, S., and Boggild, P., IEEE/RSJ International Conference on Intelligent Robots and Systems 1–9, 297 (2007).
Fahlbusch, S., Mazerolle, S., Breguet, J. M., Steinecker, A., Agnus, J., Perez, R., and Michler, J. *J. Mater. Process. Technol.* **167**, 371 (2005).
Fairbrother, R. J., and Simmons, S. J. R. *Part. Part. Syst. Char.* **15**, 16 (1998).
Fallman, E., Schedin, S., Andersson, M., Jass, J., and Axner, O., *in* "Manipulation and Analysis of Biomolecules, Cells and Tissues" (D. V. Nicolau, J. Enderlein, R. C. Leif, and D. L. Farkas Eds.), Vol. 4962, p. 206, SPIE-INT Society Optical Engineering, Bellingham, USA (2003).
Fan, X., Ten, P., Clarke, C., Bramley, A., and Zhang, Z. *Powder Technol.* **131**, 105 (2003).
Feng, W. W., and Yang, W. H. *Trans. Am. Soc. Mech. Eng.: J. Appl. Mech.* **40**, 209 (1973).
Feng, S. S., Mu, L., Chen, B. H., and Pack, D. *Mat. Sci. Eng. C - Bio. S.* **20**, 85 (2002).
Gaboriaud, F., and Dufrene, Y. F. *Colloids Surf. B* **54**, 10 (2007).
Garrett, T. R., Bhakoo, M., and Zhang, Z. *Biotechnol. Lett.* **30**, 427 (2008).
Gibbs, B. F., Kermasha, S., Alli, I., and Mulligan, C. N. *Int. J. Food. Sci. Nutr.* **50**, 213 (1999).
Goldman, W. H. *Biotechnol. Lett.* **22**, 431 (2000).
Gonzalez-Rodriguez, M. L., Holgado, M. A., Sanchez-Lafuente, C., Rabasco, A. M., and Fini, A. *Int. J. Pharm.* **132**, 225 (2002).
Grigorescu, G., Rosinski, S., Lewinska, D., Ritzén, L. G., Viernstein, H., Teunou, E., Poncelet, D., Zhang, Z., Fan, X., Serp, D., Marison, I., and Hunkeler, D. *J. Microencapsul.* **19**, 641 (2002).
Haddad, Y. M., "Viscoelasticity of Engineering Materials". Chapman and Hall, London (1995).
He, J. H., Xu, W., and Zhu, L. *Appl. Phys. Lett.* **90**, 90 (2007).
Hertz, H. *J. Reine Angew. Math.* **92**, 156 (1882).
Hiller, S., Bruce, D. M., and Jeronimidis, G. *J. Texture Stud.* **27**, 559 (1996).
Hiramoto, Y. *Exp. Cell Res.* **32**, 59 (1963).
Hochmuth, R. M. *J. Biomech.* **33**, 15 (2000).
Israelachvili, J. N., "Intermolecular and Surface Forces". 2nd ed. Academic Press, London (1992).
Jiang, Q., and Logan, B. E. *Environ. Sci. Technol.* **25**, 2031 (1991).
Jivraj, M., Martini, L. G., and Thomson, C. M. *Pharm. Sci. Technol. Today* **3**, 58 (2000).
Johnson, K. L., "Contact Mechanics". Cambridge University Press, Cambridge (1985).
Jones, W. R., Ting-Beall, H. P., Lee, G. M., Kelley, S. S., Hochmuth, R. M., and Guilak, F. *J. Biomech.* **32**, 119 (1999).
Jones, R., Pollock, H. M., Cleaver, J. A. S., and Hodges, C. S. *Langmuir* **18**, 8045 (2002).
Keller, M. W., and Sottos, N. R. *Exp. Mech.* **46**, 725 (2006).
Kikuchi, A., Kawabuchi, M., Sugihara, M., Sakurai, Y., and Okano, T. *J. Control. Release* **47**, 21 (1997).
Kleinig, A. R., Cell Disruption Mechanics, Ph.D. thesis, University of Adelaide, Australia (1997).
Klis, F. M., Boorsma, A., and de Groot, P. W. J. *Yeast* **23**, 185 (2006).
Kotte, M. K., and Rudnic, E. M., Tablet dosage forms, *in* "Modern Pharmaceutics" (G. S. Banker Ed.), Informa Healthcare, New York (1995).
Kuen, Y. L., Martin, C. P., and David, J. M. *Adv. Mater.* **13**, 837 (2001).
Kuo, S. C., and Sheetz, M. P. *Science* **260**, 232 (1993).
Kuznetsova, T. G., Starodubtseva, M. N., Yegorenkov, N. I., Chizhik, S. A., and Zhdanov, R. I. *Micron* **38**, 824 (2007).
Lardner, T. J., and Pujara, P., Compression of Spherical Cells, *in* "Mechanics today" (S. Nemat-Nasser, Ed.), Vol. 5, p. 161. Pergamon, New York (1980).
Le Meste, M., Champion, D., Roudaut, G., Blond, G., and Simatos, D. *J. Food Sci.* **67**, 2444 (2002).

Lee, K. Y., Peters, M. C., and Mooney, D. J. *Adv. Mater.* **13**, 837 (2001).
Leipzig, Nic. D., and Athanasiou, K. A. *J. Biomech.* **38**, 77 (2005).
Lim, C. T., Zhou, E. H., Li, A., Vedula, S. R. K., and Fu, H. X. *Mater. Sci. Eng.* **26**, 1278 (2006).
Liu, K. K., Williams, D. R., and Briscoe, B. J. *Phys. Rev. E* **54**, 6673 (1996).
Liu, K. K., Wang, H. G., Wan, K. T., Liu, T., and Zhang, Z. *Colloids Surf. B* **25**, 293 (2002a).
Liu, K. K., Chan, V., and Zhang, Z. *Med. Biol. Eng. Comput.* **40**, 491 (2002b).
Liu, W., Christian, G. K., Zhang, Z., and Fryer, P. *Trans. IChemE, Part C* **80**, 286 (2002c).
Liu, T., Donald, A. M., and Zhang, Z. *Mater. Sci. Technol. Ser.* **21**, 289 (2005).
Liu, W., Christian, G. K., Zhang, Z., and Fryer, P. J. *Int. Dairy J.* **16**, 164 (2006a).
Liu, W., Fryer, P. J., Zhang, Z., Zhao, Q., and Liu, Y. *Innovative Food Sci. Eng. Technol.* **7**, 263 (2006b).
Liu, W., Aziz, N. Ab., Zhang, Z., and Fryer, P. J. *J. Food Eng.* **78**, 217 (2007).
Logan, B. E., and Kilps, J. R. *Water Res.* **29**, 443 (1995).
Lu, G. Z., Thompson, F. G., and Gray, M. R. *Biotechnol. Bioeng.* **40**, 1277 (1992).
Lulevich, V. V., Radtchenko, I. L., Sukhorukov, G. B., and Vinogradova, O. I. *J. Phys. Chem. B* **107**, 2735 (2003).
Lulevich, V., Zink, T., Chen, H. Y., Liu, F. T., and Liu, G. Y. *Langmuir* **22**, 8151 (2006).
Madene, A., Jacquot, M., Scher, J., and Desobry, S. *Int. J. Food. Sci. Technol.* **41**, 1 (2006).
Madgar, l., Seidman, D. S., Levran, D., Yonish, M., Augarten, A., Yemini, Z., Mashiach, S., and Dor, J. *Hum. Reprod.* **11**, 2151 (1996).
Marszalek, P. E., Oberhauser, A. F., Pang, Y. P., and Fernandez, J. M. *Nature* **396**, 661 (1998).
Martinsen, A., Skjakbraek, G., and Smidsrod, O. *Biotechnol. Bioeng.* **33**, 79 (1989).
Mashmoushy, H., Zhang, Z., and Thomas, C. R. *Biotechnol. Tech.* **12**, 925 (1998).
Masuda, T., Ingram, A., Zhang, Z., and Seville, J. P. K., 13th World Clean Air and Environmental Protection Congress and Exhibition, London, UK, paper number 461 (2004).
Mattice, J. M., Lau, A. G., Oyen, M. L., and Kent, R. W. *J. Mater. Res.* **21**, 2003 (2006).
Mendelson, N. H., Sarlls, J. E., Wolgemuth, C. W., and Goldstein, R. E. *Phys. Rev. Lett.* **84**, 1627 (2000).
Michaels, J. D., Petersen, J. F., McIntire, L. V., and Papoutsakis, E. T. *Biotechnol. Bioeng.* **38**, 169 (1991).
Middelberg, A. P. J. *Biotechnol. Adv.* **13**, 491 (1995).
Mills, J. P., Qie, L., Dao, M., Lim, C. T., and Suresh, S. *Mol. Cell Biol.* **1**, 169 (2004).
Mow, V. C., Kuei, S. C., Lai, W. M., and Amstrong, C. G. *ASME J. Biomech. Eng.* **102**, 73 (1980).
Müller, E., Chung, J. T., Zhang, Z., and Sprauer, A. *J. Chromatogr. A* **1097**, 116 (2005).
Murhammer, D. W., and Goochee, C. F. *Biotechnol. Prog.* **6**, 391 (1990).
Nakabayashi, D., and Silva, P. C. *Int. J. Nanotechnol.* **4**, 609 (2007).
Nguyen, V. B., Wang, C. X., Thomas, C. R., and Zhang, Z. *Chem. Eng. Sci.* **64**, 821 (2009).
Nishizaka, T., Miyata, H., Yoshikawa, H., Ishiwata, S., and Kinosita, K. *Nature* **377**, 251 (1995).
Ohtsubo, T., Tsuda, S., and Tsuji, K. *Polymer* **32**, 2395 (1991).
Papoutsakis, E. T. *Trends Biotechnol.* **9**, 427 (1991).
Pelling, A. E., Sehati, S., Gralla, E. B., Valentine, J. S., and Gimzewski, J. K. *Science* **305**, 1147 (2004).
Poncelet, D., and Neufeld, R. T. *Biotechnol. Bioeng.* **33**, 95 (1989).
Read, T., Stensvaag, V., Vindenes, H., Ulvestad, E., Bjerkvig, R., and Thorsen, F. *Int. J. Neurosci.* **7**, 653 (1999).
Rehor, A., Canaple, L., Zhang, Z., and Hunkeler, D. *J. Biomater. Sci. Polym.* **12**, 157 (2001).
Ren, Y., Donald, A. M., and Zhang, Z. *Mater. Sci. Technol. Ser.* **23**, 857 (2007).
Ren, Y. L., Donald, A. M., and Zhang, Z., *Scanning* **30**, 435 (2008).
Ribeiro, A. J., Neufeld, R. J., Arnaud, R. J., and Chaumeil, J. C. *Int. J. Pharm.* **187**, 115 (1999).
Roberts, R. J., and Rowe, R. C. *Chem. Eng. Sci.* **42**, 903 (1987).

Rubinstein, M. H., "Pharmaceutics, The Science of Dosage Form Design". Churchill Livingstone, Edinburgh (2000).
Schuldt, U., and Hunkeler, D. *Minerva Biotecnol.* **12**, 249 (2000).
Shamlou, P. A., and Titchener-Hooker, N., Turbulent aggregation and breakup of particles in liquids in liquids in stirred vessels, *in* "Processing of Solid–Liquid Suspensions" (P. A. Shamlou Ed.), pp. 1–25. Butterworth-Heinemann Ltd, Oxford (1993).
Sharp, J. M., Duran, R. S., and Dickinson, R. B. *J. Colloid Interface Sci.* **299**, 180 (2006).
Shieh, A. C., and Athanasiou, K. A. *J. Biomech.* **39**, 1595 (2006).
Shiu, C., Zhang, Z., and Thomas, C. R. *Biotechnol. Tech.* **13**, 707 (1999).
Sindel, U., and Zimmermann, I. *Powder Technol.* **117**, 247 (2001).
Smith, S. B., Cui, Y., and Bustamante, C. *Science* **271**, 795 (1996).
Smith, A. E., Moxham, K. E., and Middelberg, A. P. *J. Chem. Eng. Sci.* **53**, 3913 (1998).
Smith, A. E., Zhang, Z., and Thomas, C. R. *Chem. Eng. Sci.* **55**, 2031 (2000a).
Smith, A. E., Moxham, K. E., and Middelberg, A. P. *J. Chem. Eng. Sci.* **55**, 2043 (2000b).
Sorensen, B. F., Horsewell, A., Jorgensen, O., and Kumar, A. N. *J. Am. Ceram. Soc.* **81**, 661 (2001).
Stenekes, R. J. H., De Smedt, S. C., Demeester, J., Sun, G. Z., Zhang, Z. B., and Hennink, W. E. *Biomacromolecules* **1**, 696 (2000).
Stocks, S. M., and Thomas, C. R. *Biotechnol. Bioeng.* **75**, 702 (2001).
Stokes, D. J., and Donald, A. M. *J. Mater. Sci.* **35**, 599 (2000).
Strand, B. L., Morch, Y. A., and Skjak-braek, G. S. *Minerva Biotecnol.* **12**, 223 (2000).
Sugimoto, T., Takahashi, T., Itoh, H., Sato, S., and Muramatsu, A. *Langmuir* **13**, 5528 (1997).
Sun, G., and Zhang, Z. *J. Microencapsul.* **18**, 593 (2001).
Sun, G., and Zhang, Z. *Int. J. Pharm.* **242**, 307 (2002).
Tang, S., Preece, J. M., McFarlane, C. M., and Zhang, Z. *J Colloid Interface Sci.* **221**, 114 (2000).
Tatara, Y. *J. Eng. Mater. - Trans. ASME* **113**, 285 (1991).
Tatara, Y. *J. Eng. Mater. - Trans. ASME* **36**, 190 (1993).
Thiel, B. L., and Donald, A. M. *Ann. Bot. - Lond.* **82**, 727 (1998).
Thomas, C. R., Al-Rubeai, M., and Zhang, Z. *Cytotechnology* **15**, 329 (1994).
Thomas, C. R., Zhang, Z., and Cowen, C. *Biotechnol. Lett.* **22**, 531 (2000).
Tomos, D. *Biotechnol. Lett.* **22**, 437 (2000).
Tomos, A. D., and Leigh, R. A. *Annu. Rev. Plant Phys.* **50**, 447 (1999).
Torre, M. L., Maggi, L., Vigo, D., Galli, A., Bornaghi, V., Maffeo, G., and Conte, U. *Biomaterials* **21**, 1493 (2000).
Touhami, A., Nysten, B., and Dufrene, Y. F. *Langmuir* **19**, 4539 (2003).
Van Raamsdonk, J. M., and Chang, P. L. *J. Biomed. Mater. Res.* **54**, 264 (2001).
Vandenberg, G. W., Drolet, C., Scott, S. L., and Noue, J. de la. *J. Control. Release* **77**, 297 (2001).
Wang, C. X., Wang, L., and Thomas, C. R. *Ann. Bot. - Lond.* **93**, 443 (2004).
Wang, C. X., Cowen, C., Zhang, Z., and Thomas, C. R. *Chem. Eng. Sci.* **60**, 6649 (2005).
Wang, C. X., Pritchard, C. R., and Thomas, C. R. *J. Texture Stud.* **37**, 597 (2006a).
Wang, L., Hukin, D., Pritchard, J., and Thomas, C. R. *Biotechnol. Lett.* **28**, 1147 (2006b).
Ward, I. M., and Hadley, D. W., "An Introduction to the Mechanical Properties of Solid Polymers". Wiley, Chichester, UK (1993).
Watkinson, A. P., Müller-Steinhagen, H., and Malayeri, M. R., Heat Exchanger Fouling and Cleaning: Fundamentals and Applications, ECI Symposium Series, Vol. RP1, Bypress Publications, Berkeley, USA (2004).
Weiss, L. *Exp. Cell Res.* **8**, 141 (1961).
Welsh, J. P., Relationships between Hybridoma Cell Mechanical Properties and Physiology, Ph.D. thesis, The University of Birmingham (1998).
Willett, C. D., Zhang, Z., and Seville, J. P. K., "The 1997 IChemE Research Event", Vol. 1, p. 401. Chameleon Press Ltd., London (1997).

Wilson, D. I., Fryer, P. J., and Hasting, A. P. M., "Fouling, Cleaning and Disinfection in Food Processing". University of Cambridge, Cambridge, UK, p. 263 (2002).
Wu, H., Spence, R. D., Sharpe, P. J. H., and Goeschl, J. D. *Plant Cell Environ.* **8**, 563 (1985).
Xie, C., and Li, Y. *J. Appl. Phys.* **93**, 2982 (2003).
Xue, J., Novel Encapsulation of Active Ingredients for Dental Care, Ph.D. thesis, University of Birmingham, UK (2008).
Xue, J., and Zhang, Z. *J. Microencapsul.* **25**, 523 (2008).
Yap, S. F., Adams, M., Seville, J., and Zhang, Z. *China Particuology* **4**, 35 (2006).
Yap, S. F., Adams, M. J., Seville, J. P. K., and Zhang, Z. *Powder Technol.* **185**, 1 (2008).
Yeung, A. K. C., and Pelton, R. *J. Colloid Interface Sci.* **184**, 579 (1996).
Yoneda, M. *J. Exp. Biol.* **41**, 893 (1964).
Yoneda, M. *Adv. Biopys.* **4**, 153 (1973).
Yu, M. F., Dyer, M. J., Skidmore, G. D., Rohrs, H. W., Lu, X. K., Ausman, K. D., Von Her, J. R., and Ruoff, R. S. *Nanotechnology* **10**, 244 (1999).
Zhang, J., and Buffle, J. *Colloids Surf. A* **107**, 175 (1996).
Zhang, Z., Ferenczi, M. A., and Thomas, C. R. *Chem. Eng. Sci.* **47**, 1347 (1992a).
Zhang, Z., Al-Rubeai, M., and Thomas, C. R. *Enzyme Microb. Technol.* **14**, 980 (1992b).
Zhang, Z., Saunders, R., and Thomas, C. R. *J. Microencapsul.* **16**, 117 (1999).
Zhao, L., and Zhang, Z. *Artif. Cell Blood Sub.* **32**, 25 (2004).
Zhao, Q., Wang, S., and Müller-Steinhagen, H. *Appl. Surf. Sci.* **230**, 371 (2004).
Zhao, L., Schaefer, D., Xu, H. J., Modi, S. J., Lacourse, W. R., and Marten, M. R. *Biotechnol. Prog.* **21**, 292 (2005).

CHAPTER 3

Particle Image Velocimetry Techniques and its Applications in Multiphase Systems

Feng-Chen Li[1] and **Koichi Hishida**[2,*]

Contents

1.	Introduction	88
2.	Fundamentals of Particle Image Velocimetry	90
	2.1 Seeding the flow	91
	2.2 Illumination and image recording	92
	2.3 PIV analysis	95
	2.4 Post-processing of velocity vectors	101
3.	Various Types of Particle Image Velocimetry	103
	3.1 2D-2C PIV techniques	103
	3.2 2D-3C PIV techniques	105
	3.3 3D-3C PIV techniques	109
	3.4 Others	115
4.	Measurement of Multiphase Flow Using Particle Image Velocimetry	118
	4.1 Liquid–liquid two-fluid flows	119
	4.2 Gas–liquid two-phase flows	121
	4.3 Particle-laden multiphase flows	137
5.	Summary and Outlook	140
	Notation	141
	References	142

---

1 School of Energy Science and Engineering, Harbin Institute of Technology, Harbin 150001, China
2 Department of System Design Engineering, Keio University, Yokohama 223-8522, Japan

*Corresponding author
E-mail address: hishida@sd.keio.ac.jp

Abstract	This chapter is devoted to the methodology of particle image velocimetry (PIV) techniques and its applications to multiphase flow systems. It reviews, first, the fundamental issues of a conventional PIV with considerations of improvements of spatial resolution and accuracy; second, the state of the art in various types of PIV techniques from the viewpoint of dimensions and velocity components of the measurement, the flow passage scales, and the hardware components of the system; third, the state of the art in some issues about the measurement of multiphase flow systems using PIV techniques. The multiphase flows to which the applications of PIV techniques are discussed include liquid–liquid two fluid flows, gas–liquid two-phase flows, and particle-laden multiphase flow systems.
	The emphasis in this chapter is on the fruitful methodology of PIV techniques that emerge in the recent publications instead of the detailed discussions on any individual research topic of the measurement target of PIV. The purpose is to provide an overall instructive introduction and guidance to the PIV techniques and its applications particularly in the research field of multiphase flows. To this end, fruitful examples of PIV measurements of free-surface liquid flows, bubbly flows, particle-laden multiphase flows, etc., are elucidated.

1. INTRODUCTION

Particle image velocimetry (PIV) is one of the nonintrusive flow diagnostics tools, which provide quantitative measurements with high spatial and high temporal resolution of whole field velocity profiles in liquids, gases, multiphase flows, and even the fluid-like flows of solid particles or granular flows. Since its invention for more than two decades, PIV, as a nonintrusive, accurate, reliable, convenient, and very powerful experimental technique for flow diagnostics, has been playing a highly important role in the fundamental studies on fluid dynamics, a wide range of industrial applications, understanding a variety of natural phenomena, and even exploring some flow characteristics in the body of human beings. The widely demonstrated applications of PIV technique may range from very low speed flows to supersonic flows (e.g., Ganapathisubramani et al., 2006a; Ganapathisubramani, 2007), from flows in a nano/micro-scaled passage to flows at a large scale as a river (e.g., Ettema et al., 1997), from single-phase to multiphase flows, from nonreactive flows to reactive flows as combustion, from cryogenic (e.g., Van Sciver et al., 2007; Harada et al., 2006; Zhang and Van Sciver, 2005) to very high temperature flows (e.g., Balakumar and Adrian, 2004), and from inanimate flows to flows around living beings (e.g., Stamhuis, 2006).

For summarization of the development of PIV, there have timely appeared many volumes of books or monographs or pieces of review papers on the topic (e.g., Adrian, 1991; Raffel et al., 2007; Westerweel, 1997, among others), which have well-documented the PIV technique in common senses.

In lieu of its great importance, widely spreading communities of PIV techniques, including the development of PIV and application of PIV, have emerged all over the world. There are always a large part of participations giving presentations of PIV techniques to the relevant international conferences or symposiums or subsessions of them. Sought from the "ISI Web of Knowledge" (http://apps.isiknowledge.com) using the keyword of PIV or its full spelling, it can be seen that the number of published papers about PIV in the peer-reviewed journals increases more than 10 times from 39 papers in the year of 1994 to 464 papers in 2007, as shown in Figure 1. If sought from "Google" website, the number of PIV issues, which include journal papers, conference papers, and other issues, is from 263 items in 1994 to 2270 items in 2007. After the continuously increasing development and application, the PIV techniques have partly become one of the standard diagnostics tools of flow measurement. Nevertheless, PIVs are still partly under development such as the holographic micro-scaled PIV technique for three-dimensional (3D) flow measurement in microchannels.

Multiphase flow systems appear almost everywhere in the world, including the natural phenomena, industrial applications, and daily life of human beings. The application of PIV techniques to multiphase flow measurement has also been flourished. Among the flow measurement

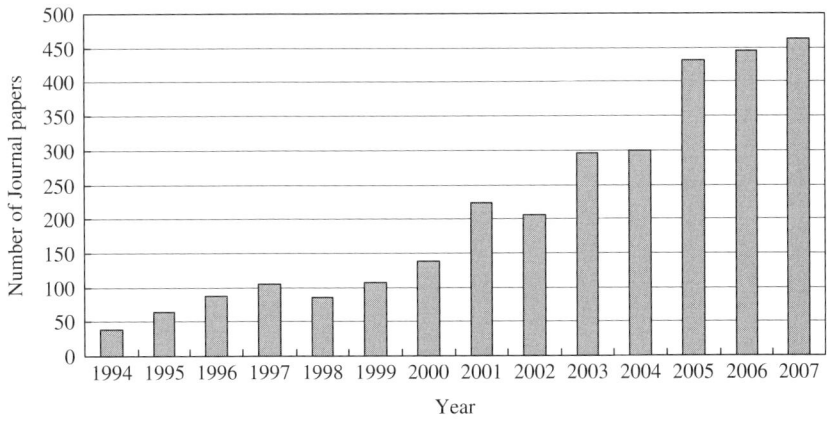

Figure 1 A survey of the number of published journal papers about development or application of PIV technique from 1994 to 2007 based on the database from "ISI Web of Knowledge" (http://apps.isiknowledge.com).

using PIV techniques, multiphase flows may have taken the most complex and difficult supplemental considerations to the experimentalists. The complexity emerges from that the multiphase systems usually involve a variety of operating modes of gas, liquid, and solid phases, including those with solid particles and liquid droplets in dispersed states, which will severely influence a normal PIV measurement without any particular treatment. The summarization of PIV techniques used for flow measurement of multiphase systems is, however, relatively scarce compared with those of PIV technique in common senses.

This chapter is therefore intended to describe the fundamental issues on PIV technique in brief, the versatility of PIV techniques from the viewpoint of its hardware components as well as the resolved dimensions. The chapter is also intended to discuss the state of the art in the special PIV techniques (sometime with the aid of other supplementary technique) applied to the measurement of multiphase flows. Note that, the emphasis in this chapter is on the methodology of PIV techniques instead of details of the individual study on any research topic.

2. FUNDAMENTALS OF PARTICLE IMAGE VELOCIMETRY

The principle of conventional two-dimensional (2D) digital PIV technique is to illuminate (for transparent flow media) a particle-seeded flow field with two laser sheet pulses separated by a time delay and capture the image of the particles with a charge-coupled device (CCD) camera. The captured images are then subdivided into an array of small size interrogation windows. In each interrogation window, all particles are assumed to have essentially the same velocity. The overall displacement of particles within each interrogation window is then calculated with a numerical correlation algorithm. Finally, the velocity vectors of the whole illuminated flow field are obtained with dividing the displacement by the time delay between the illuminating laser sheet pulses. For any kind of fluid flow that is transparent to enable imaging of the seeded particles mixed in the fluid, including gas flow, liquid flow, gas–liquid two-phase flow, and gas–liquid–solid three-phase flow, PIV can be applied to measure the whole field velocity distribution. A generalized PIV setup is shown in Figure 2. Apart from this kind of conventional PIV, that is, pulsed lasers, particle seeding, and transparency of the flow media are required, imaging procedure and velocity derivation algorithm of PIV technique has been applied to develop measurement approaches for the flow velocity distribution at the surface of some kind of fluid-like flow or opaque fluid flow, such as granular flow (e.g., Deng and Wang, 2003; Lueptow et al., 2000; Ostendorf and

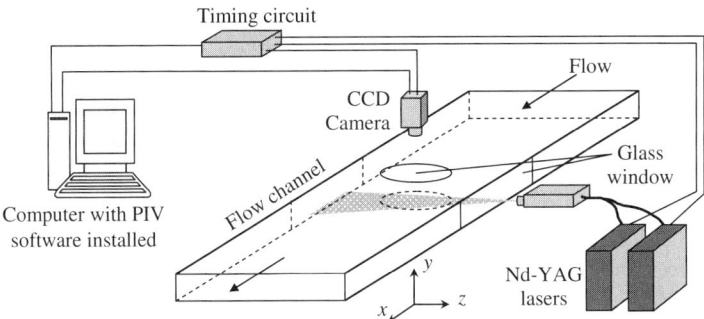

Figure 2 Diagram of a generalized 2D-PIV setup showing all major components: flow channel with the particle seeded fluid flow, laser sheet pulses illuminating one plane in the fluid, a CCD camera imaging the particles in the laser-illuminated sheet in the area of interest, a computer with PIV software installed, a timing circuit communicating with the camera and computer and generating pulses to control the double-pulsed laser. The PIV software setups and controls the major components, and analyses the images to derive a vector representation of flow field (see Plate 4 in Color Plate Section at the end of this book).

Schwedes, 2005; Sielamowicz et al., 2005; Steingart and Evans, 2005; Zhao et al., 2008), ice flow floating at the surface of a river (e.g., Ettema et al., 1997) and so on. In those cases, not all of pulsed lasers, particle seeding, and transparency of the flow are necessary. The following fundamental descriptions are based on a conventional 2D digital PIV technique. For other kinds of PIV, the principles and analysis algorithms are straightforward, which will be individually briefed in the next section.

2.1 Seeding the flow

To visualize the flow for PIV purposes, the measured fluid flow has to be seeded with particles, which need to be neutrally buoyant and small with respect to the flow phenomena studied (e.g., Raffel et al., 2007). For different sorts of visualized flow, seeding particles could be quite different.

For liquid flows, a series of polymer powders or particulate half-products for plastics with a range of densities are available and can be matched to the fluid density. In water flow for example, polystyrene, polythene, nylon, pliolite particles work well and are commercially produced in a range of diameters typically from 5 to 200 µm. Seeding gas flows, particularly at low speed, might be more difficult than seeding water flow, since particles tend to sink due to its relatively high density compared with that of gas. The seeding particles thus need to be very light and small, for example, very small polystyrene beads with diameter of 5–10 µm (Stamhuis, 2006 among others). In high speed air flow, for example, in a wind tunnel flow, water droplets or vegetable oil produced with an aerosol generator in diameters of 1–10 μm are commonly used (Raffel et al., 2007). Melling (1997) comprehensively stressed the issues on tracer particles and seeding for PIV, including the size specifications for suitable tracer particles particularly with respect to their flow tracking capability, a wide variety of tracer materials used in liquid and gases, and methods of generating seeding particles and introducing the particles into the gas flows.

In the gas–liquid two-phase flows illuminated by a laser sheet, for example, the intensity of light reflected from the gas–liquid interface (mostly the gas bubble's surface) not only saturate the CCD camera, but also overwhelm the intensity of light from the seeded tracer particles in its vicinity. Fluorescent particles are often used to realize the laser-induced fluorescence (LIF) technique together with PIV (e.g., Broder and Sommerfeld, 2002; Fujiwara et al., 2004a, b; Kitagawa et al., 2005; Liu et al., 2005; Tokuhiro et al., 1998, 1999), so that both images of gas–liquid interface (e.g., bubble's geometry) and velocity distribution in the liquid phase around the gas bubbles can be obtained. Issues on PIV measurement of gas–liquid two-phase flows will be further illustrated in the latter sections.

A summarization of different seeding particles for the PIV measurement of water and air flows, which are the most often encountered, is provided in Table 1.

2.2 Illumination and image recording

A laser light sheet is usually adopted for the illumination in PIV flow studies, since it may have strong enough brightness and almost constant thickness without aberration or diffusion due to the coherent and monochromatic character of the emitted laser light. Some exceptions, such as the conventional microscopic PIV (μ-PIV), PIV technique for granular flows and so on, will be mentioned in the later relative sections. The slightly diverging light beam produced by a laser is usually transformed into a sheet by converging it with a weak positive lens and subsequently making the beam fan out in one plane to a sheet by an additional cylindrical lens. This results in a sheet with a lightly converging thickness, enabling one to select a certain sheet thickness tuned to the certain experimental conditions. The minimum thickness of the laser sheet locates at the focal point of the outermost convex lens of the optics and is calculated with the following equation when the Rayleigh length of the light sheet is much larger than the focal length of the lens: $\delta_m = 4F\lambda/\pi d_0$, where F is the focal length of the convex lens, λ the laser wave length, and d_0 the diameter of the laser beam.

Table 1 The often used seeding particles for PIV measurement of water and air flows

	For water flow			For gas flow	
Type	Material	Diameter (μm)	Type	Material	Diameter (μm)
Solid	Polystyrene	10–100	Solid	Polystyrene	0.5–10
	Polymer fluorescent microspheres	0.1–1		Alumina Al_2O_3	0.2–5
	Glass spheres	10–100		Magnesium	2–5
	Hollow glass spheres	10–150		Titania TiO_2	0.1–5
	Hollow plastic spheres	10–50		Glass microspheres	0.2–3
	Aluminum flakes	2–7		Hollow glass spheres	10–100
	Granular particles with coating	10–500		Hollow plastic spheres	10–50
	Ion exchange resin	30–700		Granular particles with coating	10–50
Liquid	Oils	50–500		Ion exchange resin	1–30
Gas	Oxygen bubbles	50–1,000		Smoke	<1
			Liquid	Oils	0.5–10
				Water	0.5–10

For a two-dimensional two-component (2D-2C) PIV measurement, the main flow direction and the laser light sheet for illumination generally have to be aligned, otherwise the out-of-plane problem of the illuminated particles, that is, the particles illuminated and pictured in the first image disappear from the illuminated plane and being replaced by others in the second image, resulting in the loss of coherence of the two images, could be serious.

Two types of lasers are commonly used for PIV illumination: continuously emitting laser also named continuous wave (CW) lasers and pulsating lasers. Pulsed lasers, for example, Nd:YAG lasers, can produce power light with high energy per pulse with very short intervals between two pulses. For a high-speed flow to be studied, pulsed lasers are therefore highly recommended, since short illumination times and a high pulse frequency are usually necessary to image the particles seeded in the flow. For a standard 2D-2C PIV, two pulsed lasers such as Nd:YAG lasers are used. The CW laser, for example, He–Ne laser or Ar laser, produces relatively low power light of good beam quality within a short interval. Pulsed lasers, like Nd:YAG lasers, are usually more expensive and more difficult to set up due to the added timing and synchronization equipment compared with CW lasers. In some particular cases, such as PIV studies of liquid Helium (e.g., Van Sciver et al., 2007; Zhang et al., 2004; Zhang and Van Sciver, 2005), CW lasers are less desirable as they provide a steady background heat loading to the experiment. Instead, short pulse Nd:YAG solid-state lasers operating at minimum power are effective.

For 2D-2C PIV measurement, the illuminated plane in which seeded particles to the flow should be clearly visible and homogeneously distributed, is imaged by a camera with its optical axis perpendicular to the plane. High quality lenses are generally acquired because the light powers are usually not very strong and large apertures have to be used. The up-to-date PIV commonly uses electronic cameras with a CCD or complementary metal-oxide semiconductor (CMOS) pick-up device. Electronic cameras are available in a whole range of resolutions, sensitivities, shutter speeds, and frame rates. For a certain flow to be studied, such as a turbulent flow with highly frequent events, repeatability, and illumination intensity need to be evaluated for choosing a suitable camera often in combination with the illumination system. Most of the electronic camera used for PIV are available with sensor resolutions of 512×512 pixels or $1\,K \times 1\,K$ pixels or higher. The frame rate of electronic cameras for PIV system can range from 25 to 30 fps, which is for the normal PIV, to 1,000 fps at full sensor resolution and much higher rates at reduced image sizes, which is for the time-resolved or dynamic PIV (e.g., Bi et al., 2003; Fore et al., 2005; Li et al., 2007; Triep et al., 2005). Dynamic PIV can be used for flow analysis, enabling mapping of unique flow velocity distributions in both space and time.

A double-exposure electronic camera is capable of recording two frames within a very short inter-frame interval and transferring both frames to the host system, and then the camera is prepared for the next image pair. Depending on the purpose, such as for measuring high speed repeatable flow events or only taking snap-shots of a flow, adjustment of the inter-frame times of double-exposure cameras can be conducted from 0.5 μs to even 0.5 s (Raffel et al., 2007; Stamhuis, 2006). On the other hand, a double-exposure camera has to be used in combination with a pulsed laser due to the construction of the camera shutting system. In addition to the combination of a double-exposure camera with pulsed lasers, a continuous video stream and a CW laser together with an external fast shutter could be a low-cost alternative for PIV application. The shuttering may result in relatively low illumination levels, which usually requires a more powerful CW laser.

2.3 PIV analysis

Figure 3a shows an example of the image pair taken by PIV cameras. Dense particle images are clearly visible in the pictures. It is not possible to measure the displacement of each tracer particle separately. Therefore, each of the frames is divided into small square areas named interrogation windows (Figure 3b). The size of the interrogation windows depends on the desired accuracy, the resolution of the measurement and the quality of the image recording. Typically, an unambiguous result for the particle-image displacement can be obtained from a minimum of 3–4 particle images using a correlation analysis. Keane and Adrian (1992) analyzed that the probability for a good analysis result is highest when the interrogation window contains approximately 8–10 particle images. Correlation analysis is then performed to find the displacement peak, by which the average displacement of the particles within each interrogation window is abstracted and the velocity vector is calculated. Repeating the procedures, the velocity distribution in the whole field of the PIV image is then obtained.

With the technological growth and performance improvement of digital cameras, digital cross-correlation methods are now almost generalized for analyzing recorded PIV images (Willert and Gharib, 1991). Cross-correlation removes the autocorrelation peak and directional ambiguity (i.e., there is only one peak) which are present in autocorrelation algorithm at the early stage of the development of PIV. Hence, only the issues related to cross-correlation approach of PIV are briefed as follows.

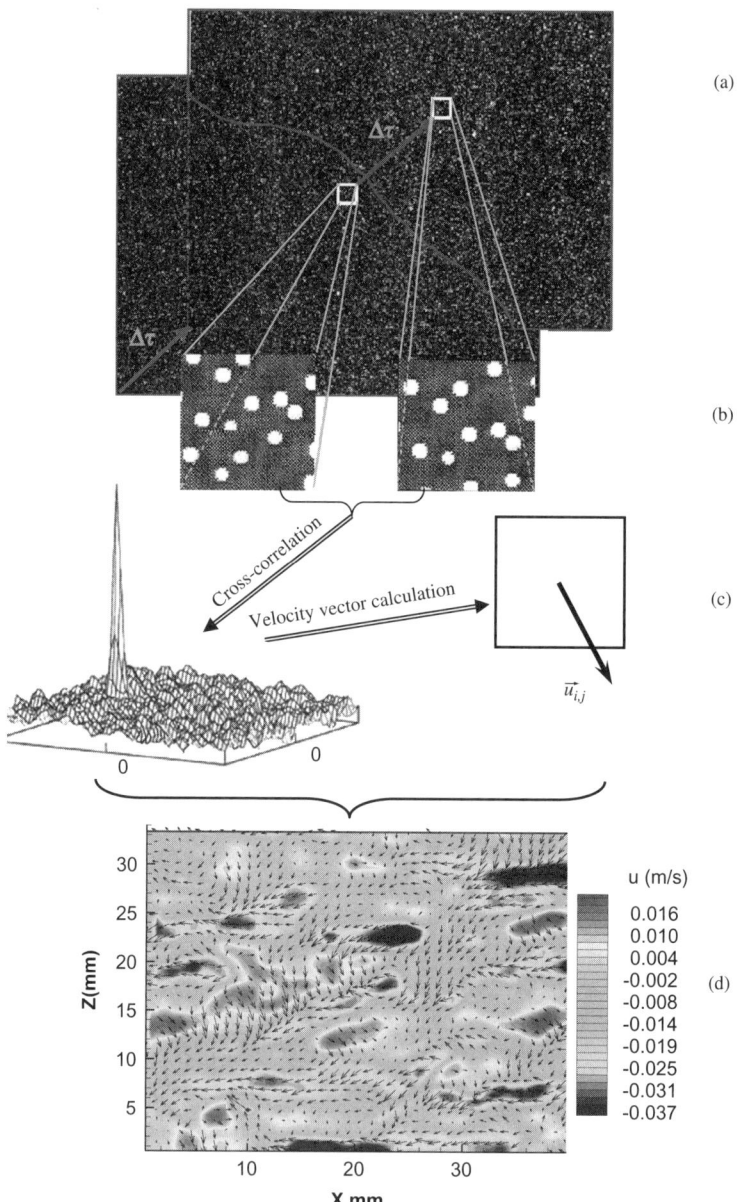

Figure 3 Diagram of the steps in PIV analysis of successively recorded particle images in a flow: (a) two successive PIV images; (b) two subimages called interrogation window from the same location of two frames; (c) a dominant peak of the cross-correlation calculation representing the most probable average displacement of particles containing in the interrogation window and through which the velocity vector is calculated; and (d) a whole field 2D velocity field obtained from the pair of PIV image.

2.3.1 Cross-correlation analysis for PIV

Traditional forms of digital PIV analysis algorithms have relied on autocorrelation and cross-correlation schemes. Cross-correlation implementations are typically favored over autocorrelation approaches since the former yielded directionally unambiguous displacements and have been found to provide more robust and superior performance, for example, lower correlation noise, larger dynamic range (the maximum velocity range that can be measured with a fixed set of instrumental parameters divided by the minimum resolvable velocity measurement; Adrian, 1997), and less gradient bias (Keane and Adrian, 1992; McKenna and McGillis, 2002). Supposing an interrogation window with size of $N \times N$ pixel, the discrete cross-correlation function is written as $c_{fg}(\Delta x, \Delta y) = \sum_{i=1}^{N}\sum_{j=1}^{N} f(x_i, y_j) g(x_i + \Delta x, y_j + \Delta y)$, where f and g are the variables (e.g., intensity values) extracted from both the images, (x_i, y_j) is the current position of calculation, and $(\Delta x, \Delta y)$ the sample shift. The normalized cross-correlation, that is, cross-correlation coefficient given by the following equation, is often used:

$$R_{fg}(\Delta x, \Delta y) = \frac{\sum_{i=1}^{N}\sum_{j=1}^{N}\left\{f(x_i, y_j) - f_m\right\} g\left\{(x_i + \Delta x, y_j + \Delta y) - g_m\right\}}{\sqrt{\sum_{i=1}^{N}\sum_{j=1}^{N}\left\{f(x_i, y_j) - f_m\right\}^2 \sum_{i=1}^{N}\sum_{j=1}^{N} g\left\{(x_i + \Delta x, y_j + \Delta y) - g_m\right\}^2}}$$

where f_m and g_m are the average values within the $N \times N$ pixel subareas in both PIV images, respectively. Early PIV cross-correlation analyses were performed in the Fourier domain, as this provided the most efficient means of processing the data (e.g., McKenna and McGillis, 2002; Willert and Gharib, 1991). With the development of faster computer CPUs, the need for transformation into frequency space is becoming less imperative and some methods of correlation are performed mandatorily in the spatial domain directly.

To get the most probable displacement of the particle pattern in the interrogation window, a mathematical correlation procedure is applied. One can imagine this procedure as "moving window 1 over window 2 until the best matching is found." "Best matching" is used here since in practice 100% matching can never be reached due to particles that have left or entered the interrogation window in the second image compared with the first. Mathematically, there are two methods used for image correlation analysis: Fourier transformations and convolution filtering. The results of both algorithms are comparable. Details on the mathematics of both methods have been well documented (e.g., Raffel et al. 2007).

Performance of cross-correlation leads to the correlation plane as seen in Figure 3c, which depicts the correlations versus the displacement in x

and y direction of a 2D coordinate system. Figure 3c shows a typical result of the calculation of the cross-correlation, with many small peaks and only one dominant peak. The displacement that belongs to this dominant peak is the most probable displacement because the dominant peak indicates the "best matching" of particle images between the two interrogation windows. With the abstracted most probable displacement, the (mean) velocity vector of the interrogated subimage area is then calculated by dividing the time interval between the two investigated PIV images, $\Delta\tau$, and is usually located at the center of the present interrogation window (Figure 3c). Repeating the preceding procedures, a whole field 2D velocity distribution is yielded finally (Figure 3d).

2.3.2 Resolution improvement

Spatial resolution is one of the most important points of the performance of a PIV system. For conventional correlation analysis of PIV, the spatial resolution is bounded by the size of the measurement volume, which is determined by the intersection of the illuminated light sheet with the interrogation window density distribution (Keane et al, 1995). In contrast with the limiting role of the interrogation window size in the resolution of conventional correlation PIV methods, an iterative multipass correlation PIV analysis approach is developed and elaborated by Nogueira et al. (1999, 2001a, b, 2005a, b), which yields resolutions below the window size using a particular weighting function (Huang et al., 1993; Jambunathan et al., 1995). The multigrid analysis obtained by progressively refining the interrogation window size and deformation during iterative interrogation is proved to increase the accuracy as well as spatial resolution of the PIV technique (Fincham and Delerce, 2000; Scarano and Riethmuller, 2000; Theunissen et al., 2007). Several well-documented approaches for improving spatial resolution are briefed as follows.

Particle tracking velocimetry (PTV) algorithms are based on the tracking of individual particle images and allow in principle the highest spatial resolution, that is, one vector for each detected particle (Keane et al., 1995; Theunissen et al., 2007). The super-resolution approach is then developed, which incorporates conventional cross-correlation PIV followed by subgrid particle tracking within the interrogation window (Bastiaans et al., 2002; Keane et al., 1995; Stitou and Riethmuller, 2001; Susset et al., 2006). The essence of this method is to enhance particle tracking by means of the PIV information. The technique is a hybrid algorithm that starts by statistical evaluation of a tracer's displacement (PIV correlation) and further refines the resolution of measurement with the tracking of individual particles (PTV) within the interrogation window. This technique improves not only the spatial resolution but also the accuracy. The velocity field obtained is unstructured due to the random position of the particles inside the flow field. For more convenience, for example, to compare with other databases or data analysis in a structured form, a data redistribution method can be applied on a structured grid.

Window displacement iterative multigrid (WIDIM) interrogation method has been proposed and developed to improve the resolution for PIV (Fincham and Delerce, 2000; Scarano, 2002, 2004; Scarano and Riethmuller, 1999, 2000; Soria, 1996; Theunissen et al., 2007). The essence of WIDIM is to compensate for the loss-of-pairs of particles due to in-plane motion. To do this, the local displacement of each interrogation window is made on the basis of a flow pattern prediction. The predicted displacement is obtained by a previous interrogation of the set of two PIV images, therefore, it is necessary to adopt an iterative procedure (Scarano and Riethmuller, 1999). At the start of the process, no *a priori* information on the flow pattern is available and the first predictor is set uniformly to zero. After the first interrogation, the coarse result will be used as a predictor. A finer windowing is then made halving the windows in both directions and the predictor is applied to the window offset by means of simple substitution of the previous iteration result. As a consequence, in the subsequent steps, the one quarter rule (Raffel et al., 2007) related to the in-plane displacement does not limit anymore the size of the windows and the PIV images can then be interrogated with a better resolution. Involving similar consideration, Hart (2000a) achieved an increased spatial resolution by recursively correlating the image frames at finer grid sizes down to the size of an individual particle image. Correlation search length is also reduced iteratively to the smallest meaningful scale parallel with the decrease of interrogation window size.

Local-field correlation particle image velocimetry (LFCPIV) is another method to resolve flow structure smaller than the interrogation window size (Nogueira et al., 1999, 2001a, 2005a, b). It is also said that LFCPIV is the only correlation PIV method being able to yield super resolution (Nogueira et al., 2001a). The interrogation windows for LFCPIV are fixed in size and location. After each iteration step, however, the image is redefined through compensation of the particle pattern deformation caused by the velocity gradient in the flow field. This implies both displacement and deformation of the image. It is performed using the displacement field from the previous evaluation. As a result, this method presents the ability to resolve small structures with large interrogation windows. In LFCPIV, a proper weighting function is defined to avoid the instability related to high spatial frequency. The combination of the two capabilities, that is, the ability to cope with large velocity gradients and to resolve small structures in the flow, results in a very robust high-resolution technique (Nogueira et al., 2001a, b).

There are also several other alternatives to resolution improvement methods for PIV, such as hierarchical processing method, reverse hierarchical processing method and so on. For those, one can refer to the relative references (Kumar and Banerjee, 1998; Rohaly et al., 2002; Susset et al., 2006; Westerweel et al., 1997, among others).

2.3.3 Inherent error elimination and accuracy improvement

The accuracy of PIV measurement (before post-processing) depends on several factors, including the properties of target flow fields, the properties and concentration of seeding particles, the optical setup, the data acquisition system, the image interrogation technique, etc. (Chen and Katz, 2005). Among them, the error introduced during image interrogation has received the most attention since it provides the widest latitude for development of optimization tools (Chen and Katz, 2005; Huang et al., 1997; Huang, 1998; Keane and Adrian, 1990; Lecordier et al., 2001; Westerweel, 1997). The uncertainties associated with PIV measurements can be classified into two categories: random error and bias error. The random errors in PIV are most often associated with electronic noise in the cameras, shot noise, and random errors associated with properly identifying the subpixel displacement (Christensen, 2004). Owing to its random nature, the influence of random errors can be reduced by statistical analysis using a sufficiently large ensemble set. Bias errors are not random in space and time and can degrade not only the accuracy of instantaneous PIV results, but also any statistic computed from biased PIV ensembles. Several bias errors can exist in a PIV measurement, including uncertainties associated with the fill ratio of the CCD camera and the algorithm used to interrogate the images (Christensen, 2004). The most significant bias error is the so-called peak-locking error (e.g., Christensen, 2004; Fincham and Spedding, 1997; Hart, 2000b; Raffel et al., 2007), that is, the biasing of particle displacements toward integer pixel values, which inherently stems from the choice of subpixel finding algorithm, underresolved optical sampling of the particle images, and the truncation of particle images by the borders of the interrogation window (Nogueira et al., 2001a). Here, some of the techniques coping with the peak-locking error are summarized briefly as follows.

As mentioned previously, the location of the peak in the cross-correlation plane of a pair of interrogation windows yields the mean particle displacement within the first window. To achieve subpixel accuracy, a smooth curve is typically fitted through 3–4 points in the vicinity of the discrete correlation peak. This subpixel curve fitting causes a bias toward discrete values of displacement (i.e., peak-locking error). To remedy or eliminate the peak-locking effect, several methods have been proposed. Westerweel (1997) proved that a Gaussian subpixel estimator is superior to both centroid and quadratic fits in terms of mitigating peak-locking effects. A *Sinc* function is recommended by Roesgen (2003) to suppress the spurious spectral side lobes in the correlation resulting in minimal peak-locking effects for adequately resolved particle images. On the basis of the fact that there should be no bias error when the true displacement is an integer pixel, a solution to reduce peak-locking is to apply the continuous window shifting technique (Gui and Merzkirch, 2000; Gui and Wereley, 2002; Liao and Cowen, 2005; Nogueira et al., 2001a), that is, iteratively shift the interrogation window by fractional displacements until the final determined subpixel displacement is driven to zero.

Since the peak-locking error is due to the subpixel curve fitting, more advanced error-elimination methods have been proposed by bypassing the subpixel curve fitting. Fincham and Delerce (2000) developed a peak anti-aliasing, spline transformed interrogation scheme involving interpolation of pixel values but without involving subpixel fitting to the correlations. Chen and Katz (2005) proposed a correlation mapping method to eliminate the peak-locking error in PIV analysis, which bypasses the subpixel curve fitting and so eliminates the peak-locking effect, but does not require iterations as the continuous window shifting technique does. Using subpixel interpolation, this method expresses the second exposure of an interrogation window as a polynomial function with unknown displacement, whose coefficients are determined by the grayscale distribution of the first image (Chen and Katz, 2005). Thus, the correlation between this function and the first exposure is also a polynomial of the displacement. This virtual correlation function can be matched with the exact correlation value at every point in the correlation map. A least-squares method is used to find the optimal displacement components that minimize the difference between the real and virtual correlation values in an (e.g., 5×5 pixel) area surrounding the discrete correlation peak. The correlation mapping method returns superior results compared to the Gaussian subpixel interpolation.

2.4 Post-processing of velocity vectors

The resultant velocity vector set of automated PIV analysis often includes a certain number of incorrect vectors that are usually obvious in the vector diagram. Such spurious vectors are usually due to imperfections in the input PIV images, caused by local variations in seeding density, local over-illumination due to an object or wall in the light sheet, strong out-of-plane flow, local low illumination close to the image borders, or crippled interrogation windows next to the image border. These problems cause lack of correlation in the normal way, and a background (noise) peak is then recognized as if it were the displacement peak, resulting in a spurious vector.

Elimination of the spurious vectors is usually performed by comparing them with their neighbors. Errors are recognized if they are inconsistent with neighbors in some statistical or physical sense. False vectors can also be identified by the ensemble method, in which outliers are removed from the ensemble-averaged velocity field based on the standard deviation of the velocity at each grid (Wernet, 2000). The simplest and commonly used post-processing technique is the local-median method. The median is the middle value (nth element) of a sequence of $2n+1$ scalar elements (Westerweel, 1994), that has been rearranged in increasing or decreasing order (for even number of elements, the median is equal to the mean of the two middle elements). Westerweel (1994) showed that the local-median method is more efficient than the global- and local-mean methods. The local-median method with constant user-adjustable thresholds is usually adequate to detect the spurious vectors when a proper threshold is set for a specified flow field. However, a single constant threshold is generally not applicable in complicated flows such as inhomogeneous gradient flows or vortical flows. For this, Liu et al. (2008) proposed a flow-adaptive data validation scheme to avoid the selection of the appropriate thresholds for specified flow fields, which shows superior performance to the local-median method.

Once the false or spurious vectors are recognized and removed, empty grids should be filled by new vectors, which should be representative of the local flow velocity as close as possible, for continuous calculation of local derivatives and gradient parameters. To do this, 2D interpolation is commonly performed. Although simple linear or polynomial interpolation can obtain reasonable results, the best and also principally the most reliable results have been yielded with 2D cubic natural spline interpolation (Spedding and Rignot, 1993). Cubic natural spline fits leave the original neighboring data unaffected and find the missing points according to "the most fluent line or plane," that is, with minimization of the local curvature. This algorithm works well for shock waves in transonic flow as well as in a noncompressible or subsonic compressible fluid flows. In some cases more often than not, spurious vectors may still exist in a PIV measured 2D velocity field after performing data-validation procedures with any methods. Afanasyev and Demirov (2005), thus, designed a variational filtration and interpolation technique to reduce the effect of spurious vectors, which cannot be removed by other methods or to reconstruct the velocity field in the areas where the outliers are removed. This method is based on the application of dynamical constraints such as continuity, smoothness, and matching to the original data, rather than a simple statistical approach. After the treatment of spurious vectors, that is, detection, elimination, and filling the holes, final analysis of the acceptable velocity vectors can be finally conducted with physical meaning.

3. VARIOUS TYPES OF PARTICLE IMAGE VELOCIMETRY

At the very early stage, the light scattered by the tracer particles for a PIV system is recorded via a high quality lens on a film camera and the PIV measurement is of course only 2D-2C (C means component of velocity). After the development of the photographical PIV, recording is digitized by means of a scanner and the output of the digital sensor is transferred to the memory of a computer directly, and such PIV system is the conventional 2D digital PIV, which is introduced in last section. The development of PIV techniques during the past 20 years has brought about a variety of PIV system. In this section, the up-to-date various types of PIV are summarized from the viewpoint of the hardware components of the system, from 2D-2C to 3D-3C PIV, and to time-resolved dynamic PIV, etc., for measurement of both macro and micro-scaled velocity field.

3.1 2D-2C PIV techniques

3.1.1 Macro scale

A typical 2D-2C PIV system is shown in Figure 1, composing with two pulsed lasers, optical lenses for laser-sheet formation, a CCD camera, a computer with PIV software installed, a timing circuit communicating with the camera and computer, and generating pulses to control the double-pulsed laser. Detailed information for the conventional macro-scaled 2D-2C PIV have been introduced in Section 2.

3.1.2 Micro scale

The fundamental difference between 2D-2C micro-scaled PIV (μ-PIV) and traditional PIV lies in the method of particle illumination (Bourdon et al., 2004a, b; Santiago et al., 1998). In traditional PIV, a thin region of the flow field is illuminated by a laser sheet, and the measurement depth is determined by the laser sheet thickness and intensity distribution. Instead of illuminating particles with a laser sheet, the entire volume of the microfluidic device of interest is illuminated, and the measurement volume depth is defined by such factors as the optics of the imaging system, wave length of emitted light, particle size (Meinhart et al., 2000; Olsen and Adrian, 2000a), the magnitude of Brownian motion (Olsen and Adrian, 2000b), and the out-of-plane velocity component (Bourdon et al., 2004a, b; Olsen and Bourdon, 2003). Figure 4 shows a typical 2D-2C μ-PIV setup, assembled about an inverted microscope system. The microfluidic device to be investigated is placed above the objective of an epifluorescent-inverted microscope. Light from the light source (mercury lamp as shown in Figure 4 or an Nd:YAG laser) enters the microscope through an aperture and is focused onto a small region of the

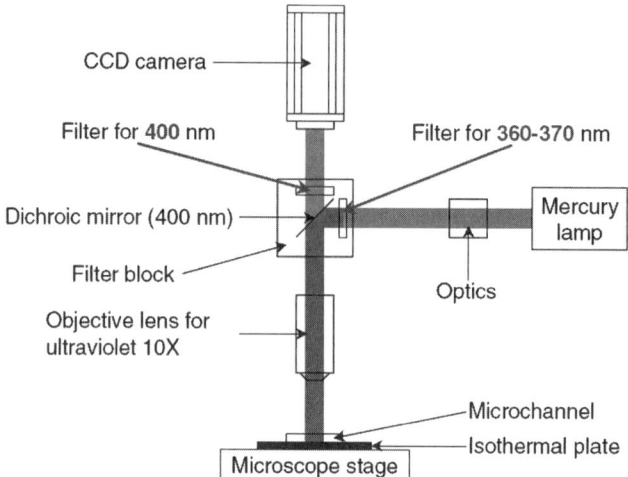

Figure 4 Schematic diagram of a typical μ-PIV (Sato et al., 2003). The light source can be mercury lamp or double-pulsed ND:YAG laser (see Plate 5 in Color Plate Section at the end of this book).

microfluidic device by the imaging objective, illuminating the entire depth of the fluid. The flow within the microfluidic device is seeded with small, fluorescent particles (order of 1 micron in diameter in general) that absorb the illuminating light and emit light at a different (usually longer) wavelength. The emitted light from the particles, as well as scattered and reflected light from light source, is long-pass filtered and imaged by a CCD camera. These filtered images containing only the light emitted by the fluorescent particles are then analyzed using PIV techniques (as described in Section 2.3) to yield the velocity vectors data. Since the camera-sampled images consist not only of in-focus particle images due to particles located at the object plane, but also out-of-focus particle images of various dimensions and intensities due to particles located away from the focal plane. Discussions on the issues associated with this major difficulty, the so-called "depth of correlation" (Olsen and Adrian, 2000a), have been well documented (Meinhart et al., 2000; Olsen and Adrian, 2000a, b). To overcome the abovementioned "depth of correlation" problem of μ-PIV, two techniques are developed, that is, selective seeding for μ-PIV and confocal μ-PIV technique.

Mielnik and Saetran (2006) designed the selective seeding technique for μ-PIV. The principles of this technique are based on selectively seeding a thin sheet of liquid within the flow in a microfluidic device. To do this, a three-layer-flow structure is generated in a microfluidic device to be investigated, by setting three inlet ports for the flow passage. Only the sheathed flow (in the center of the three) is seeded with fluorescent particles. By changing the flow rate of the other two flow streams (without seeding and keeping the same flow rate for each), the thickness of the seeded sheathed flow can be controlled. In this manner, the measurement plane is completely defined by the particle sheet, and in principle, a measurement depth corresponding to the diameter of the tracer particles may be achieved (Mielnik and Saetran, 2006). This technique is only applicable to the flow with inherent laminar flow characteristics.

Confocal µ-PIV technique is more advanced (but expensive), and emerges after the considerable progress in the development of confocal microscopy and the advantages of this technique over conventional microscopy. This method combines the conventional PIV system with a spinning disc confocal microscope (Kinoshita et al., 2007; Lima et al., 2006, 2007, 2008; Park and Kihm, 2006; Park et al., 2004). By combining its outstanding spatial filtering technique with a multipoint illumination system, the confocal microscope has the ability to obtain in-focus images with optical thickness less than 1 µm, which is extremely difficult to achieve using conventional microscopy (Lima et al., 2006). It is therefore possible to achieve a confocal µ-PIV system with an extremely high spatial resolution, that is, true depth-wise resolved µ-PIV vector field mapping (Park et al., 2004).

3.2 2D-3C PIV techniques

2D-3C PIV techniques generally refer to the stereoscopic PIVs, which allows for the measurement of three-component velocity vectors in a 2D illuminated plane by utilizing simultaneous viewing with two cameras from two directions to obtain depth-perception. Such twin camera systems mimic the binocular vision that enables human beings to distinguish between objects near and far (Prasad, 2000). In addition to the availability to measure the third velocity component in the direction perpendicular to the laser sheet, stereoscopic PIV also adds the advantage of eliminating perspective error which exists in conventional 2D-2C PIV in the presence of out-of-plane component of velocity (e.g., Gaydon et al., 1997).

3.2.1 Macro scale

There are two basic stereoscopic configurations for the macro-scaled stereoscopic PIV: the translational method and the angular method. In the translation method, the optical axes of the camera lenses are parallel, so the image magnification is constant. A disadvantage is that the stereoscopic viewing angle must remain small (less than 30°) to avoid image distortion. The angular method can be used with much larger stereoscopic viewing angles up to the maximum of 90°. Thus, it can achieve a higher precision, and is generally preferred over the translation

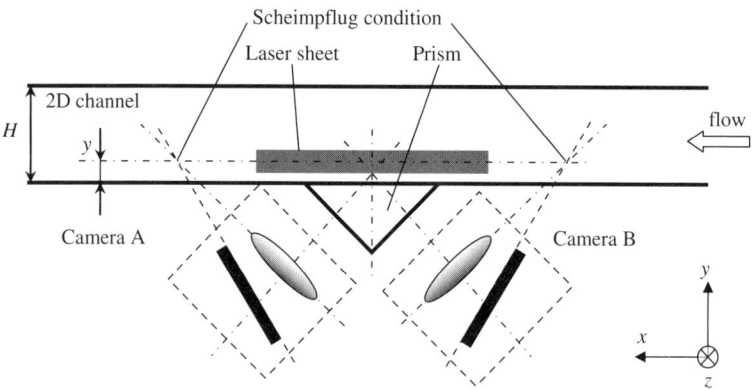

Figure 5 Schematic diagram of optical configurations for stereoscopic PIV measurement (measurement of velocity field in a x–z plane in a 2D channel flow), viewing for stereoscopic imaging with the Scheimpflug condition (Li et al., 2006a).

method (e.g., Coudert and Schon, 2001). Figure 5 shows an example of the schematic diagram of a stereoscopic PIV system with angular method.

For a stereoscopic PIV system with angular method, the image plane has to be tilted with respect to the optical axis to match the so-called Scheimpflug condition, that is, the image plane, lens plane, and object plane for each of the cameras intersect in a common line (Hinsch, 1995; Prasad and Jensen, 1995). Thus, the image magnification is no longer constant, that is, the images have a perspective distortion, and so a careful calibration procedure is necessary before any stereoscopic PIV measurement. The calibration procedure will output a space and calibration coefficient map across the image where the calibration coefficients are related to the spatial distortion of the image (Lawson and Wu, 1997). This calibration coefficient map bridges the back-projection of the data from image plane to the object plane for the reconstruction of the three velocity components. For details of stereoscopic PIV system one can refer to Soloff et al. (1997), Lawson and Wu (1997), and Prasad (2000), Raffel et al. (2007).

The calibration is a key issue before stereoscopic PIV measurement. Almost all calibrations for the reported stereoscopic PIV study utilize a target, which consists of a discrete number of markers placed on a regular Cartesian grid. However, not all the flow environment to be investigated is convenient to place a calibration target in the measurement region. To overcome this difficulty, three novel calibration methods for stereoscopic PIV are reported recently. Akedo et al. (2005) carried out a stereoscopic PIV study on flow characteristics in a cerebral aneurysm model. Since the flow passage geometry is quite irregular and it is impossible to place and move a target plate for calibration purpose. In that reason, Akedo et al. (2005) proposed a new calibration technique using lasers. The positions of calibration points are indicated as crosspoints of horizontal and perpendicular laser beams from the precisely positioned optical axis. After recording the positions of those crosspoints of laser beams (each cross-point is actually reconstructed after recording two perpendicular laser beams crossing that point) by two cameras (with angular method), the calibration procedure can be performed as usual. Wieneke (2005) designed a stereoscopic PIV technique using self-calibration on particle images. This stereoscopic PIV calibration procedure is developed based on fitting a camera pinhole model to the two cameras using single or multiple views of a 3D calibration plate. The key feature of this technique is that it is possible to derive accurate mapping functions even if the calibration plate is quite far away from the light sheet, making the calibration procedure much easier. Hence, this method allows stereoscopic PIV measurements to be taken inside closed measurement volumes (Wieneke, 2005). Fouras et al. (2007, 2008) developed a novel, accurate, and simple calibration-target-free stereoscopic PIV technique utilizing three cameras. The key feature of this technique is that there is no need of a separate calibration phase but utilizes a third camera placed in paraxial (normal to the laser light sheet) position. This calibration-target-free technique offers the advantages of the calibration-target-based stereoscopic PIV, with even greater improvements in reconstruction accuracy and without the requirement of the practitioner to conduct a distinct calibration phase, and is greatest utility when the paraxial view has minimal distortion or when it is not convenient to place a calibration target in the measurement region (Fouras et al., 2007, 2008).

There are also other types of stereoscopic PIV configurations. Gaydon et al. (1997) proposed a hybrid stereo-camera combining features of translational and angular systems. Essentially, this hybrid system uses a small translation between the camera axes as well as a small inward rotation of the axes, which is for increasing the off-axis angle beyond that in an equivalent translation method, but without increasing the nonuniformity in magnification to the level of an equivalent angular method. Grant et al. (1995) introduced an in-line stereoscopic PIV system, in which the two cameras are in-line arranged and both perpendicular to the object plane. The in-line arrangement is facilitated with the use of a semi-silvered mirror placed along the common optical axis. The requirement for differing views is satisfied by using differing magnifications and object-distances for each camera. One advantage of this in-line stereoscopic system is that each particle forms images at the same angular location on each film, which makes the particle matching easier for their particle-tracking algorithm. Arroyo and Greated (1991) devised a setup to take two stereoscopic images of the flow to be investigated simultaneously with only one camera, by means of a mirrors system. The idea is to take into account that correlating two photographs taken by two cameras can be a source of error and time consuming. Using such system, the correlation between the two photographs taken by only one camera is very straightforward and can be done automatically once everything is set up (Arroyo and Greated, 1991).

3.2.2 Micro scale

The stereoscopic μ-PIV is another version of 2D-3C PIV technique, which is based on several common components and procedures (Lindken et al., 2006) including the 2D-2Cμ-PIV method as first introduced by Santiago et al. (1998), the stereo-PIV method (e.g., Prasad, 2000), and stereomicroscopy. Figure 6 shows the schematic diagram of a stereoscopic μ-PIV system. The calibration procedure named "self-calibration on particle images" (Wieneke, 2005) has to be applied for the stereoscopic μ-PIV due to the very small confinement of the measurement target. Although preliminary results have been published from stereoscopic μ-PIV systems (Bown et al., 2006; Lindken et al., 2006), there are several limitations to this technique. A limitation is that stereoscopic μ-PIV requires a stereo-objective that typically has a low numerical aperture (NA = 0.14–0.28) and large depth-of-focus as opposed to high NA lenses (up to 0.95 without immersion) for standard μ-PIV (Lindken et al., 2006). It is also found that the accuracy of the correlation-based PIV technique is limited by the degree of overlap of the two focal planes in the stereomicroscope (Bown et al., 2006).

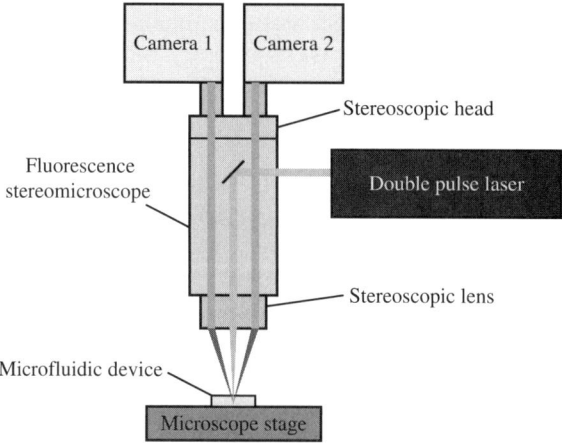

Figure 6 Schematic diagram of a stereoscopic μ-PIV system.

3.3 3D-3C PIV techniques

3.3.1 Macro scale

There have been several different 3D-3C (or volumetric) PIV techniques for macro-scaled flow measurement. Among them, holographic PIV (HPIV), defocusing PIV (DPIV), and tomographic PIV (TPIV) are the three that inherently measure all the three components of velocity field in a 3D volume. Other PIV techniques available to obtain a 3D velocity field, such as scanning PIV, dynamic PIV, that are based on some indirect algorithms to reconstruct a 3D velocity field, will be mentioned in the later sections.

3.3.1.1 Holographic particle image velocimetry. HPIV technique utilizes the interference pattern of a reference light beam with light scattered by a particle, which is recorded on a hologram, to determine the particle location in depth (Hinsch, 1995, 2002; Meng et al., 2004; Royer, 1997; Svizher and Cohen, 2006a, b), and then calculates the instantaneous 3D velocities of those particles in the illuminated volume by finding the 3D displacements of the particles between two exposures separated by a short time delay. Figure 7 depicts the principle of HPIV (Meng et al., 2004). The early HPIV systems are based on film recording of the images and so limited to double-exposure single-frame measurement. Digital HPIV emerges from about the beginning of the 21st century, which directly records a time series of holograms using a CCD camera and reconstructs the 3D velocity field through correlation-based analysis of the digital images. The evaluation method (correlation method) of the digitized data of HPIV is similar to those used in the classical 2D PIV. Although the wide application of film-based HPIV technique has been hampered by its formidable cost and technical complexity involved, the digital HPIV, with the relatively simple hardware components and the

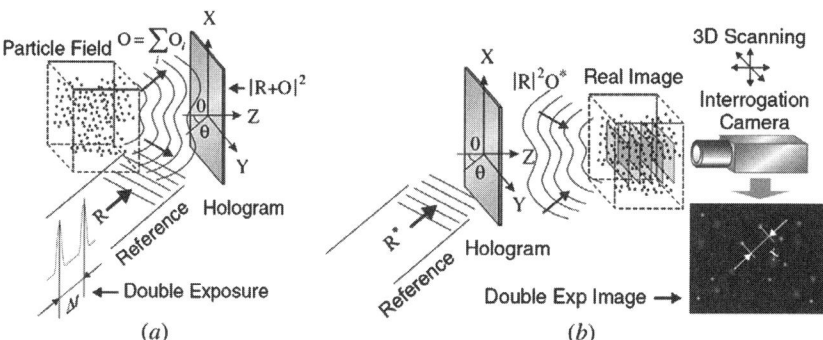

Figure 7 Principle of HPIV with traditional setup: (a) holographic recording of particle ensembles and (b) holographic reconstruction of particle images and their interrogation (Meng et al., 2004).

relative ease of operation, is hopefully to be developed and expanded as a robust, reliable, and commercialized four-dimensional (3D in space and 1D in time) flow field measurement technique.

HPIV configurations can be broadly categorized into two types based on the nature of the holographic scheme: in-line HPIV, where only one beam is employed to produce both the object wave and the reference wave, and off-axis HPIV, where separate object beam and reference beam(s) are introduced (Pu and Meng, 2000). For off-axis HPIV measurement, the seeding particles in the measured flow are illuminated with a properly expanded laser light wave, and the scattered light field from particles is stored by superposing a reference wave from the same laser source and having an oblique angle relative to the object light and recording the interference pattern on a file or a CCD sensor. The off-axis HPIV inherently requires good coherence and relies on an accurate reproduction of the reference wave for reconstruction of the image of an object, and is also easy to be suffered from vibrational noise. The in-line HPIV can provide better stability but has relatively simple arrangement and requires lower laser coherence and energy. The conventional in-line HPIV usually encounters with difficulty when the seeding particles have large density (e.g., at a level for a normal PIV measurement), since excessive speckle noise stemming from the superposition of the real image, virtual image, and reference waves may be produced to interfere with the reproduction of particle images. Another problem with in-line HPIV associated with the measurement accuracy is the large depth-of-focus in the reconstruction of particle images. Among the approaches proposed for the abovementioned problems, the in-line recording off-axis viewing technique is proved to be an efficient method to suppress the noise and improve the accuracy.

HPIV is rather complicated (and expensive) compared to 2D PIV, since it uses coherent optics for imaging, and it records, transmits, and processes information in not just two but three dimensions (Meng et al., 2004). The key issues of HPIV system are reduction of speckle noise, handling of huge quantities of data, extraction of 3D velocity in presence of large gradients and fluctuations and system complexity (Pu and Meng, 2000). Studies toward improvement of data analysis techniques for HPIV are still under development (e.g., Ooms et al., 2006, 2008). A "universal" HPIV scheme, that can provide successful measurements for any given flow, has not yet been proposed, and so the HPIV system requirements imposed by the flow phenomenon to be studied have to be first estimated before choosing a suitable approach (Svizher and Cohen, 2006a, b). Nevertheless, the HPIV technique has been successfully applied to measure a variety of flow phenomena, for example, 3D characteristics and statistics of a turbulent water flow in a square channel (Tao et al., 2000, 2002; van der Bos et al., 2002; Zhang et al., 1997); a cylinder wake flow in air and a free air nozzle flow (Herrmann et al., 2000); acoustically excited air jet and the wake of a surface-mounted tab in water channel flow (Pu and Meng, 2000); large wind-tunnel flow (Herrmann and Hinsch, 2004); instantaneous topology and 3D structures of hairpin vortices in a turbulent air channel flow (Svizher and Cohen, 2006a, b); instantaneous flow fields within a motored Diesel engine (Coupland et al., 2006), and has shown its great powerfulness.

3.3.1.2 Defocusing particle image velocimetry. DPIV technique utilizes a defocusing or blurring concept to obtain information regarding a particle's position in space, as shown in Figure 8 (Kajitani and Dabiri, 2005). This concept is applied to 3D particle tracking in a flow of a vortex ring by Willert and Gharib (1992), in which a single camera with a

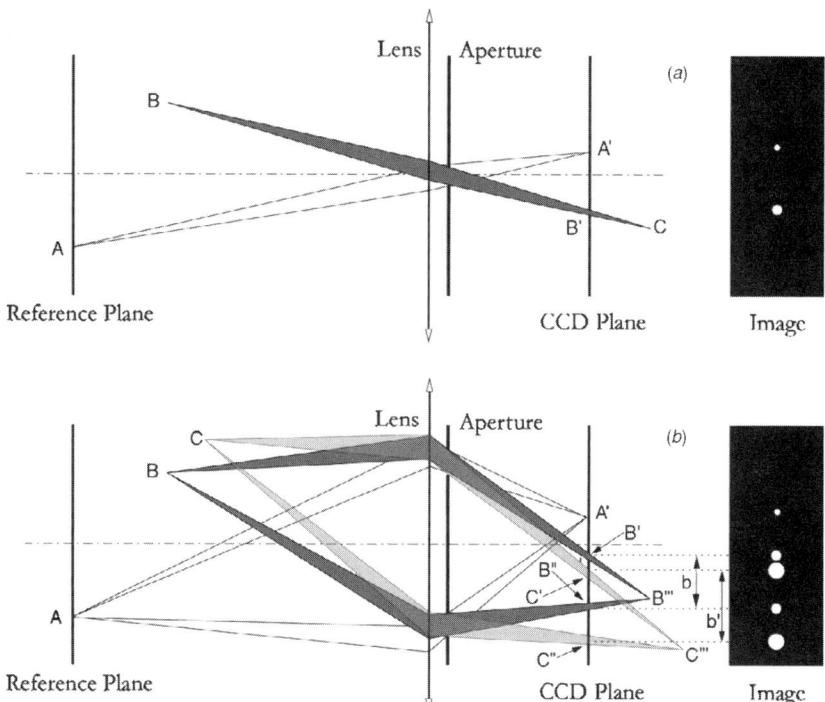

Figure 8 Defocusing concepts: (a) a standard defocusing setup with aperture on-axis; (b) defocusing setup with two off-axis apertures. Point A focuses from the reference plane onto the CCD plane; point B focuses behind the CCD plane at point B''', leaving two slightly blurred images on the CCD plane (B' and B') at a distance b apart; point C focuses further behind the CCD plane at point C''', leaving two slightly larger blurred images on the CCD plane (C' and C'') at a larger distance b' apart (Kajitani and Dabiri, 2005).

modified three-hole aperture is used. Lately, three individual cameras are used instead of a single one to improve accuracy with large three-hole separation (Kajitani and Dabiri, 2005; Pereira and Gharib, 2002; Pereira et al., 2000), which combines the three camera images into one image for processing to obtain the triangle-image of each particle. Pereira and Gharib (2002) introduced a DPIV system and its geometric analysis, and derive fundamental equations to estimate 3D particle locations in the measured flow field. Kajitani and Dabiri (2005) described full 3D derivations and modified equations for a DPIV system. The processing of the DPIV technique using three cameras generally includes the detection of all particle images, 2D Gaussian functions fit of the particles, identification of triplets of detected particle images forming equilateral triangles, and a 3D cross-correlation displacement estimation. The DPIV technique has been used to map the bubbly flow field in a two-blade model boat propeller (Pereira et al., 2000) and in the wake of a hydrofoil (Jeon et al., 2003), and may potentially be used to study transient flow phenomena and time-averaged statistical behavior due to its ability to acquire sequences of particle images (Kajitani and Dabiri, 2005).

3.3.1.3 Tomographic particle image velocimetry. In TPIV (as shown in Figure 9 for the principle), the pulsed light source illuminating fluid volume is viewed and recorded simultaneously from several directions by at least three digital cameras (Elsinga et al., 2006a). The setting of digital cameras is similar to stereoscopic PIV, that is, Scheimpflug condition between the image plane, lens plane, and the mid-object plane is satisfied, and so the recorded images are distorted, which requires a calibration procedure common to stereoscopic PIV to establish the relation between image coordinates and the physical space (Elsinga et al., 2006a). The cameras record 2D image projections of the 3D particle distribution in the viewed fluid volume. The 3D particle distribution is then recovered as a 3D light intensity distribution from its 2D projections, which does not rely on particle identification techniques as performed in photogrammetric PTV scheme (e.g., Maas et al., 1993). The reconstruction is an inverse problem and its solution is not straightforward: a single set of projections can result from many different 3D objects. Thus, recovering the most likely 3D distribution of particles is the topic of tomography. After reconstruction of the 3D distribution, the particle displacement (and in consequence velocity) within an interrogation volume is then obtained by the 3D cross-correlation at the two exposures, which can be an extension of any matured 2D cross-correlation algorithm to 3D. TPIV is of great attractiveness because the optical configuration is similar to a stereoscopic PIV system. The differences are only the number of cameras used and novel software for data processing. Elsinga et al. (2006a, b)

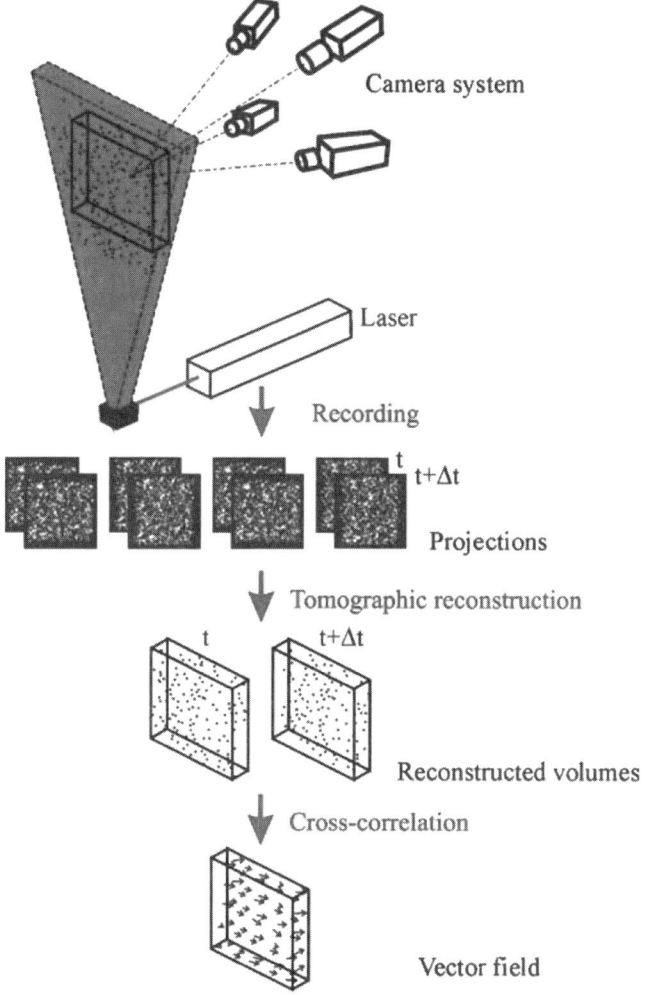

Figure 9 Principle of tomographic PIV (Elsinga et al., 2006a).

validated the TPIV technique through measurement of a circular cylinder wake flow. The results show a good agreement of the returned flow statistics compared with a stereoscopic PIV measurement (Elsinga et al., 2006b). In another feasibility study, the TPIV technique is used by Schroder et al. (2008) to time-resolved PIV recordings for the study of the growth of a turbulent spot in a laminar flat plate boundary layer and to visualize the topology of coherent flow structures within a tripped turbulent boundary layer flow. The complete time-dependent 3D velocity gradient tensor (VGT) within the measurement volume, a model of the connection of the turbulence producing Qudrant-2 and Qudrant-4 events in a spatio-temporal flow topology, and some important aspects of Lagrangian fluid dynamics of the turbulent flow, have been successfully derived from this TPIV study (Schroder et al., 2008).

3.3.2 Micro scale

For micro scale flow measurements, there are very few papers mentioning inherent 3D-3C PIV techniques. Yang et al. (2004) introduced a digital μ-HPIV method using holographic principles, which is similar to the macro-scale HPIV technique in principle. However, holographic-based approaches need complicated system and involve difficulties with analysis and noise (Yang et al., 2004; Yoon and Kim, 2006). The μ-HPIV technique is still under development. Very recently, contrast enhancing techniques in digital holographic microscopy are discussed by Lobera and Coupland (2008). The implementation of contrast enhancing techniques as a post-processing tool increases the analysis' possibilities. Development in holographic microscopy is bound to enhance the improvement in μ-HPIV technique.

μ-DPIV is another PIV technique that inherently obtains a 3D-3C velocity field at micro scale, which goes a little further than μ-HPIV technique. Yoon and Kim (2006) and Pereira et al. (2007) described a μ-DPIV system for detecting 3D particle positions and conducting 3D velocity field measurement at micro scale via a three-pinhole defocusing concept. The basic concept of defocusing at micro scale is the same as at macro scale. In the reported μ-DPIV systems (Pereira et al., 2007; Yoon and Kim, 2006), a mask with three pinholes arranged at the vertices of an equilateral triangle is positioned on an objective lens of a microscope. The light from a seeded particle in the illuminated flow passes through each pinhole and then reaches three different positions on the image plane. The distance between the triangle vertices of the images increases with depth-wise distance between the particle position and the focusing plane, from which the particle positions in the depth direction (hence the depth-wise velocity component) can be estimated from the dimension of the triangular pattern by the defocusing concept. This μ-DPIV technique is validated through measuring the 3D flow field in a 50-μm deep micro backward-facing step (Yoon and Kim, 2006) and tracking the seeded fluorescent particles inside an evaporating water droplet at micro scale (Pereira et al., 2007). For the μ-DPIV technique, calibration procedure before the measurement is necessary (Pereira et al., 2007; Yoon and Kim, 2006).

There are also some other techniques available to obtain a 3D-3C velocity field in a micro-scaled flow passage in an indirect way, for example, using the continuity equation for a laminar flow (Bown et al., 2007; Kinoshita et al., 2007), high-speed scanning μ-PIV system using a rotating disc (Angele et al., 2006) as briefed in the later section, and so on.

3.4 Others

Apart from the abovementioned PIV techniques, there are still some other PIV systems designed for the particular purposes. Among them, the time-resolved or dynamic PIV, high-speed scanning PIV, dual-plane PIV, and simultaneous orthogonal-plane PIV are briefly introduced as follows.

3.4.1 Dynamic PIV

For a turbulent or transient flow, resolved temporal details of the flow characteristics are usually as important as the spatial structures. The appearance of kilohertz frame rate PIV, named time-resolved or dynamic PIV, allows the experimentalists to obtain the temporal and spatial information of the measured flow simultaneously. The particular parts of a dynamic PIV system different from a conventional 2D PIV includes an illumination system, which can be either pulsed lasers with a high repetition rate ranging from several to several tens kHz or a CW laser, an imaging system, which is usually a high-speed CMOS camera (or two CMOS cameras for a time-resolved stereoscopic PIV) and an accurate timing circuit (e.g., Bi et al., 2003; Burgmann et al., 2008; Li et al., 2007; Shinohara et al., 2004; Triep et al., 2005). At micro scale, a mercury lamp assembled to an epi-fluorescent microscope can also be used as light source for a time-resolved μ-PIV (Sugii et al., 2005). Other aspects other than the issues associated with the temporal resolution of the dynamic PIV, such as calibration procedures, data analysis algorithms, etc., are the same as those of a conventional 2D-2C PIV or a normal stereoscopic PIV.

3.4.2 Scanning PIV

A high-speed scanning stereoscopic PIV system is developed by Hori and Sakakibara (2004). This scanning PIV system is composed of a high-repetition-rate pulsed Nd:YLF laser, an optical scanner, and two high-speed CMOS cameras. The optical arrangement is schematically plotted in Figure 10 (Hori and Sakakibara, 2004). The laser sheet produced by the high-repetition-rate laser is scanned by an optical scanner in the direction normal to the sheet, forming an illuminated volume by collecting all the scanned laser sheets at all instants. The two CMOS cameras capture the illuminated particle images, and the stereoscopic PIV approach is adopted to construct the 3D-3C velocity distribution in the measured flow. This scanning stereoscopic PIV technique is validated with measurement of 3D vorticity field in a round water jet flow. Another type of scanning PIV technique is developed by Brucker (1997). Other than a high-speed CMOS camera for imaging at high frequency, the scanning illumination system consists of ten adjustable laser diodes in continuous mode, which can be adjusted and positioned independently,

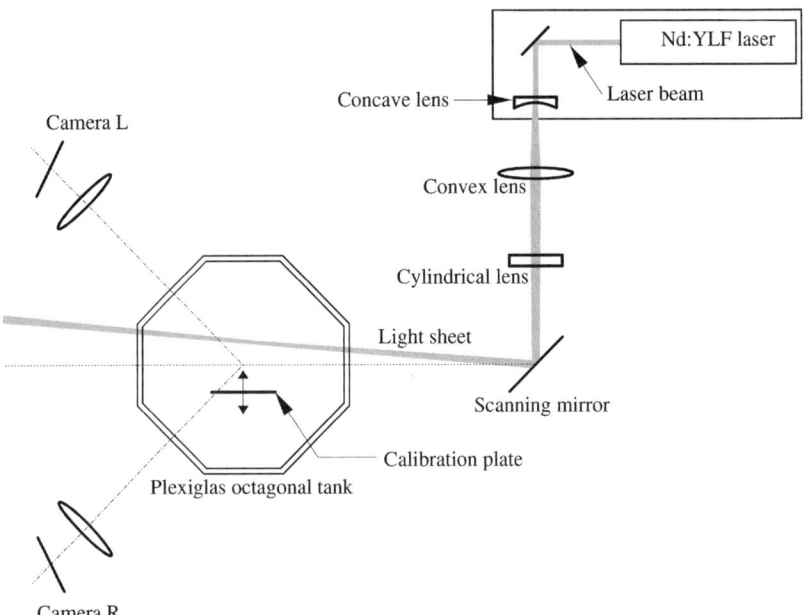

Figure 10 Optical arrangement for a high-speed scanning stereoscopic PIV (Hori and Sakakibara, 2004).

and a special electronic control allowed to pulse the laser diodes in any desired sequence or simultaneously. In this way, it is possible to form an illuminated 3D volume. This scanning PIV technique has been successfully applied to measure a laminar bubble and a transitional separation bubble on an airfoil (Burgmann et al., 2006, 2008).

At micro scale, two types of scanning μ-PIV techniques are worthy to be mentioned: a 3D scanning μ-PIV system using a piezo actuator and a high-speed scanning μ-PIV system using a rotating disc. Shinohara et al. (2005) reported the 3D scanning μ-PIV technique combining a high-speed camera, an epifluorescent microscope, a CW laser, and a piezo actuator. The piezo actuator is the unique part of this system, which is for moving the objective lens in the depthwise direction of the microscope. By inputting a current signal from a function generator, the piezo actuator equipped on the objective lens can be displaced in the depthwise direction, and so the objective lens is also moved and the information on the vertical axis (and thus a 3D velocity field) can be obtained. Angele et al. (2006) developed the high-speed scanning μ-PIV system, where the focal plane of the microscope is moved rapidly in the depthwise direction by changing the optical length using a rotating disc having glass windows with different thickness. By means of this scanning μ-PIV system, time-resolved (up to 100 Hz), pseudo-3D flow field information can be achieved.

3.4.3 Dual-plane PIV

Until now, basically two types of dual-plane PIV techniques emerged in the literature: polarization-based technique (Ganapathisubramani et al., 2005a, 2006b; Hu et al., 2001, 2002; Kahler, 2004; Kahler and Kompenhans, 1999, 2000; Liberzon et al., 2003) and frequency-based technique (Mullin and Dahm, 2005, 2006a, b). Kahler and Kompenhans (1999) first reported a dual-plane stereoscopic PIV measurement, in which two separate stereoscopic PIV systems are used to provide 3C velocity distribution in two parallel laser light sheets. This dual-plane PIV technique uses a polarization-based approach, where the two laser light sheets are arranged with orthogonal polarizations so that each stereoscopic camera pair records the scattered light from only one of the sheets. The realization of separation of the scattered light from two illuminating laser sheets with orthogonal polarization direction is based on the polarization conservation characteristic of Mie scattering, that is, in the Mie scattering regime, the scattering has a dominant forward direction and nonuniform lobed scattering towards the side directions. The polarization-based dual-plane stereoscopic PIV system consists of a four-pulse laser system delivering orthogonally polarized light, two pairs of CCD cameras set in a regular way for normal stereoscopic PIV, two high-reflectivity mirrors, and a pair of polarizing beam-splitter cubes (Kahler and Kompenhans, 2000). This technique has been successfully applied to measure all the nine components of velocity gradients in different turbulent flow fields to be investigated (Ganapathisubramani et al., 2005a, 2006b; Hu et al., 2001, 2002; Kahler, 2004).

Mullin and Dahm (2005) pointed out several drawbacks of the polarization-based dual-plane PIV technique: first, the reported experimental studies using polarization-based dual-plane PIV only measure large-scale features of turbulent flows, whereas the velocity gradients on the quasi-universal intermediate and small scales are not resolved, indicating the coarse spatial resolution of the measurement; maintaining the orthogonal polarization in the Mie scattered light requires the scattering particles to be spherical, resulting in a sever limitation to the seeding particles (fine liquid droplets are used for gas flow) and the flow environments to be investigated (only nonreacting flow is suitable to be measured). To overcome the abovementioned drawbacks, a frequency-based dual-plane stereoscopic PIV technique is developed by Mullin and Dahm (2005). The reported frequency-based dual-plane PIV system consists of four Nd:YAG lasers, two pulsed dye lasers, four CCD cameras, and an onboard timing circuit. The technique is based on two essentially independent stereoscopic PIV systems that simultaneously provide measurements in two differentially spaced data planes, by means of two different laser frequencies in conjunction with filters to separate the light scattered from the seeded particles onto the individual stereoscopic camera pairs. Hence, the traditional seeding particles can be used as the seed, and the measurement of any reacting flows is also permitted. Furthermore, it is shown that this technique provides for significantly higher spatial resolution than the polarization-based dual-plane PIV. Mullin and Dahm (2005) stressed that the spatial resolution achieved in their measurement is a factor of 3–4 times finer than the local viscous length scale of the turbulent flow. The powerfulness of frequency-based dual-plane stereoscopic PIV is testified in the measurements of VGT fields in turbulent shear flow (Mullin and Dahm, 2006a, 2006b).

3.4.4 Orthogonal-plane PIV

The orthogonal-plane PIV technique is recently proposed for investigating the 3D characteristics of the coherent structures in a turbulent boundary layer flow (Hambleton et al., 2006; Kim et al., 2006). The hardware components and principle of this technique are the same as polarization-based dual-plane PIV. The only difference is to set up both laser sheets mutually perpendicular to each other instead of parallel to each other in the dual-plane PIV system. This allows for measuring velocity distributions in both streamwise-spanwise and streamwise-wall-normal planes simultaneously, so that the salient features of the coherent structures in a turbulent boundary layer flow as the legs and the head of the hairpin vortices can be detected (Hambleton et al., 2006; Kim et al., 2006).

4. MEASUREMENT OF MULTIPHASE FLOW USING PARTICLE IMAGE VELOCIMETRY

Since the term "particle image velocimetry" being first proposed in 1984 (Adrian, 1984; Adrian, 2005; Pickering and Halliwell, 1984), the development and application of PIV techniques have been flourished in the research fields of measurement science and technology, turbulence, fluid mechanics, fluid engineering, aerodynamics, multiphase flow, combustion, granular flow, microfluidics, cryogenics, biological fluid dynamics, etc. Among the published experimental studies using PIV, one of the most successful application areas of this technique could be, for example, the detection of the coherent vortical structures and investigation of its statistical characteristics of wall-bounded turbulent flows without or with control (Adrian et al., 2000a, b; Adrian, 2007; Carlier and Stanislas, 2005; Christensen and Adrian, 2001; Ganapathisubramani et al., 2003, 2005a, b, 2006a, b; Hambleton et al., 2006; Hou et al., 2008; Hutchins et al., 2005; Kahler, 2004; Kim et al., 2006; Liu et al., 2001; Li et al., 2005a, b, 2006a, b, 2008; Longmire et al., 2003; Natrajan et al., 2007;

Svizher and Cohen, 2006a, b; Tao et al., 2000, 2002; Tomkins and Adrian, 2003; Warholic et al., 2001; White et al., 2004, among others). Experimental investigation on a multiphase flow by means of PIV technique might have some additional complexity and difficulty as compared with its single-phase counterpart. Besides that images taken by a PIV system are statistically analyzed to construct the velocity vectors, the patterns or intensity distribution of the images taken for a multiphase flow can also be used for phase-resolution, providing further information for multiphase flows in addition to the velocity field. The existence of different phases in a multiphase flow may bring about additional problems to a normal PIV measurement, such as strong reflection at the interface between two phases (particularly gas–liquid interface), problems associated with different refractive index of two phases, precisely distinguishing two phases, and so forth. Compared with PIV techniques for the single-phase flow, multiphase flow PIV measurement usually needs additional analytical algorithms or even supplementary technique(s), for example, LIF technique is often used together with PIV to resolve gas and liquid phases as well as the velocity field. Hence, some issues associated with measurement of multiphase flows using PIV technique are introduced in this section. Note here that the "multiphase" here holds extensive meaning that the multiphase flow to be investigated cannot only include different phases (states of matter) such as a gas–liquid two-phase flow, but also include same phases but different fluids such as a liquid–liquid two-fluid flow.

4.1 Liquid–liquid two-fluid flows

Although the liquid–liquid two-fluid flow has the same phase in the viewpoint of state of matter, some problems might be encountered when utilizing PIV to measure the velocity field(s) in one or both of fluid flows, such as mismatching of refractive indices of the two fluids, physical properties that influence the mixing of seeding particles, resolving the two fluids from each other, etc.

It is usually necessary to match the refractive indices of two fluids (and the transparent wall of flow passage in some cases particularly for microchannel flow). For example, in an experimental study on the self-preserving structure of steady round buoyant turbulent plumes in cross flow (Diez et al., 2005), planar-LIF (PLIF) and PIV techniques are utilized to measure the mean concentration of source fluid and mean velocity fields simultaneously. Both PLIF and PIV measurements in this study necessitate matching the indices of refraction of the source (water solution of potassium phosphate, monobasic KH_2PO_4, containing Rhodamine 6G dye) and ambient fluids (ethyl alcohol/water) to avoid scattering the laser beam away from the buoyant flow. Visual inspection of different instantaneous PLIF images taken with and without index matching does show obvious effect when the indices of the source and ambient fluids are not matched: blurred islands appear on the PLIF image without index matching (Diez et al., 2005). Resolving the source and ambient fluids from each other can be straightforwardly realized by detection of the PLIF images: color part represents the source fluid containing dye and colorless part represents the ambient fluid.

Creating staggered flow in a microchannel to generate 3D internal circulation within liquid droplets or liquid slugs is one of the hot topics in the research field of microfluidics with various application backgrounds such as creation of diverse and individually addressable sample stacks for serial processing, narrow residence time distributions, enhancement of mixing capabilities, and so on (Gunther et al., 2005; Kashid et al., 2005; Kinoshita et al., 2007; Malsch et al., 2008; Sarrazin et al., 2006). To explore the 3D flow characteristics μ-PIV techniques are widely used as well as numerical simulations. A comprehensive 3D measurement and visualization of internal flow features of a moving droplet in a polydimethylsiloxane (PDMS) microchannel are carried out by Kinoshita et al. (2007) using high-speed confocal μ-PIV technique. Water solution of glycerol is pumped into a microchannel flow of silicone oil, which is immiscible with glycerol solution, through a T-shaped branch, and torn off at the outlet of the branch by the oil-phase flow to form droplets. The most important advantage of confocal μ-PIV is the high resolution in the depthwise direction, which is realized by the fact that the light from out-of-focus particles is cut off optically and only the illuminated particles in the focal plane are imaged at high contrast. To accomplish the full advantage of confocal μ-PIV measurement, however, it needs to eliminate any negative factors including the refractive index mismatching problem. To minimize the refraction and reflection of light at the liquid–liquid and liquid–solid interfaces, the refractive indices of the working fluids (water solution of glycerol and silicon oil) and the channel material of PDMS have to be matched. The silicon oil has a refractive index of 1.414, which is only slightly different from that of PDMS 1.412. A mixture of 45% water and 55% glycerol with a refractive index of 1.414 is adopted to be the droplet liquid. By such relatively precise treatment, the 3D velocity field (with the aid of continuity equation for the out-of-plane velocity component) and hence the 3D recirculation internal flow characteristic is clarified (Kinoshita et al., 2007). Velocity distribution in the oil phase is usually difficult to measure by means of μ-PIV because of the lack of suitable seeding particles (Kinoshita et al., 2007; Malsch et al., 2008). In some other experimental studies on liquid–liquid two fluid flows in a microchannel by means of μ-PIV, however, issues about the matching of refractive indices are not mentioned (Gunther et al., 2005; Kashid et al., 2005; Malsch et al., 2008;

Sarrazin et al., 2006, among others). Note that these studies use a normal µ-PIV system instead of confocal µ-PIV. The negligence of mismatching of refractive indices might be due to the fact that the normal µ-PIV is relatively insensitive to the effect of refractive indices on the depthwise resolution compared with its confocal counterpart, and the channel wall is flat with zero curvature.

4.2 Gas–liquid two-phase flows

Among all the researches on multiphase flows in the published literature, those on gas–liquid two-phase flows holding the largest volume might be pertinent. It might also be the case for PIV measurement of multiphase flow. The PIV investigations on gas–liquid two-phase flows that simultaneously measure velocity field in one or both of the two phases and resolve the gas–liquid interface are discussed in this section. The key issue associated with such kind of PIV measurement of gas–liquid two-phase flows is how to overcome the complexity and difficulty stemming from the influence of gas–liquid interface on the illuminating and imaging of PIV. Experimental studies on gas–liquid two-phase flows using PIV techniques (in some cases supplemental techniques like LIF are needed) to both measure the velocity distribution and distinguish different phases can be largely categorized into three groups: liquid free-surface flow, bubbly flow, and gas–liquid two-phase flows in micro-channels.

4.2.1 Free-surface flow

Free-surface flow with interfacial transport processes is a subject of great interest since its effects can be seen both in nature and practical devices, such as the air–sea interface, ship wakes, and chemical processes like gas-absorption equipment. In many cases, it is necessary to investigate the interaction of the flow and the free surface or correlate the free-surface deformation with the flow characteristics beneath the liquid surface. To this end, PIV technique can be applied to some free-surface flows as a powerful experimental tool.

Li et al. (2005c) reported an approach for simultaneously measuring the turbulent velocity field and surface wave amplitude in an open-channel flow by means of a conventional PIV system only. The experimental facility is shown in Figure 11. PIV measurement is performed in the x–y plane following the coordinate system shown in Figure 11. The construction of velocity field in the illuminated liquid phase is according to the well-documented procedures as introduced in Section 2. The determination of free-surface level is based on such a fact that the gas and liquid phases are shown by different patterns on a PIV image, that is, a relatively bright region for the gas phase and a dark

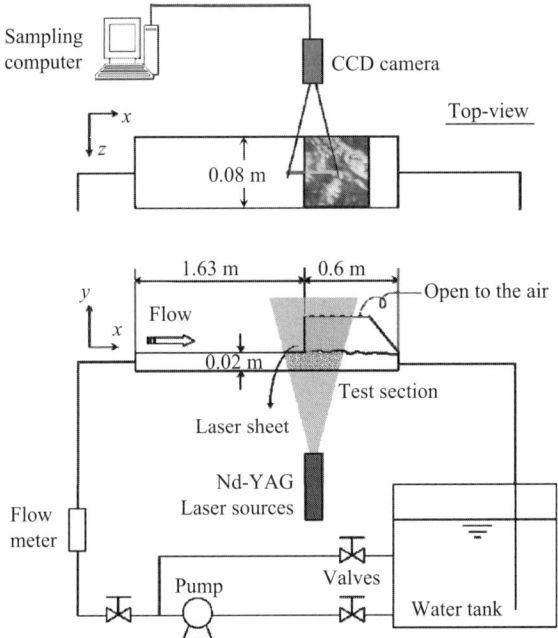

Figure 11 Schematic diagram of the experimental facility for simultaneous measurement of turbulent velocity field and free-surface wave amplitude in an open channel flow using PIV (Li et al., 2005c).

region with dense bright points (illuminated tracing particles) for the liquid phase, which allowed for the free-surface level to be automatically detected by an algorithm.

Figure 12 shows an example of a PIV image and the free-surface-level tracking procedure. First, the intensity of contrast of each pixel on a PIV image is digitized. The programmed surface-tracking algorithm (developed using the MATLAB language) is then applied to the digitized matrix of contrast intensity. Instead of inspecting the intensity of contrast at each pixel, the local standard deviation, $\tilde{p}(x_i, y_j)$, of the contrast intensity in a window of 3×3 pixels is calculated for each pixel to obtain a new matrix of $\tilde{p}(x_i, y_j)$. The value of $\tilde{p}(x_i, y_j)$ is low at the gas-phase region on a PIV image and high at the liquid-phase region. The matrix of $\tilde{p}(x_i, y_j)$ is then filtered to screen out the spikes of low $\tilde{p}(x_i, y_j)$ that would hinder the determination of the free-surface based on a threshold value of $\tilde{p}(x_i, y_j)$. This procedure was done by extracting and recording the local maximum $\tilde{p}(x_i, y_j)$ in a window of 3×3 pixels. Figure 12b shows an example of the filtered map of $\tilde{p}(x_i, y_j)$ corresponding to the rectangular area marked in Figure 12a. To determine the free-surface level, two threshold values are used, as shown in Figure 11c plotting the profile of

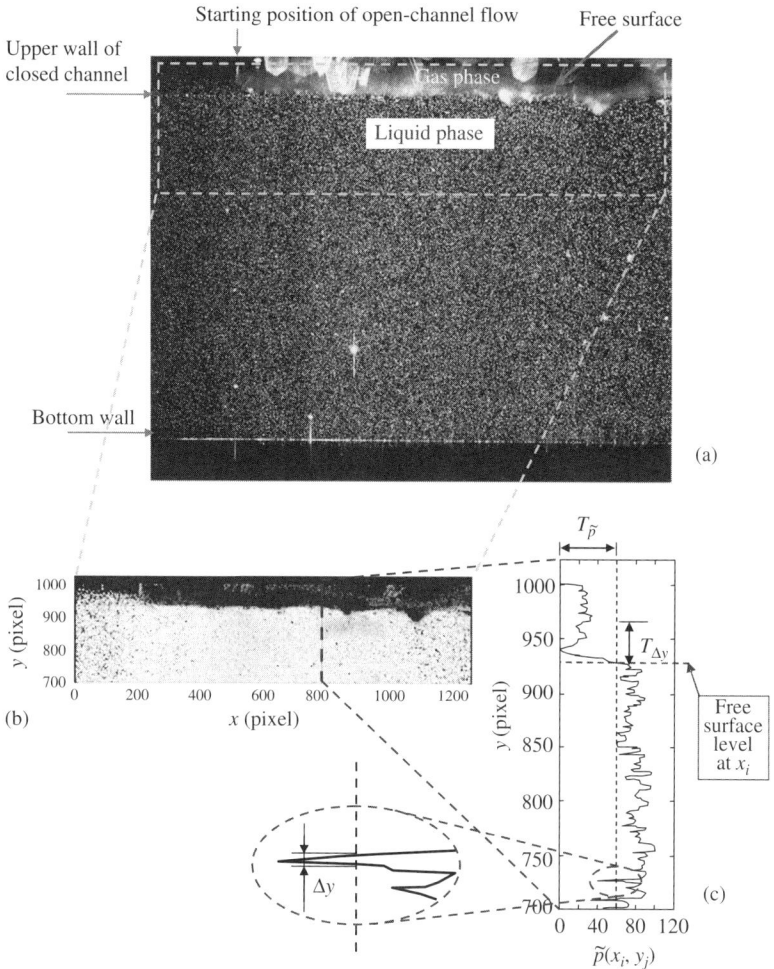

Figure 12 An example of a PIV image and the free-surface level tracking procedure: (a) PIV image showing interfaces and different phases; (b) 2D distribution of filtered local standard deviation of the contrast intensity; (c) distribution of filtered local standard deviation of the contrast intensity at $x_i = 800$ pixels showing the determination of free-surface level (Li et al., 2005c).

$\tilde{p}(x_i, y_j)$ at $x_i = 800$ pixels. The first threshold value $T_{\tilde{p}}$ was set so that $\tilde{p}(x_i, y_j)$ generally below $T_{\tilde{p}}$ at the gas-phase region and mostly above $T_{\tilde{p}}$ at the liquid-phase region. Although filtering has been applied to $\tilde{p}(x_i, y_j)$ to screen out the low-value spikes, points at which $\tilde{p}(x_i, y_j)$ is lower than $T_{\tilde{p}}$ may still remain, which calls for the second threshold value, $T_{\Delta y}$, as shown in Figure 12c. Δy is the interval in which the points have the $\tilde{p}(x_i, y_j)$ value lower than $T_{\tilde{p}}$ and out of which $\tilde{p}(x_i, y_j)$ values are larger than $T_{\tilde{p}}$. On the basis of these two threshold values, the free-surface level at x_i satisfies the following constraints: $\tilde{p}(x_i, y_{j-1}) > T_{\tilde{p}}$ and $\tilde{p}(x_i, y_{j+k}) < T_{\tilde{p}}, k = 1, 2, \ldots, T_{\Delta y}$. The obtained profile of free-surface level at all the streamwise locations is then smoothed by using the zero-phase filter. The position of the bottom wall of the channel, upper wall of the closed channel, and the starting point of the open-channel flow can be directly determined from the PIV image, as shown in Figure 12a.

Figure 13 plots an example of the processed PIV frame. The turbulent velocity field and its boundaries, solid wall, and liquid-free surface are simultaneously shown in Figure 13. The turbulence structures such as the coherent vortical structure near the bottom wall and its modification after release from the no-slip boundary condition near the free surface of the open-channel flow, and the evolvement of the free-surface wave can be seen in Figure 13. This simultaneous measurement technique for free-surface level and velocity field of the liquid phase using PIV has been successfully applied to the investigation of wave–turbulence interaction of a low-speed plane liquid wall-jet flow (Li et al., 2005d), and the characteristics of a swirling flow of viscoelastic fluid with deformed free surface in a cylindrical container driven by the constantly rotating bottom wall (Li et al., 2006c).

Several alternative algorithms for detection of free-surface level in PIV images have also been reported in the literature. Zarruk (2005)

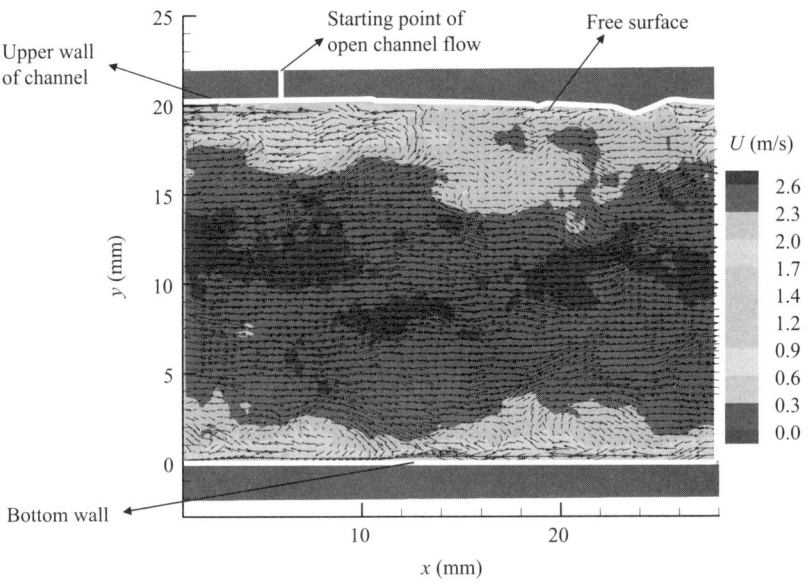

Figure 13 An example of the processed PIV frame, including turbulent velocity field and the position of the wavy free surface (Li et al., 2005c) (see Plate 6 in Color Plate Section at the end of this book).

proposed a measurement method for the free-surface deformation in PIV images, in which the gradient of the image intensity in each column, $\partial I/\partial y$, is calculated, and then the location of the maximum gradient is labeled as the potential free-surface location. If the free surface is well illuminated, it is sufficient to obtain the location of the maximum intensity gradient or the location of the maximum intensity to identify the location of the free surface (Zarruk, 2005). Misra et al. (2006) described an estimation algorithm based on gray-level co-occurrence matrix (GLCM) and "snakes" (parametric in conjunction with geometric active contours) to calculate the air–water interface from PIV images of an air-entraining laboratory hydraulic jump. Active contours minimizing energy functionals based on the internal (controlling the shape of the contour) and external (characterizing image information) energies of the contour are used to increase the accuracy of the predicted interface. This method is relatively complicated, but has the ability to give a first approximation to the location of the complex turbulent water-free surface. In the turbulent hydraulic jump experiment, the free air–water interface is compared to the visually interpreted interface and the deviation is found to be within the typical measurement resolution for the fluid velocities (Misra et al., 2006).

Another aspect of PIV investigation of a free-surface flow is to measure the gas side velocity over a liquid surface, in which identification of the gas–liquid interface is also necessary. Reul et al. (1999, 2008) reported the velocity measurements in the crest–trough region above the water surface using PIV technique. The dynamics of instantaneous velocity and vorticity structures in the near-surface flow are discussed in detail. Shaikh and Siddiqui (2008) reported an experimental investigation of the airflow structure in the immediate vicinity of the wind-sheared air–water interface using PIV technique. To obtain reliable estimates of the near-surface velocity, it is necessary to locate the water surface accurately. In the experiment (Shaikh and Siddiqui, 2008), the contrast between the air and water is improved by dissolving dark blue food color in the water before PIV measurement, resulting in the uniform grayscale value in the waterside regions of the PIV images. This allows for the implementation of a simple image processing technique to identify the air–water interface. In this way, both location of air–water interface and velocity distribution in the vicinity of the interface are obtained precisely (Shaikh and Siddiqui, 2008).

4.2.2 Bubbly flow

In the industry of chemical process, nuclear engineering, maritime engineering, environmental engineering, etc., there is a vast interest in the behavior of gas–liquid two-phase flows with bubbles involved. Although computational fluid dynamics (CFD) is a powerful tool that is being used to simulate the multiphase flow behavior, there is still a lot of dispute about the formulation and closure of the fluid dynamical equations of theoretical and empirical models. This calls for the need for experimental validation of CFD. PIV technique is one of the measurement approaches and has received a lot of attention for this purpose in the past two decades.

Lindken et al. (1999) reported a PIV measurement of velocity field in multiphase flows, where a digital mask in combination with the minimum quadratic difference (MQD) method (Gui and Merzkirch, 1996) is adopted to enhance the evaluation of velocity field in the contact area between the gas and liquid phases. The application of digital mask technique is allowed by the fact that the sizes of gas bubbles and tracer particles seeded to liquid phase appearing on the PIV images have significant differences. Lindken and Merzkirch (2000) extended the combination technique of MQD-based PIV and digital mask to simultaneously measure the 3D shape, position, and velocity of the bubbles and the 2D pseudo-turbulence velocity distribution induced by the bubbles in the liquid phase. To obtain the 3D information of the bubbles, a second, independent illumination and recording system, which provides a laser light sheet perpendicular to the PIV light sheet and records the in-plane images by a high-speed camera, is used simultaneously in addition to a normal PIV system as applied by Lindken et al. (1999). From a time series of the images of the recorded bubbles taken with the additional system and the rise velocities measured by PIV, a 3D picture of the system of bubbles including an estimate of the bubbles' size and shape can be reconstructed. Hence, full information on the instantaneous 3D bubble positions and velocities, and the 2D-2C velocity field of the water flow is achieved. Hassan et al. (1998) carried out an investigation of 3D two-phase flow structure in a bubbly pipe flow using a PIV system together with other three digital cameras. In addition to the measurement of velocity field in the liquid phase, the bubble's shape is reconstructed based on the dynamic generalized Hough transform algorithm. The DPIV technique was also successfully applied to the measurement of bubbly flow (Jeon et al., 2003; Pereira and Gharib, 2002; Pereira et al., 2000), in which high resolution images of the bubble field were recorded and analyzed to provide both bubble size, bubble location, and bubble velocity field within a cubic foot volume. The velocity vectors of the bubbles were computed from the volumetric cross-correlation of consecutive 3D sets of bubble locations and the bubble size information was abstracted from the bubble-scattered peak intensity.

Although some experimental studies on multiphase flow using PIV-only technique have been reported, the presence of the dispersed gas bubbles or columns usually introduces problems to the PIV measurement. Deen et al. (2002) provided an example of PIV image taken in a bubbly flow at a void fraction of approximately 1%. It indicates at least the following problems: in the subregion on the PIV image where the bubble concentration is rather high, there is little space left for the tracer particles, resulting in low valid detection probability; the dispersed phase can introduce shadows, which together with bubbles in front of the light plane reduce the amount of information present in the PIV images; and the deformation of gas bubbles during the time delay between the two recordings of the flow may deteriorate the precision of the PIV measurement (Deen et al., 2002). To overcome those problems emerged in PIV measurement of bubbly flow, several advanced techniques have been developed, such as DPIV technique as introduced in Section 3.3, combination of PIV with one or both of LIF technique and shadow image technique (SIT), etc. Nevertheless, PIV measurement technique used for bubbly flow is limited to relatively low volume fractions of the dispersed phase.

When merely a PIV system is used for the measurement of a bubbly flow, the intensity of light reflected from the bubbles' surface not only saturates the CCD camera but also overwhelms the intensity of light refracted from the seeding particles for PIV in its vicinity. Philip et al. (1994), Broder and Sommerfeld (2002), and Liu et al. (2005) hence used the hybrid PIV–LIF technique (wavelength discrimination), that is, fluorescent seeds and appropriate filters (LIF technique) are adopted to distinguish between the light reflected by the liquid–gas interface and the light refracted by the surrounding tracer particles, in their experimental studies on the characteristics of steam bubble collapse, a laboratory bubble at higher void fractions, or bubble-induced flow structure.

To clarify the mutual interactions between the gas bubbles and its surrounding liquid flow (mostly turbulent) in a bubbly flow, information of bubble's shape and motion is one of the key issues as well as the surrounding liquid velocity distribution. Tokuhiro et al. (1998, 1999) enhanced the PIV/LIF combination technique proposed by Philip et al. (1994) with supplementation of SIT to simultaneously measure the turbulent flow velocity distribution in liquid phase around the gas bubble(s) and the bubble's shape and motion in a downward flow in a vertical square channel. The typical experimental setup of the combination of PIV, LIF, and SIT is shown in Figure 14. The hybrid measurement system consists of two CCD cameras; one for PIV/LIF (rear camera) and the other for SIT (front). The fluorescent particles are Rhodamine-B impregnated, nominally 1–10 μm in diameter with specific density of 1.02, and illuminated in a light sheet of approximately 1 mm thickness (Tokuhiro et al., 1998, 1999). The fluorescence is recorded through a color filter (to cut reflections) by the rear camera. A shadow of the gas bubble is produced from infrared LEDs located behind the gas bubble. A square "window" set within the array of LEDs provides optical access for

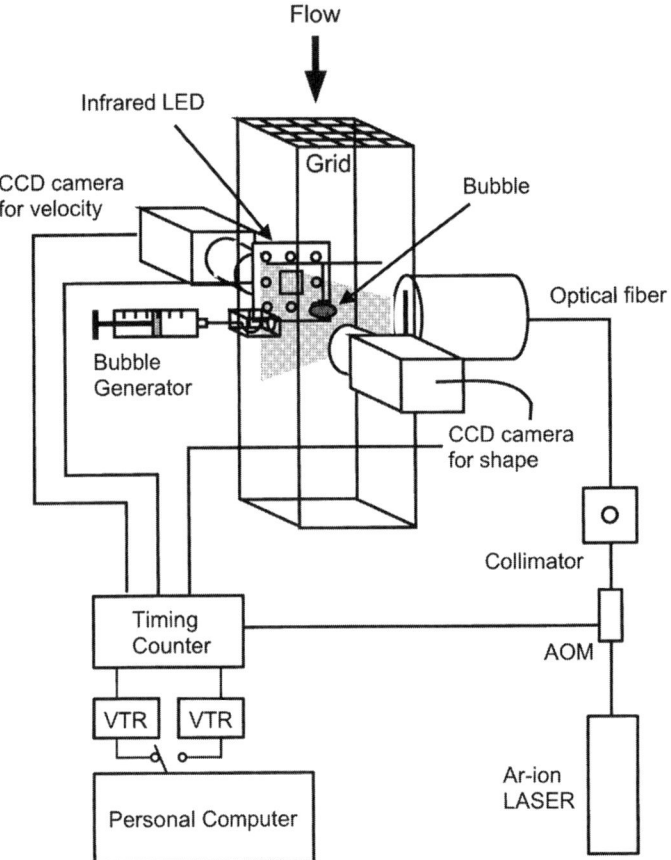

Figure 14 Schematic diagram of a hybrid PIV/LIF/SIT system used for measuring liquid-phase velocity distribution and bubble's shape and motion (Tokuhiro et al., 1998).

PIV/LIF. The emitted light is filtered through a translucent cover sheet and produces a shadow of the bubble, which is then captured by the front CCD camera. To capture the bubble's shape and the surrounding liquid flow field simultaneously, the triggering of the laser, the LEDs, and the two CCD cameras are synchronized by a timing circuit.

Figure 15 shows an example of typical images of the illuminated tracer particles for PIV/LIF and the SIT images of two bubbles in three representative positions. It is clearly seen that the tracer particles in the wake and a partial outline of the bubble boundaries are visible on the PIV image. The number of tracer particles is sufficient enough to calculate velocity vectors through the normal cross-correlation-based PIV algorithm.

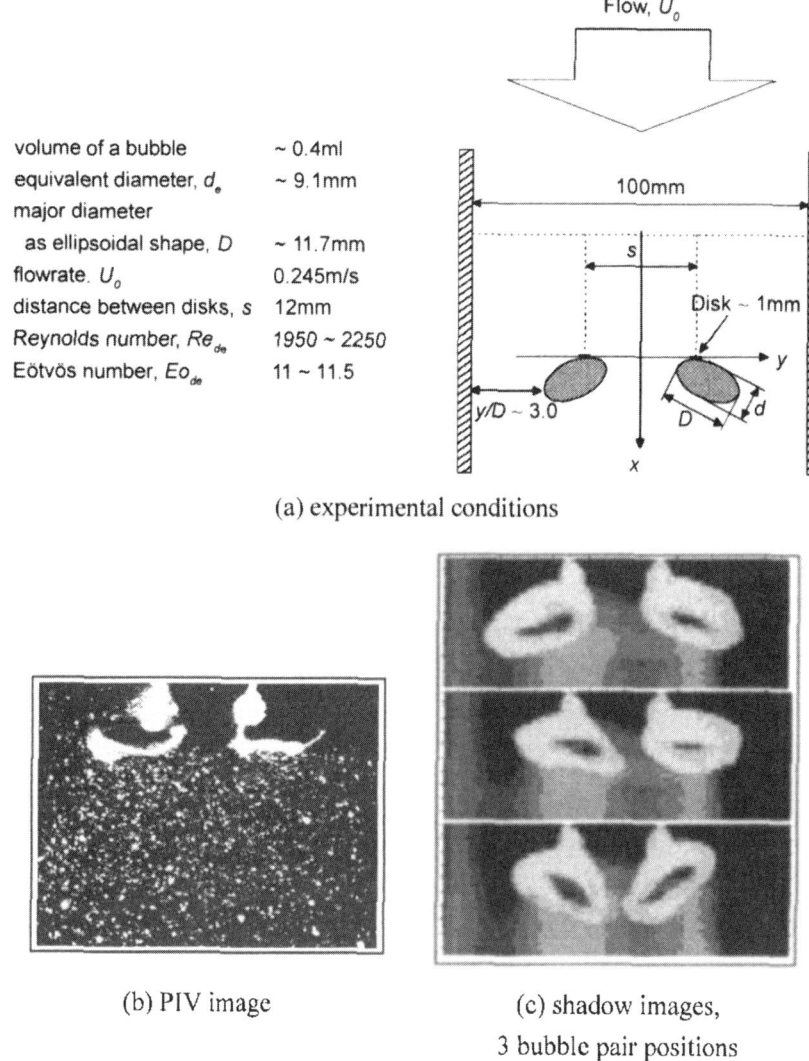

Figure 15 An example of PIV/LIF/SIT taken images: (a) schematic of experimental conditions and ranges; (b) typical PIV image showing illuminated tracer particles around the two bubbles; (c) typical shadow images of the two bubbles at three instants (Tokuhiro et al., 1999).

A representative time sequence of four PIV/LIF-derived velocity vector distributions together with the SIT-derived bubble shadows is plotted in Figure 16 as a typical result obtained by the PIV/LIF/SIT system. Note that even with LIF technique used, there are also "white-out" regions (intensity saturation), and the laser sheet entering from the

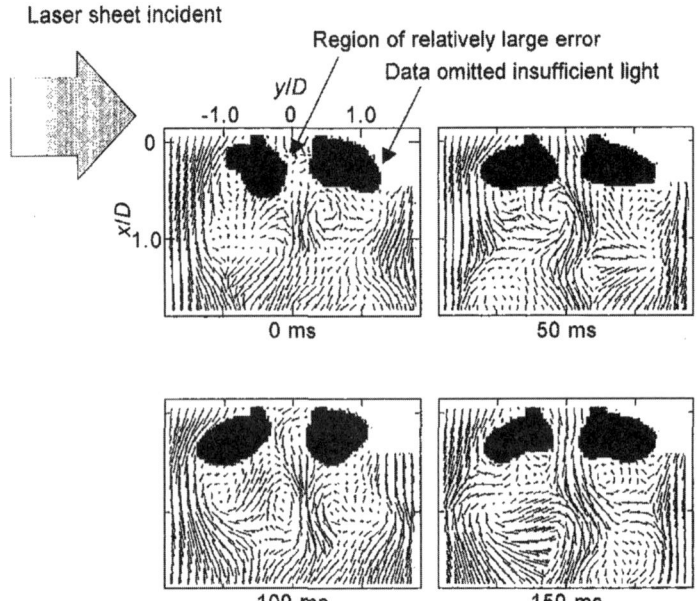

Figure 16 An example of PIV/LIF/SIT taken turbulent velocity fields in the wake region of two bubbles together with the bubble shadows at four instants (Tokuhiro et al., 1999).

left cannot illuminate the regions in between the two bubbles and to the far right, resulting in either relatively large error of or lack of velocity vectors locally, as can be seen from Figures 15b and 16. Nevertheless, the detailed behaviors of gas bubble(s) subjected to a liquid flow, the characteristics of the wake flow behind the bubble(s), the interactions between the two bubbles, and between bubble and surrounding liquid-phase flow have been explored successfully by means of the hybrid PIV/LIF/SIT technique (Tokuhiro et al., 1998, 1999).

Tokuhiro et al. (1998, 1999) applied the hybrid techniques of PIV/LIF/SIT to a relative simple bubbly flow case, that is, with one or two bubbles involved. The application of this technique to a normal bubbly flow with many bubbles but at low void fraction can be extended straightforwardly. An experimental study on the effect of bubble diameter on modification of turbulence in an upward bubbly flow in pipe is carried out by Fujiwara et al. (2004b) using the PIV/LIF/SIT system as measurement technique. Since many bubbles are involved in this bubbly flow, some additional procedures for the detection of bubble shapes and positions are provided, which include (1) obtaining the background image from the average of more than 1,000 SIT images in a time series; (2) detecting the bubble-like objects by subtracting the background image from the original images; (3) changing the bubbles in focal plane images to be binary (black–white) by using a certain threshold level of light intensity; and (4) adopting the roundness, given by $L^2/4\pi A$, where L and A are the circumference length and area of the object, to prevent miss-recognition of overlapped bubbles as single bubbles and remove irregular images. Figure 17 gives an example of both PIV/LIF and SIT images and the final reconstruction of velocity field together with the bubble images (Fujiwara et al., 2004b).

Furthermore, Fujiwara et al. (2004a) performed an experimental study using PIV/LIF combining with double-SIT to construct approximated 3D shape deformation of bubbles as well as to investigate quantitatively the 3D wake flow structures behind bubbles in a simple shear flow. The

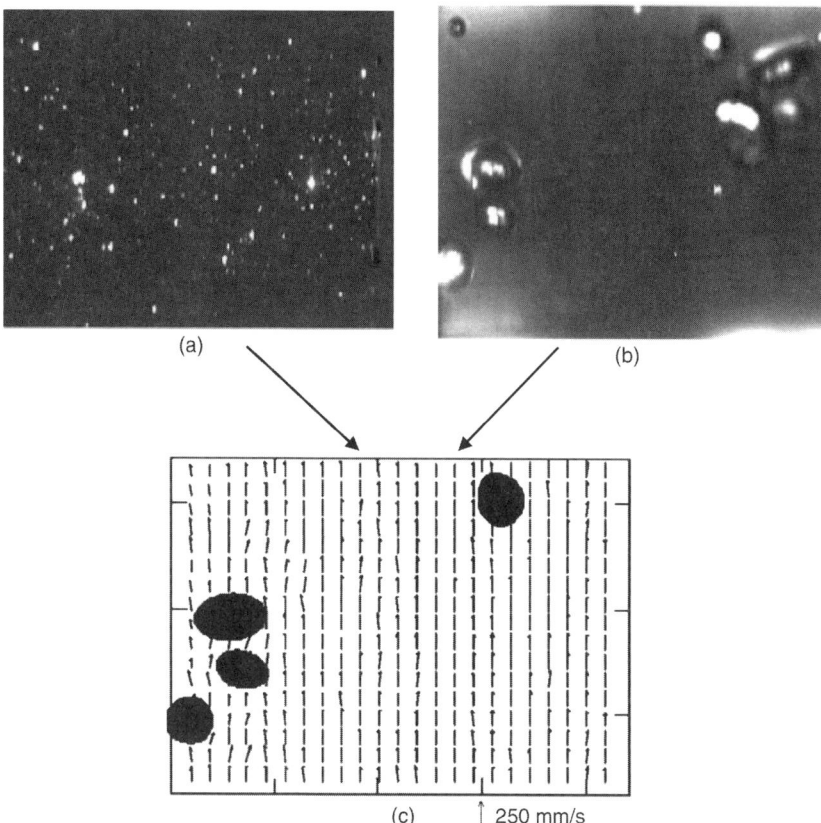

Figure 17 An example of PIV/LIF/SIT-measured velocity field together with the bubble shapes and positions in an upward bubbly flow: (a) PIV image with fluorescent tracer particles; (b) SIT snapshot with bubble shadow images at the same instant as (a); and (c) velocity vector field together with detected bubbles reconstructed from (a) and (b) (Fujiwara et al., 2004b).

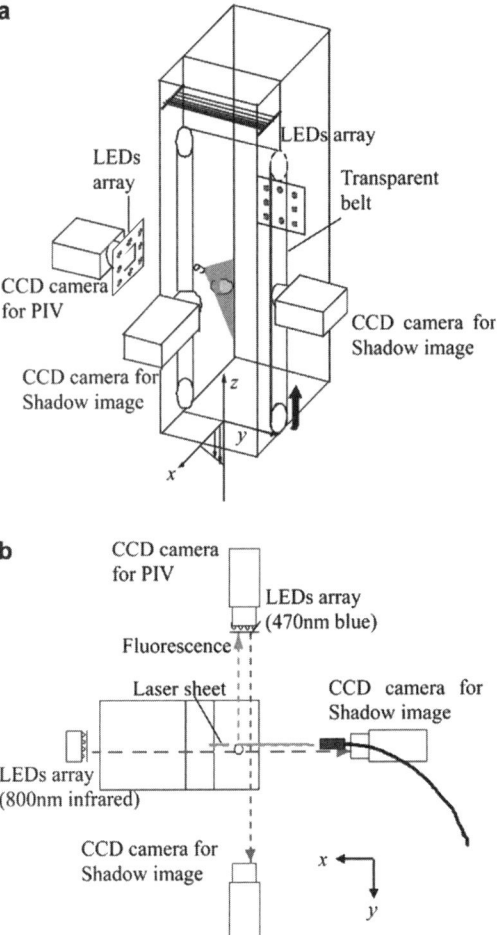

Figure 18 Schematic of experimental setup for measurement of 3D bubble deformation and flow structure in the wake using the combination of PIV/LIF and double-SIT: (a) schematic of the measurement system; and (b) top view of the experimental facility (Fujiwara et al., 2004a).

combination system of PIV/LIF and double-SIT is schematically shown in Figure 18 (Fujiwara et al., 2004a). The vertical shear flow is produced between the looped transparent belt driven by a variable-speed motor and the channel wall. The measurement system consists of three CCD cameras: one for PIV/LIF and the other two set perpendicularly to each other for detecting the bubble shape in two planes, from which the approximated 3D bubble shape can be reconstructed. A timing circuit is built for simultaneously synchronizing the triggering of the laser, the LEDs, and the three CCD cameras.

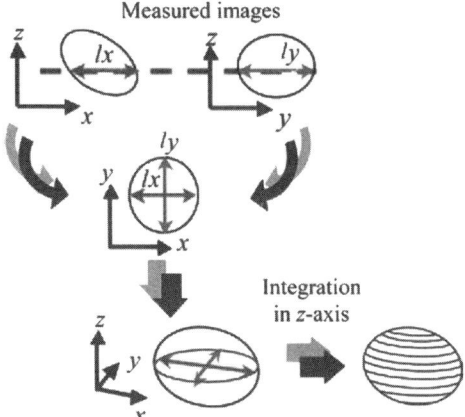

Figure 19 Approximation of 3D bubble shape from the bubble shadow images taken in two perpendicular planes (Fujiwara et al., 2004a).

The post-processing procedure is almost the same as that in PIV/LIF/SIT measurements (Fujiwara et al., 2004b; Tokuhiro et al., 1998, 1999) except for the approximation of the 3D bubble shape and tracing its trajectory. The reconstruction of 3D bubble shape from the recorded SIT images in two planes perpendicular to each other is realized by an approximation method. Figure 19 depicts the schematic of the reconstruction procedure. First, the bubble's x–y cross-section (following the coordinate provided in Figure 19) is assumed to be ellipsoid. Then, the SIT images of bubble in both the x–z and y–z planes allow the estimation of the length of the major and minor axes of the ellipsoidal x–y cross-section at each position in the z direction. Finally, the instantaneous 3D bubble shape can be reconstructed by integrating each x–y cross-section in time series. The approximated 3D bubble shape, the trajectory of the bubble in time series, and the corresponding flow structures calculated by PIV technique are plotted in Figure 20 as an example. With the full information, the detailed characteristics of the bubble transition path, the interactions between the deformed bubble, and the 3D wake flow structure, etc. have been investigated (Fujiwara et al., 2004a).

An extended version of the hybrid technique of PIV/LIF/SIT is reported by Kitagawa et al. (2005), in which the PTV technique is employed to measure the velocity field in liquid phase and track the velocity distribution of dispersed bubbles, in addition to the SIT measurement of bubbles' shape and location in a microbubble-laden turbulent channel flow. It is well known that microbubbles injected into the turbulent boundary layer developing on a solid wall have a significant skin friction reduction effect. To investigate the interactions between the injected microbubbles (the void fraction is actually low but

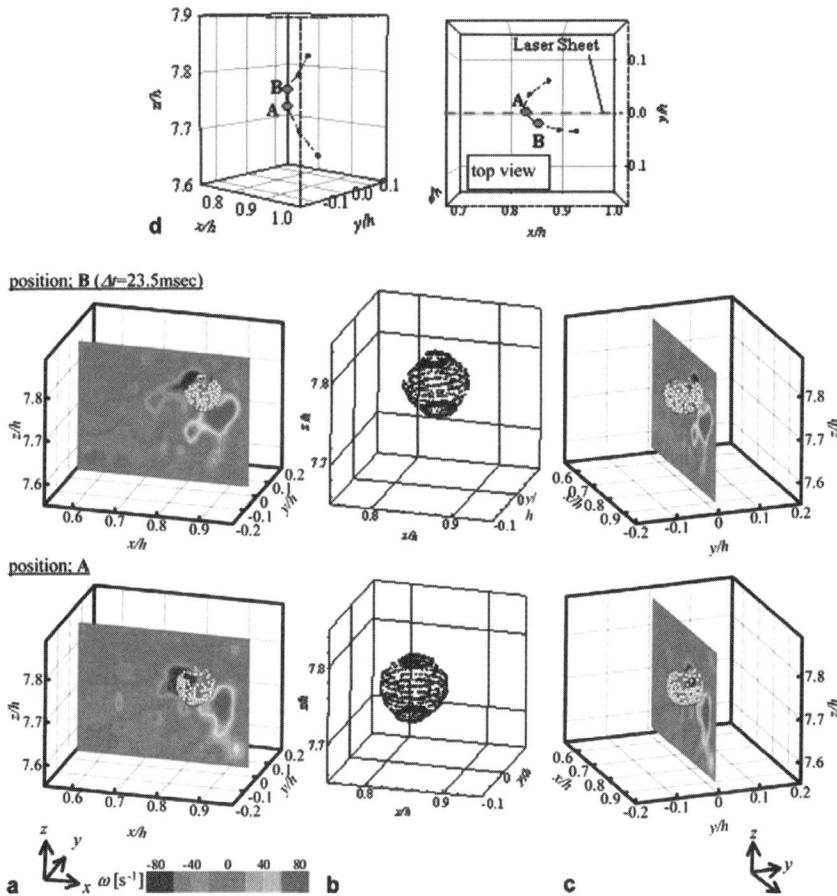

Figure 20 An example of the approximated 3D bubble shape and corresponding flow structure estimated from the measurement using PIV/LIF combining with double-SIT system: (a) characteristic vorticity structure around the bubble (bubble moves in the y–z plane); (b) reconstructed 3D bubble shape; (c) relation between bubble location and measured plane for PIV; and (d) 3D bubble trajectory (Fujiwara et al., 2004a) (see Plate 7 in Color Plate Section at the end of this book).

the number of microbubbles is huge) and turbulence structures in a turbulent channel flow of water, which is the key point of understanding the drag-reduction mechanism of microbubble-injection method, an experimental study using the combination system of PIV/LIF/SIT is thus designed by Kitagawa et al. (2005). For the measurement of liquid turbulent flow with large number of bubbles involved, the PTV technique instead of cross-correlation-based PIV algorithm is employed. This is because liquid velocity vectors in high accuracy are usually difficult to be estimated with increase of void fraction with PIV

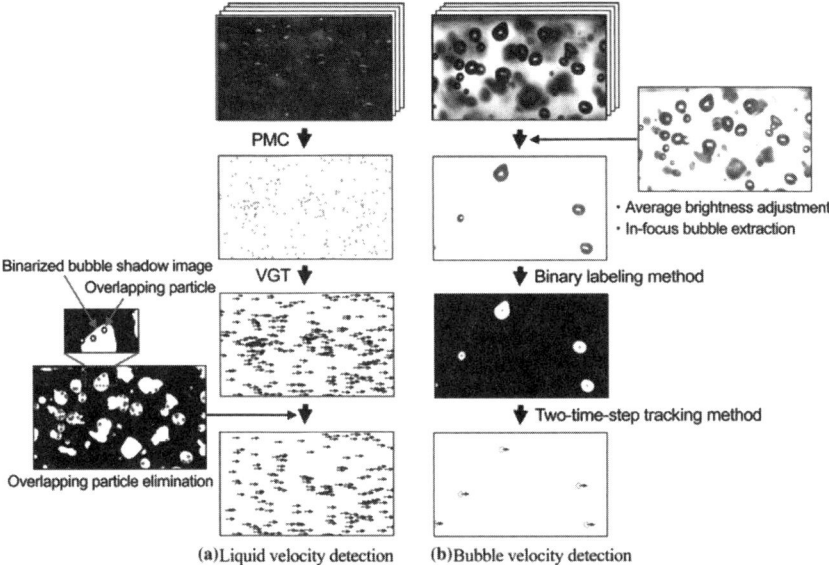

Figure 21 Flow chart of the processing procedure of PIV/LIF/SIT-taken images for liquid phase and bubble velocity estimation by means of PTV technique (Kitagawa et al., 2005).

technique, which originates from the decrease of inter-bubble gaps in relation to the interrogation window dimensions and the increase of liquid fluctuation velocities induced in the bubble wake.

The data-processing procedures of both PIV/LIF and SIT images for estimation of velocity distribution in liquid phase and of bubbles, respectively, using the PTV technique, are illustrated in Figure 21. The PTV procedures for measuring the liquid-phase velocity field are twofold: first, by using the particle mask correlation (PMC) approach (Takehara and Etoh, 1999), the centroid of each fluorescent tracer particle is calculated from the original PIV images; then, the velocity vectors of individual fluorescent tracer particles are calculated by the VGT method (Ishikawa et al., 2000) with a time interval of 26 μs between pairs of images. The procedure for measuring the bubbles' velocity vectors, without considering the in-focus overlapping bubbles, is as follows: first, the average brightness of the original SIT images is adjusted to emphasize the edges of the bubbles and to remove the noise, and then only in-focus bubbles are extracted; then, the centroids of the bubbles are estimated from the images with invert-brightness and binary treatment using the binary-labeling method (e.g., Yamamoto et al., 1996); at last, the velocity vectors of the bubbles are simply calculated from bubble positions in two frames. To accurately obtain the velocity of the liquid phase, tracer particles that overlap with the bubble SIT images are regarded as bad points and need to be eliminated in the post-processing. This elimination process is accomplished in two steps: first, the original SIT images are binarized, that is, the binary value is 1 for bubble and 0 for nonbubble; then, the tracer particle positions obtained by the PMC method are collated with the binary data. When the binary value at the tracer particle position is 1, it is defined as particle overlapping on a bubble shadow and is then eliminated.

The velocity vectors of the liquid phase and bubbles are accurately detected after the abovementioned data-processing procedures, so that the characteristics of turbulence in a channel flow modified by microbubbles injection and the bubble–turbulence interactions are able to be explored statistically (Kitagawa et al., 2005).

Lindken and Merzkirch (2002) also reported a measurement technique for two-phase bubbly flows using the combination of PIV/LIF/SIT, in which only one CCD camera is used for recording both PIV/LIF and SIT images. Nd:YAG laser at 532 nm wavelength is used to illuminate a 2D light sheet in the bubbly flow. The emit light from the applied fluorescent tracer particles has a wavelength of 555–585 nm with an emission peak at 566 nm. The LED array for SIT emits light at 675 nm. An optical high pass filter with a steep transmission edge at 570 nm $\pm$ 5 nm is then used to completely block the reflected light at 532 nm, whereas the shadow image with light of wavelength 675 nm and most of the fluorescent light passes the filter and are imaged by the same CCD camera. Thus, both information of liquid-phase velocity field and the gas bubble's shape and position can be determined by properly designed algorithms. In an experimental study on single bubble dynamics in a small diameter pipe (Hassan et al., 2001), stereoscopic PIV combined with SIT, without LIF, however, are used as the measurement approaches. The recording system consists of four CCD cameras. One camera together with optical filter records SIT images in the direction parallel to the laser sheet; one camera is set perpendicular to the laser sheet for PIV and records PIV images for analysis of bubble information together with the SIT images. The other two cameras are for stereoscopic PIV images recording. The centroid, rotation angle, and semiaxes of the ellipse that best fits the edge points of the bubble image are computed using the direct generalized Hough transform algorithm, which only applied to images from the orthogonally positioned cameras. In this way, the velocity field generated by the passage of a single air bubble rising in quiescent water in conjunction with the bubble size and shape can be obtained simultaneously.

4.2.3 Gas–liquid two-phase flows in microchannels

Research topics on gas–liquid two-phase flows in microchannels are receiving increasing notice recently due to its importance in such areas as two-phase microchannel heat sinks, identification of micro-scaled two-phase regime, multiphase microfluidics, etc. Experiments on gas–liquid two-phase microchannel flows using PIV technique (μ-PIV) may not encounter sever difficulty resulted from the strong reflection of illumination light from the gas–liquid interface, as possibly happens to a PIV (only) measurement of gas–liquid two-phase flow in a normal-sized geometry, since LIF technique is actually always utilized to any μ-PIV system, either a normal μ-PIV or a confocal PIV system. The data-processing procedure is thus relatively simple. Only two examples of PIV techniques used for measurement of gas–liquid two-phase flows in microchannels are described as follows.

Wang et al. (2006) reported a hybrid experimental and computational method for reconstructing 3D bubble geometry as well as providing other critical information associated with nucleating bubbles in gas–liquid two-phase microchannel flows. A normal μ-PIV system is used to measure the water phase 2D-2C velocity field surrounding a nucleating bubble. This 2D-2C velocity field locates in a 2D slice of the 3D flow. To yield the 3D flow field as well as bubble geometry, a variety of 3D flow solutions for bubbles with varying diameters and center locations but identical bubble cross-sections are then numerically predicted, and then the in-plane projection of the predicted 3D velocity vectors at the focal plane are compared to the respective μ-PIV measurements; finally, the best match between the predicted and μ-PIV-measured velocity fields gives an estimate of the actual bubble geometry (Wang et al., 2006). In the μ-PIV study of the formation of segmented flow in microfluid T-junctions (Van Steijn et al., 2007), the digital mask technique is utilized to eliminate spurious vectors due to background noise in the gas phase. To build up a mask, the so-called Q-ratio, that is, the ratio between the largest and second-largest peak in the correlation field of PIV images is computed, resulting in grayscale images. These images are then binarized and used as a mask for the PIV vector fields. The shape and movement of the interface between the liquid and gas phases are measured by means of a high-speed camera. Hence, the characteristics of such transient flow field during the formation of bubbles in a microfluidic T-junction can be investigated.

4.3 Particle-laden multiphase flows

Particle-laden multiphase flows, usually turbulent, cover a wide range of applications, such as pollution control, sediment transport, combustion processes, erosion effects in gas turbines, and so on. One of the most important aspects of particle-laden turbulent flows is the mutual interactions between particles and turbulence. PIV techniques, as a powerful tool other than numerical simulation method and theoretical analysis, have been applied to this research field of particle-laden multiphase flows. Note that, dispersed-phase particles in particle-laden multiphase flows usually do not refer to the seeding particles for PIV, the former are normally much larger than the latter. In the PIV investigation of particle-laden multiphase flows, the key issue is to not only measure velocity field in the continuous phase, but also distinguish the dispersed particles or droplets from its environment and detect the dispersed particle path location within the illuminating laser sheet. Several typical examples of treatment method for PIV measurement of particle-laden multiphase flows are described later.

In an experimental study on a downward particle-laden water channel flow by Sato and Hishida (1996), a special function for particle-laden two-phase flows is supplemented to the conventional PIV system: a two-control volume method comprising an Ar-laser sheet and two infrared laser diode sheets located along both sides of the Ar-laser sheet is used. In this way, the dispersed particles are correctly recognized and its velocities and water-phase velocity field are obtained simultaneously, so that the interactions between dispersed particles and fluid turbulence have been studied (Sato and Hishida, 1996). A further study of the effect of inter-particle spacing on turbulence modulation in a downward particle-laden water channel flow is carried out by Sato et al. (2000) using a Lagrangian PIV technique. A high-speed CCD camera for recording PIV images and a cylindrical lens for introducing a laser sheet to the flow field to be investigated are mounted on a moving shuttle, which moves from top to the bottom, parallel to the water channel wall, with mean streamwise velocity of the laden particles on which it focuses, to realize the Lagrangian measurement technique. By using the PIV technique combined with the moving shuttle, it is possible to measure the Lagrangian velocity field following some targeted particles, so that the distortion of turbulence in the presence of particles and the small-scale structure of turbulence modulation in terms of the inter-particle spacing can be investigated.

Kiger and Pan (2000) reported a PIV technique for the simultaneous measurement of dilute solid–liquid two-phase flows, in which a 2D median filter is employed to separate the larger dispersed phase particle image from the continuous phase tracer particle image. This phase separation technique is based on two aspects that a median filter is a nonlinear signal processing technique that has been found effective in reducing random noise and periodic interference patterns without severely degrading the signal, and that for a two-phase image with both small tracer particles for PIV and large dispersed particles, the small tracer particles can be regarded as noise scattered over a uniform background. After phase separation, the velocity vectors of the dispersed large particles are then calculated using a correlation tracking algorithm and the velocity field of the continuous phase is estimated through a standard cross-correlation-based method (Kiger and Pan, 2000).

Borowsky and Wei (2007) proposed a hybrid PIV/LIF/PIA (particle image accelerometry) technique to measure the kinematic and dynamic parameters of a liquid–solid pipe flow. In this technique, two CCD cameras are utilized and positioned 180° apart such that both focuses on the same briefly illuminated region of flow passing vertically through the transparent tube test section made of acrylic resin. The dispersed phase solid particles are silver-coated glass spheres and the seeding particles for PIV are fluorescent acrylic particles. Optical filters are employed to each camera so that the cameras recorded the trajectories of either the solid particles for PIA or the neutrally buoyant fluorescent fluid tracer particles for PIV. Simultaneous two-color PIV/PIA measurements are then realized. Both velocity and acceleration fields are calculated using the correlation-based PIV algorithm. The kinematic terms and dynamic parameters as the second and third central moments of temporal acceleration for both water and solid phases are then investigated.

PIV techniques have also been applied to the measurement of gas–liquid–solid three-phase flows. Chen and Fan (1992) and Chen et al. (1994) applied a 2D PIV system to simultaneously measure the instantaneous full-field flow properties of all components in a gas–liquid and gas–liquid–solid fluidization system. Reese et al. (1995) extend the 2D PIV system developed by Chen and Fan (1992) to a 3D PIV and measure the 3D local flow properties of three-phase fluidization systems. The three-phase flow to be investigated consists of the continuous water phase, neutrally buoyant Pliolite particles of 200–500 μm in diameter for PIV tracer particles and solid phase as well, and dispersed gas bubbles injected from the bottom of water pipe flow. The 3D PIV system consists of three major components: an Ar-CW laser with optics providing laser sheet with 1.0 cm in thickness, one CCD camera for image recording, and image-processing system. An optical arrangement using six pieces of mirror is utilized for the one-camera-based 3D PIV technique. The left and right sides (half-half) of each PIV image recorded by the only one camera with the aid of the optical arrangement contains the displacement of particles in two orthogonally different planes, respectively. The 3D PIV algorithms (and calibration procedures) are then developed to identify particle (where "particle" is used in a broad sense to include gas bubbles, solid particles, and liquid seeding particles) images, locate centroids of particle images, distinguish particle images between different phases, compute displacements between image pairs, match image pairs from the orthogonal views of the flow field, and reconstruct the 3D spatial coordinates of particle images and thus the 3D instantaneous velocities. This 3D PIV system has been applied to the 3D gas–liquid–solid fluidization system operating in the dispersed bubble flow regime and testified that it is capable of providing the 3D full-field instantaneous velocities, accelerations, and holdups (concentrations) of different phases (Reese et al., 1995).

5. SUMMARY AND OUTLOOK

The conventional 2D PIV technique has almost matured and become a standard nonintrusive diagnostic tool of flow measurement. 3D PIV measurement, which can achieve the most comprehensive information of the flow characteristics to be studied, is usually either difficult to accomplish or expensive, motivating experimentalists to develop more and more convenient, inexpensive and robust 3D PIV techniques. Indeed, several techniques such as the 3D μ-PIV, particularly for highly unsteady flow measurement in microchannels, are still at its infant stage. The development of PIV techniques relies on the development of its hardware components as well as designing ideas of the arrangement of illumination, optics, image recording system, and the analysis algorithms.

Applications of PIV techniques, ranging from 2D-2C to 3D-3C, from parallel and orthogonal dual-plane PIV to time-resolved PIV, to the measurement of single-phase flows have obtained vast achievement in understanding the fluid dynamics at both macro and micro scales. PIV measurement of multiphase flow systems normally need to achieve additional information like the interactions between phases as well as the velocity field in the continuous phase flow. This requirement needs more skills in addition to the PIV measurement of single-phase flow, including the particularly designed data analysis methods for phase-distinguishing and estimation of velocity vectors of dispersed phase(s), and sometime additional measurement techniques such as LIF and SIT. It is reasonable to comment that, with careful and detailed considerations, PIV techniques are generally applicable to the measurement of multiphase flow systems as far as the PIV measurement can be performed on a single-phase flow flowing in the same environment, but lots of knowhow is necessary for these additional considerations.

The essence of PIV (including PTV) technique, that is, yielding the velocity vectors from analyses of the viewed and recorded image patterns, can be quoted to establish novel estimation techniques for wider field of researches. The estimation of velocity distribution of the floating ice pieces on a river based on PIV algorithm is one of the best examples (Ettema et al., 1997). Freeing one's imagination, the multiphase flow systems can also be extended to a much broader scope, which may involve, for example, cherry-blossom petals falling down in the wind, clouds floating in the sky, a group of snow finches flying in the air, rocks or other solid pieces rolling with the earthquake or heavy rain-induced landslide, pollution materials floating and collecting at the water surface, cars running in a big city, and so on. PIV algorithms are suitable to estimate those movement behaviors as far as the viewed and recorded scope can be large enough to cover them. PIV might be able to measure the flowing or moving or floating velocities for everything in the imagination of human beings, which, on the other hand, requires the endless development of this technique.

NOTATION

A	area of the investigated object
C_{fg}	cross-correlation function
d_0	diameter of laser beam
f	variable extracted from the PIV image
f_m	average value of f
g	variable extracted from the PIV image
g_m	average value of g
H	height of flow channel
I	image intensity
i	ith location
j	jth location
L	circumference length of the investigated object
N	pixel number in one direction of an interrogation window
n	counted number
$\tilde{p}(x_i, y_j)$	local standard deviation of the contrast intensity of an image
R_{fg}	cross-correlation coefficient
$T_{\tilde{p}}$	a threshold limiting the value of $\tilde{p}(x_i, y_j)$
$T_{\Delta y}$	A threshold limiting the spreading distance of the peak of $\tilde{p}(x_i, y_j)$ in the y direction
t	time
u	velocity component in the x direction
$\vec{u}_{i,j}$	velocity vector at grid (i, j)
x	x component in the Cartesian coordinate
y	y component in the Cartesian coordinate
z	z component in the Cartesian coordinate
Δx	sample shift in the x direction
Δy	sample shift in the y direction

GREEK LETTERS

δ_m	minimum thickness of laser sheet
λ	wavelength of laser light
π	ration of the circumference of a circle to its diameter

ABBREVIATIONS

2D-2C	two-dimensional two-component
2D-3C	two-dimensional three-component
3D-3C	three-dimensional three-component
CCD	charge-coupled device
CFD	computational fluid dynamics
CMOS	complementary metal-oxide semiconductor
CW	continuous wave
DPIV	defocusing particle image velocimetry
GLCM	gray-level co-occurrence matrix
HPIV	holographic particle image velocimetry
LFCPIV	local-field correlation particle image velocimetry
LIF	laser-induced fluorescence
NA	numerical aperture
PDMS	polydimethylsiloxane
PIA	particle image accelerometer
PIV	particle image velocimetry
PLIF	planar laser-induced fluorescence
PMC	particle mask correlation
PTV	particle tracking velocimetry
SIT	shadow image technique
TPIV	tomographic particle image velocimetry
VGT	velocity gradient tensor
WIDIM	window displacement iterative multigrid
μ-PIV	micro-scaled particle image velocimetry
μ-DPIV	micro-scaled defocusing particle image velocimetry
μ-HPIV	micro-scaled holographic particle image velocimetry

REFERENCES

Adrian, R. J. *Appl. Opt.* **23**, 1690–1691 (1984).
Adrian, R. J. *Annu. Rev. Fluid Mech.* **23**, 261–304 (1991).
Adrian, R. J. *Meas. Sci. Technol.* **8**, 1393–1398 (1997).
Adrian, R. J. *Exp. Fluids* **39**, 159–169 (2005).
Adrian, R. J. *Phys. Fluids* **19**, 041301 (2007).
Adrian, R. J., Christensen, K. T., and Liu, Z.-C. *Exp. Fluids* **29**, 275–290 (2000a).
Adrian, R. J., Meinhart, C. D., and Tomkins, C. D. *J. Fluid Mech.* **422**, 1–54 (2000b).
Afanasyev, Y. D., and Demirov, E. K. *Exp. Fluids* **39**, 828–835 (2005).
Akedo, Y., Oshima, M., Oishi, M., and Saga, T.,"Visualization of flow structure in cerebral aneurysm model", The 8th Asian Symposium on Visualization, Chiangmai, Thailand, ID46 (2005).
Angele, K. P., Suzuki, Y., Miwa, J., and Kasagi, N. *Meas. Sci. Technol.* **17**, 1639–1646 (2006).
Arroyo, M. P., and Greated, C. A. *Meas. Sci. Technol.* **2**, 1181–1186 (1991).
Balakumar, B. J., and Adrian, R. J. *Exp. Fluids* **36**, 166–175 (2004).

Bastiaans, R. J. M., Plas, G. A. J., and Kieft, R. N. *Exp. Fluids* **32**, 346–356 (2002).
Bi, W. T., Sugii, Y., Okamoto, K., and Madarame, H. *Meas. Sci. Technol.* **14**, L1–L5 (2003).
Borowsky, J., and Wei, T. *ASME J. Fluid Eng.* **129**, 1415–1421 (2007).
Bourdon, C. J., Olsen, M. G., and Gorby, A. D. *Exp. Fluids* **37**, 263–271 (2004a).
Bourdon, C. J., Olsen, M. G., and Gorby, A. D. *Meas. Sci. Technol.* **15**, 318–327 (2004b).
Bown, M. R., MacInnes, J. M., and Allen, R. W. K. *Exp. Fluids* **42**, 197–205 (2007).
Bown, M. R., MacInnes, J. M., Allen, R. W. K., and Zimmerman, W. B. J. *Meas. Sci. Technol.* **17**, 2175–2185 (2006).
Broder, D., and Sommerfeld, M. *Exp. Fluids* **33**, 826–837 (2002).
Brucker, C. *Meas. Sci. Technol.* **8**, 1480–1492 (1997).
Burgmann, S., Brucker, C., and Schroder, W. *Exp. Fluids* **41**, 319–326 (2006).
Burgmann, S., Dannemann, J., and Schroder, W. *Exp. Fluids* **44**, 609–622 (2008).
Carlier, J., and Stanislas, M. *J. Fluid Mech.* **535**, 143–188 (2005).
Chen, R. C., and Fan, L.-S. *Chem. Eng. Sci.* **47**, 3615–3622 (1992).
Chen, J., and Katz, J. *Meas. Sci. Technol.* **16**, 1605–1618 (2005).
Chen, R. C., Reese, J., and Fan, L.-S. *AIChE. J.* **40**(1093), 1104 (1994).
Christensen, K. T. *Exp. Fluids* **36**, 484–497 (2004).
Christensen, K. T., and Adrian, R. J. *J. Fluid Mech.* **431**, 433–443 (2001).
Coudert, S. J. M., and Schon, J. P. *Meas. Sci. Technol.* **12**, 1371–1381 (2001).
Coupland, J. M., Garner, C. P., Alcock, R. D., and Halliwell, N. A. *J. Phys. Conf. Ser.* **45**, 29–37 (2006).
Deen, N. G., Westerweel, J., and Delnoij, E. *Chem. Eng. Technol.* **25**, 97–101 (2002).
Deng, R.-S., and Wang, C.-H. *Phys. Fluids* **15**, 3718–3729 (2003).
Diez, F. J., Bernal, L. P., and Faeth, G. M. *Int. J. Heat Fluid Flow* **26**, 873–882 (2005).
Elsinga, G. E., Scarano, F., Wieneke, B., and Van Oudheusden, B. W. *Exp. Fluids* **41**, 933–947 (2006a).
Elsinga, G. E., Van Oudheusden, B. W., and Scarano, F., Experimental assessment of tomographic-PIV accuracy, Proceedings of the 13th International Symposium Application of Laser Techniques to Fluid Mechanics, Lisbon, Portugal (2006b).
Ettema, R., Fujita, I., Muste, M., and Kruger, A. *Cold Reg. Sci. Technol.* **26**, 97–112 (1997).
Fincham, A. M., and Delerce, G. *Exp. Fluids* **29**, S13–S22 (2000).
Fincham, A. M., and Spedding, G. R. *Exp. Fluids* **23**, 449–462 (1997).
Fore, L. B., Tung, A. T., Buchanan, J. R., and Welch, J. W. *Exp. Fluids* **39**, 22–31 (2005).
Fouras, A., Dusting, J., and Hourigan, K. *Exp. Fluids* **42**, 799–819 (2007).
Fouras, A., Jacono, D. L., and Hourigan, K. *Exp. Fluids* **44**, 317–329 (2008).
Fujiwara, A., Danmoto, Y., Hishida, K., and Maeda, M. *Exp. Fluids* **36**, 157–165 (2004a).
Fujiwara, A., Minato, D., and Hishida, K. *Int. J. Heat Fluid Flow* **25**, 481–488 (2004b).
Ganapathisubramani, B. *Phys. Fluids* **19**, 098108 (2007).
Ganapathisubramani, B., Clemens, N. T., and Dolling, D. S. *J. Fluid Mech.* **556**, 271–282 (2006a).
Ganapathisubramani, B., Hutchins, N., Hambleton, W. T., Longmire, E. K., and Marusic, I. *J. Fluid Mech.* **524**, 57–80 (2005b).
Ganapathisubramani, B., Longmire, E. K., and Marusic, I. *J. Fluid Mech.* **478**, 35–46 (2003).
Ganapathisubramani, B., Longmire, E. K., and Marusic, I. *Phys. Fluids* **18**, 055105 (2006b).
Ganapathisubramani, B., Longmire, E. K., Marusic, I., and Pothos, S. *Exp. Fluids* **39**, 222–231 (2005a).
Gaydon, M., Raffel, M., Willert, C., Rosengarten, M., and Kompenhans, J. *Exp. Fluids* **23**, 331–334 (1997).
Grant, I., Fu, S., Pan, X., and Wang, X. *Exp. Fluids* **19**, 214–221 (1995).
Gui, L., and Merzkirch, W. *Exp. Fluids (Suppl.)* S465–S468 (1996).
Gui, L., Merzkirch, W., and Fei, R. *Exp. Fluids* **29**, 30–35 (2000).
Gui, L., and Wereley, S. T. *Exp. Fluids* **32**, 506–517 (2002).

Gunther, A., Jhunjhunwala, M., Thalmann, M., Schmidt, M. A., and Jensen, K. F. *Langmuir* **21**, 1547–1555 (2005).
Hambleton, W. T., Hutchins, N., and Marusic, I. *J. Fluid Mech.* **560**, 53–64 (2006).
Harada, K., Murakami, M., and Ishii, T. *Cryogenics* **46**, 648–657 (2006).
Hart, D. P. *J. Vis.* **3**, 187–194 (2000a).
Hart, D. P. *Exp. Fluids* **29**, 13–22 (2000b).
Hassan, Y. A., Ortiz-Villafuerte, J., and Schmidl, W. D. *Int. J. Multiphase Flow* **27**, 817–842 (2001).
Hassan, Y. A., Schmidl, W., and Ortiz-Villafuerte, J. *Meas. Sci. Technol.* **9**, 309–326 (1998).
Herrmann, S. F., Hinrichs, H., Hinsch, K. D., and Surmann, C. *Exp. Fluids* **29**, S108–S116 (2000).
Herrmann, S. F., and Hinsch, K. D. *Meas. Sci. Technol.* **15**, 613–621 (2004).
Hinsch, K. D. *Meas. Sci. Technol.* **6**, 742–753 (1995).
Hinsch, K. D. *Meas. Sci. Technol.* **13**, R61–R72 (2002).
Hori, T., and Sakakibara, J. *Meas. Sci. Technol.* **15**, 1067–1078 (2004).
Hou, Y. X., Somandepalli, V. S. R., and Mungal, M. G. *J. Fluid Mech.* **597**, 31–66 (2008).
Hu, H., Saga, T., Kobayashi, T., and Taniguchi, N. *Phys. Fluids* **14**, 2128–2138 (2002).
Hu, H., Saga, T., Kobayashi, T., Taniguchi, N., and Yasuki, M. *Exp. Fluids* **31**, 277–293 (2001).
Huang, H. *Exp. Fluids* **24**, 364–372 (1998).
Huang, H., Dabiri, D., and Gharib, M. *Meas. Sci. Technol.* **8**, 1427–1440 (1997).
Huang, H. T., Fiedler, H. E., and Wang, J. J. *Exp. Fluids* **15**, 263–273 (1993).
Hutchins, N., Hambleton, W. T., and Marusic, I. *J. Fluid Mech.* **541**, 21–54 (2005).
Ishikawa, M., Murai, Y., Wada, A., Iguchi, M., Okamoto, K., and Yamamoto, F. *Exp. Fluids* **29**, 519–531 (2000).
Jambunathan, K., Ju, X. Y., Dobbins, B. N., and Ashforth-Frost, S. *Meas. Sci. Technol.* **6**, 507, 514 (1995).
Jeon, D., Pereira, F., and Gharib, M. *Part. Part. Syst. Charact.* **20**, 193–198 (2003).
Kahler, C. J. *Exp. Fluids* **36**, 114–130 (2004).
Kahler, C. J., and Kompenhans, J., Multiple plane stereo PIV: Technical realization and fluid-mechanical significance. Proceedings of the 3rd International Workshop on Particle Image Velocimetry, Santa Barbara, CA, USA (1999).
Kahler, C. J., and Kompenhans, J. *Exp. Fluids (Suppl.)* S70–S77 (2000).
Kajitani, L., and Dabiri, D. *Meas. Sci. Technol.* **16**, 790–804 (2005).
Kashid, M. N., Gerlach, I., Goetz, S., Franzke, J., Acker, J. F., Platte, F., Agar, D. W., and Turek, S. *Ind. Eng. Chem. Res.* **44**, 5003–5010 (2005).
Keane, R. D., and Adrian, R. J. *Meas. Sci. Technol.* **1**, 1202–1215 (1990).
Keane, R. D., and Adrian, R. J. *J. Appl. Sci. Res.* **49**, 191–215 (1992).
Keane, R. D., Adrian, R. J., and Zhang, Y. *Meas. Sci. Technol.* **6**, 754–768 (1995).
Kiger, K. T., and Pan, C. *ASME J. Fluids Eng.* **122**, 811–818 (2000).
Kim, K. C., Yoon, S. Y., Kim, S. M., Chun, H. H., and Lee, I. *Exp. Fluids* **40**, 876–883 (2006).
Kinoshita, H., Kaneda, S., Fujii, T., and Oshima, M. *Lab Chip* **7**, 338–346 (2007).
Kitagawa, A., Hishida, K., and Kodama, Y. *Exp. Fluids* **38**, 466–475 (2005).
Kumar, S., and Banerjee, S. *Phys. Fluids* **10**, 160–177 (1998).
Lawson, N. J., and Wu, J. *Meas. Sci. Technol.* **8**, 1455–1464 (1997).
Lecordier, B., Demare, D., and Vervisch, L. M. J. *Meas. Sci. Technol.* **12**, 1382–1391 (2001).
Li, F.-C., Kawaguchi, Y., Hishida, K., and Oshima, M. *Exp. Fluids* **40**, 218–230 (2006a).
Li, F.-C., Kawaguchi, Y., Segawa, T., and Hishida, K. *Phys. Fluids* **17**, 075104 (2005a).
Li, F.-C., Kawaguchi, Y., Segawa, T., and Hishida, K. *Chin. Phys. Lett.* **22**, 624–627 (2005b).
Li, F.-C., Kawaguchi, Y., Segawa, T., and Suga, K. *Exp. Fluids* **39**, 945–953 (2005c).
Li, F.-C., Kawaguchi, Y., Segawa, T., and Suga, K. *Phys. Fluids* **17**, 082101 (2005d).
Li, F.-C., Kawaguchi, Y., Segawa, T., and Hishida, K. *Chin. Phys. Lett.* **23**, 1226–1229 (2006b).

Li, F.-C., Oishi, M., Kawaguchi, Y., Oshima, N., and Oshima, M., Experimental study of swirling flow of a viscoelastic fluid with deformed free surface, Proceedings of the ASME Joint U.S.–European Fluids Engineering Summer Meeting, July 17–20, Miami, FL, USA, Paper No. FEDSM2006-98387 (2006c).

Li, F.-C., Oishi, M., Oshima, N., Kawaguchi, Y., and Oshima, M., Statistical characteristics of elastic turbulence in a free-surface swirling flow, New Trends in Fluid Mechanics Research, Proceedings of the 5th International Conference on Fluid Mechanics, Shanghai, China, August 15–19, pp. 91–94, Tsinghua University Press and Springer (2007).

Li, F.-C., Yu, B., Wei, J.-J., Kawaguchi, Y., and Hishida, K. *Int. J. Heat Mass Transfer* **51**, 835–843 (2008).

Liao, Q., and Cowen, E. A. *Exp. Fluids* **38**, 197–208 (2005).

Liberzon, A., Gurka, R., and Hetsroni, G. *Exp. Fluids* **36**, 355–362 (2003).

Lima, R., Wada, S., Takeda, M., Tsubota, K., and Yamaguchi, T. *J. Biomech.* **40**, 2752–2757 (2007).

Lima, R., Wada, S., Tanaka, S., Takeda, M., Ishikawa, T., Tsubota, K., Imai, Y., and Yamaguchi, T. *Biomed. Microdevices* **10**, 153–167 (2008).

Lima, R., Wada, S., Tsubota, K., and Yamaguchi, T. *Meas. Sci. Technol.* **17**, 797–808 (2006).

Lindken, R., Gui, L.-C., and Merzkirch, W. *Chem. Eng. Technol.* **22**, 202–206 (1999).

Lindken, R., and Merzkirch, W. *Exp. Fluids (Suppl.)* S194–S201 (2000).

Lindken, R., and Merzkirch, W. *Exp. Fluids* **33**, 814–825 (2002).

Lindken, R., Westerweel, J., and Wieneke, B. *Exp. Fluids* **41**, 161–171 (2006).

Liu, Z., Adrian, R. J., and Hanratty, T. J. *J. Fluid Mech.* **448**, 53–80 (2001).

Liu, Z.-L., Jia, L.-F., Zheng, Y., and Zhang, Q.-K. *Chem. Eng. Sci.* **60**, 3537–3552 (2005).

Liu, Z.-L., Jia, L.-F., Zheng, Y., and Zhang, Q.-K. *Chem. Eng. Sci.* **63**, 1–11 (2008).

Lobera, L., and Coupland, J. M. *Meas. Sci. Technol.* **19**, 025501 (2008).

Longmire, E. K., Ganapathisubramani, B., Marusic, I., Urness, T., and Interrante, V. *J. Turbulence* **4**, 023 (2003).

Lueptow, R. M., Akonur, A., and Shinbrot, T. *Exp. Fluids* **28**, 183–186 (2000).

Maas, H. G., Gruen, A., and Papantoniou, D. *Exp. Fluids* **15**, 133–146 (1993).

Malsch, D., Kielpinski, M., Merthan, R., Albert, J., Mayer, G., Kohler, J. M., Susse, H., Stahl, M., and Henkel, T. *Chem. Eng. J.* **135S**, S166–S172 (2008).

McKenna, S. P., and McGillis, W. R. *Exp. Fluids* **32**, 106–115 (2002).

Meinhart, C. D., Wereley, S. T., and Gray, H. B. *Meas. Sci. Technol.* **11**, 809–814 (2000).

Melling, A. *Meas. Sci. Technol.* **8**, 1406–1416 (1997).

Meng, H., Pan, G., Pu, Y., and Woodward, S. H. *Meas. Sci. Technol.* **15**, 673–685 (2004).

Mielnik, M. M., and Saetran, L. R. *Exp. Fluids* **41**, 155–159 (2006).

Misra, S. K., Thomas, M., Kambhamettu, C., Kirby, J. T., Veron, F., and Brocchini, M. *Exp. Fluids* **40**, 764–775 (2006).

Mullin, J. A., and Dahm, W. J. A. *Exp. Fluids* **38**, 185–196 (2005).

Mullin, J. A., and Dahm, W. J. A. *Phys. Fluids* **18**, 035101 (2006a).

Mullin, J. A., and Dahm, W. J. A. *Phys. Fluids* **18**, 035102 (2006b).

Natrajan, V. K., Wu, Y., and Christensen, K. T. *J. Fluid Mech.* **574**, 155–167 (2007).

Nogueira, J., Lecuona, A., and Rodriguez, P. A. *Exp. Fluids* **27**, 107–116 (1999).

Nogueira, J., Lecuona, A., and Rodriguez, P. A. *Meas. Sci. Technol.* **12**, 1911–1921 (2001a).

Nogueira, J., Lecuona, A., and Rodriguez, P. A. *Exp. Fluids* **30**, 309–316 (2001b).

Nogueira, J., Lecuona, A., and Rodriguez, P. A. *Exp. Fluids* **39**, 305–313 (2005a).

Nogueira, J., Lecuona, A., and Rodriguez, P. A. *Exp. Fluids* **39**, 314–321 (2005b).

Olsen, M. G., and Adrian, R. J. *Exp. Fluids* **29**, S166–S174 (2000a).

Olsen, M. G., and Adrian, R. J. *Opt. Laser Technol.* **32**, 621–627 (2000b).

Olsen, M. G., and Bourdon, C. J. *J. Fluids Eng.* **125**, 895–901 (2003).

Ooms, T., Koek, W., Braat, J., and Westerweel, J. *Meas. Sci. Technol.* **17**, 304–312 (2006).

Ooms, T., Koek, W., and Westerweel, J. *Meas. Sci. Technol.* **19**, 074003 (2008).
Ostendorf, M., and Schwedes, J. *Powder Technol.* **158**, 69–75 (2005).
Park, J. S., Choi, C. K., and Kihm, K. D. *Exp. Fluids* **37**, 105–119 (2004).
Park, J. S., and Kihm, K. D. *Opt. Lasers Eng.* **44**, 208–223 (2006).
Pereira, F., and Gharib, M. *Meas. Sci. Technol.* **13**, 683–694 (2002).
Pereira, F., Gharib, M., Dabiri, D., and Modarress, D. *Exp. Fluids (Suppl.)* S78–S84 (2000).
Pereira, F., Lu, J., Castano-Graff, E., and Gharib, M. *Exp. Fluids* **42**, 589–599 (2007).
Philip, O. G., Schmidl, W. D., and Hassan, Y. A. *Nucl. Eng. Des.* **149**, 375–385 (1994).
Pickering, C. J. D., and Halliwell, N. *Appl. Opt.* **23**, 2961–2969 (1984).
Prasad, A. K. *Exp. Fluids* **29**, 103–116 (2000).
Prasad, A. K., and Jensen, K. *Appl. Opt.* **34**, 7092–7099 (1995).
Pu, Y., and Meng, H. *Exp. Fluids* **29**, 184–197 (2000).
Raffel, M., Willert, C. E., Werely, S. T., and Kompenhans, J., "Particle Image Velocimetry: A Practical Guide". 2nd Ed. Springer, Berlin (2007).
Reese, J., Chen, R. C., and Fan, L.-S. *Exp. Fluids* **19**, 367–378 (1995).
Reul, N., Branger, H., and Giovanageli, J. P. *Phys. Fluids* **11**, 1959–1961 (1999).
Reul, N., Branger, H., and Giovanageli, J. P. *Boundary Layer Meteorology* **126**, 477–505 (2008).
Roesgen, T. *Exp. Fluids* **35**, 252–256 (2003).
Rohaly, J., Frigerio, F., and Hart, D. P. *Meas. Sci. Technol.* **13**, 984–996 (2002).
Royer, H. *Meas. Sci. Technol.* **8**, 1562–1572 (1997).
Santiago, J. G., Wereley, S. T., Meinhart, C. D., Beebe, D. J., and Adrian, R. J. *Exp. Fluids* **25**, 316–319 (1998).
Sarrazin, F., Loubiere, K., Prat, L., Gourdon, C., Bonometti, T., and Magnaudet, J. *AIChE J.* **52**, 4061–4069 (2006).
Sato, Y., Fukuichi, U., and Hishida, K. *Int. J. Heat Fluid Flow* **21**, 554–561 (2000).
Sato, Y., and Hishida, K. *Int. J. Heat Fluid Flow* **17**, 202–210 (1996).
Sato, Y., Irisawa, G., Ishizuka, M., Hishida, K., and Maeda, M. *Meas. Sci. Technol.* **14**, 114–121 (2003).
Scarano, F. *Meas. Sci. Technol.* **13**, R1–R9 (2002).
Scarano, F. *Meas. Sci. Technol.* **15**, 475–486 (2004).
Scarano, F., and Riethmuller, M. L. *Exp. Fluids* **26**, 513–523 (1999).
Scarano, F., and Riethmuller, M. L. *Exp. Fluids (Suppl.)* S51–S60 (2000).
Schroder, A., Geisler, R., Elsinga, G. E., Scarano, F., and Dierksheide, U. *Exp. Fluids* **44**, 305–316 (2008).
Shaikh, N., and Siddiqui, K. *Ocean Dyn.* **58**, 65–79 (2008).
Shinohara, K., Sugii, Y., Aota, A., Hibara, A., Tokeshi, M., Kitamori, T., and Okamoto, K. *Meas. Sci. Technol.* **15**, 1965–1970 (2004).
Shinohara, K., Sugii, Y., Jeong, J. H., and Okamoto, K. *Rev. Sci. Instrum.* **76**, 106109 (2005).
Sielamowicz, I., Blonski, S., and Kowalewski, T. A. *Chem. Eng. Sci.* **60**, 589–598 (2005).
Soloff, S. M., Adrian, R. J., and Liu, Z.-C. *Meas. Sci. Technol.* **8**, 1441–1454 (1997).
Soria, J. *Exp. Therm. Fluid Sci.* **12**, 221–233 (1996).
Spedding, G. R., and Rignot, E. J. M. *Exp. Fluids* **15**, 417–430 (1993).
Stamhuis, E. J. *Aquatic Ecol.* **40**, 463–479 (2006).
Steijn, V., Kreutzer, M. T., and Kleijn, C. R. *Chem. Eng. Sci.* **62**, 7505–7514 (2007).
Steingart, D. A., and Evans, J. W. *Chem. Eng. Sci.* **60**, 1043–1051 (2005).
Stitou, A., and Riethmuller, M. L. *Meas. Sci. Technol.* **12**, 1398–1403 (2001).
Sugii, Y., Okuda, R., Okamoto, K., and Madarame, H. *Meas. Sci. Technol.* **16**, 1126–1130 (2005).
Susset, A., Most, J. M., and Honore, D. *Exp. Fluids* **40**, 70–79 (2006).
Svizher, A., and Cohen, J. *Exp. Fluids* **40**, 708–722 (2006a).
Svizher, A., and Cohen, J. *Phys. Fluids* **18**, 014105 (2006b).
Takehara, K., and Etoh, T. *J. Visual* **1**, 313–323 (1999).
Tao, B., Katz, J., and Meneveau, C. *Phys. Fluids* **12**, 941–944 (2000).

Tao, B., Katz, J., and Meneveau, C. *J. Fluid Mech.* **457**, 35–78 (2002).
Theunissen, R., Scarano, F., and Riethmuller, M. L. *Meas. Sci. Technol.* **18**, 275–287 (2007).
Tokuhiro, A., Fujiwara, A., Hishida, K., and Maeda, M. *ASME J. Fluids Eng.* **121**, 191–197 (1999).
Tokuhiro, A., Maekawa, M., Iizuka, K., Hishida, K., and Maeda, M. *Int. J. Multiphase Flow* **24**, 1383–1406 (1998).
Tomkins, C. D., and Adrian, R. J. *J. Fluid Mech.* **490**, 37–74 (2003).
Triep, M., Brucker, Ch., and Schroder, W. *Exp. Fluids* **39**, 232–245 (2005).
Van der Bos, F., Tao, B., Meneveau, C., and Katz, J. *Phys. Fluids* **14**, 2456–2474 (2002).
Van Sciver, S. W., Fuzier, S., and Xu, T. *J. Low Temp. Phys.* **148**, 225–233 (2007).
Wang, E. N., Devasenathipathy, S., Lin, H., Hidrovo, C. H., Santiago, J. G., Goodson, K. E., and Kenny, T. W. *Exp. Fluids* **40**, 847–858 (2006).
Warholic, M. D., Heist, D. K., Katcher, M., and Hanratty, T. J. *Exp. Fluids* **31**, 474–483 (2001).
Wernet, M. P. *Exp. Fluids* **28**, 97–115 (2000).
Westerweel, J. *Exp. Fluids* **16**, 236–247 (1994).
Westerweel, J. *Meas. Sci. Technol.* **8**, 1379–1392 (1997).
Westerweel, J., Dabiri, D., and Gharib, M. *Exp. Fluids* **23**, 20–28 (1997).
White, C. M., Somandepalli, V. S. R., and Mungal, M. G. *Exp. Fluids* **36**, 62–69 (2004).
Wieneke, B. *Exp. Fluids* **39**, 267–280 (2005).
Willert, C. E., and Gharib, M. *Exp. Fluids* **10**, 181–193 (1991).
Willert, C. E., and Gharib, M. *Exp. Fluids* **12**, 353–358 (1992).
Yamamoto, F., Wada, A., Iguchi, M., and Ishikawa, M. *J. Flow Visual Image Proc.* **3**, 65–78 (1996).
Yang, H., Halliwell, N. A., and Coupland, J. M., Micro holographic particle image velocimetry: Digital 3C3D measurement of free jet flow. Proceedings of the 12th International Symposium on Application of Laser Techniques to Fluid Mechanics, Lisbon, Portugal (2004).
Yoon, S. Y., and Kim, K. C. *Meas. Sci. Technol.* **17**, 2897–2905 (2006).
Zarruk, G. A. *Meas. Sci. Technol.* **16**, 1970–1975 (2005).
Zhang, J., Tao, B., and Katz, J. *Exp. Fluids* **23**, 373–381 (1997).
Zhang, T., Celik, D., and Van Sciver, S. W. *J. Low Temp. Phys.* **134**, 985–1000 (2004).
Zhang, T., and Van Sciver, S. W. *J. Low Temp. Phys.* **138**, 865–870 (2005).
Zhao, X.-L., Li, S.-Q., Liu, G.-Q., Song, Q., and Yao, Q. *Powder Technol.* **183**, 79–87 (2008).

CHAPTER 4

Positron Emission Imaging in Chemical Engineering

J.P.K. Seville[1,*], **A. Ingram**[2], **X. Fan**[3] and **D.J. Parker**[3]

Contents		
	1. Introduction	150
	2. Positron Emission Techniques and their Recent Development	151
	2.1 Positron emission particle tracking (PEPT) and positron emission tomography (PET)	151
	2.2 Positron-emitting tracers	153
	2.3 Detectors	154
	2.4 Technique development	155
	3. Applications	156
	3.1 Fluidised beds	156
	3.2 Rotating drums and kilns	162
	3.3 Solids mixing	163
	3.4 Other applications	168
	4. Portable PEPT	168
	5. Summary and Future Plans	174
	Symbols and Abbreviations	176
	Acknowledgements	177
	References	177

Abstract   Better understanding, design and operation of engineering processes demand visualisation of the material flows within them under realistic conditions. Methods based on radioactive tracers

1 School of Engineering, University of Warwick, Coventry CV4 7AL, UK
2 School of Chemical Engineering, University of Birmingham, Birmingham B15 2TT, UK
3 School of Physics and Astronomy, University of Birmingham, Birmingham B15 2TT, UK

*Corresponding author.
E-mail address: J.P.K.Seville@warwick.ac.uk

enable visualisation to be performed on real processes taking place within opaque walls. Positron emission methods rely on detecting the pairs of back-to-back gamma rays produced when a positron (emitted in radioactive decay) annihilates with an electron, and are variants of positron emission tomography (PET) which is widely used in medicine for determining the distribution in 3D of a labelled fluid. In chemical engineering applications, extensive use has been made of the alternative technique of positron emission particle tracking (PEPT), invented at the University of Birmingham, in which a single tracer particle is radioactively labelled and can be accurately tracked at high speed. This has now been developed to the point where it has the capability to track tracer particles down to approximately 60 μm in size, moving at up to 10 m/s, yielding locations to within ±1 mm at frequencies better than 100 Hz. Most importantly, gamma rays are sufficiently penetrating that good location data can be obtained within real process vessels. Applications have been extremely diverse, and include both gas-phase-continuous and liquid-phase-continuous systems. Particularly strong contributions have been made to the study of mixing processes and applications of fluidisation. Hitherto, the method has been confined to the laboratory. However, a modular transportable positron camera has now been developed and has been used for the first time on large-scale plant at an industrial site.

1. INTRODUCTION

Better understanding, design and operation of engineering processes demand visualisation of the material flows within them under realistic conditions. The range of techniques available for imaging industrial processes is vast (Chaouki et al., 1997). A fundamental distinction can be made between (a) projection-imaging techniques, exemplified most simply by conventional photography, (b) tomographic techniques, in which an image is obtained of a slice through an object, and (c) particle tracking techniques, such as those which employ a radioactive tracer. The interest here is in techniques that can be used to image processes taking place within metal-walled vessels. This rules out most of the optical and electrical methods and places the emphasis on x-ray and radioactive techniques.

Within the radiation emission tracking techniques, there are two main variants: positron emission, in which the tracer position is determined by triangulation as described in Section 2, and the "proximity" techniques, in which a gamma emitter is placed within the system of interest and its position found by measuring the relative count rates in an array of detectors. An example of the latter is computer-automated radioactive particle tracking technique (CARPT) developed at the University of Washington St. Louis and the École Polytechnique, Montréal. Both can be used to image material flows in industrial processes and both have advantages and disadvantages. One major difference between them, however, is that proximity techniques require extensive experimental calibration and pre-calculation to take account of the effect of the vessel walls, whereas positron emission methods do not require this. In both cases, a feature of particular benefit to engineering studies is the fact that the actual particles of interest may be used as tracers, rather than dissimilar materials of unknown behaviour.

2. POSITRON EMISSION TECHNIQUES AND THEIR RECENT DEVELOPMENT

2.1 Positron emission particle tracking (PEPT) and positron emission tomography (PET)

Positron emission particle tracking (PEPT) is derived from the commonly used medical diagnostic technique of positron emission tomography (PET). The main difference is that in PET a distribution of radioactivity is imaged in a relatively long time (some minutes), whereas in PEPT a single small source of radioactivity is located very frequently (normally up to approximately 100 times per second). The basis of all positron-imaging methods is that positrons emitted from the tracer annihilate with free electrons very close to their point of emission to produce pairs of "back-to-back" γ-rays, which travel along the same line in opposite directions. These are then detected using two large position-sensitive detectors (the "positron camera"), from which, in PEPT, a line can be constructed on which the tracer particle must lie (as shown in Figure 1). In theory, two such lines are sufficient to locate the tracer in three dimensions; in practice, normally approximately 100 are used. An algorithm

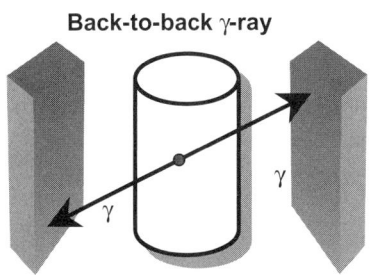
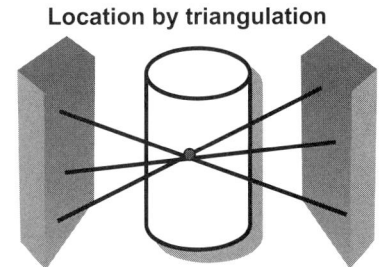

Figure 1 Principles of PEPT.

then finds the most probable position of the tracer by eliminating the outliers from the set of γ lines.

A significant number of the detected events are invalid, as either they correspond to a "random coincidence" between two unrelated γ-rays or else one or both of the γ-rays has been scattered before detection. The PEPT algorithm attempts to discard these invalid events, using an iterative procedure in which the centroid of the events is calculated, the γ-rays passing furthest from the centroid are discarded, and this process is repeated until a specified fraction f of the original events remains. The optimum value of f depends on the mass of material between the tracer and the detectors, which adds to the number of scattered events. For example, when studying flow inside a vessel with 15 mm thick steel walls, 80% of the detected events must be discarded, so that the fraction f of useful events is 0.2. The precision Δ of a PEPT location is given approximately by

$$\Delta \approx \frac{w}{\sqrt{fN}} \quad (1)$$

where w is the intrinsic spatial resolution of the positron camera (roughly 10 mm in a conventional camera), N the number of events detected during the location interval and f the fraction of these actually used for location. Assuming a data rate of $50\,\mathrm{k\,s^{-1}}$ with $f = 0.2$ one expects to be able to locate the tracer to within approximately 1 mm every 10 ms using the existing camera. This is typical of the tracking achieved for relatively slow moving tracers. During this time interval a tracer moving at 1 m/s will move 10 mm, so that for faster moving tracers it is necessary to locate more frequently and with slightly lower precision.

Features of PEPT of particular benefit to engineering studies include the fact that the actual particles of interest may be used as tracers, rather than dissimilar materials of unknown behaviour, and that γ-rays are sufficiently penetrating that location is unimpaired by the presence of metal walls, for example. In recent developments, the minimum size of particles which can be tracked has been reduced to approximately 60 μm. It is now possible to track multiple particles, to determine particle rotation and to track motion within real industrial equipment by use of a mobile modular positron camera. These developments are described later.

Figure 2 shows examples of how the data from PEPT can be treated, in this case from a spouted bed. The simplest data output is a continuous trajectory, consisting of a large file of x, y, z, time values, plus optional continuously recorded user-defined parameters such as speed or blade position. Velocity values can be obtained from this data file, in practice by a multi-point weighted averaging technique, and time averaged velocity profiles can then be derived. It is also possible to obtain values of

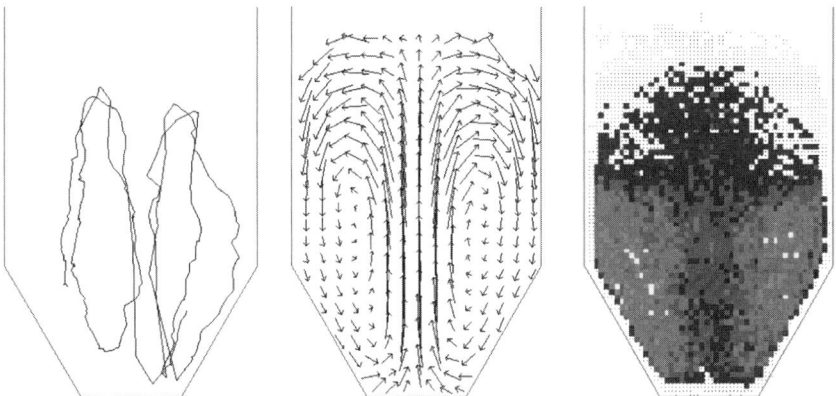

Figure 2 Typical PEPT output for a spouted bed (Left to Right: single trajectory; time-averaged velocity vectors; time-averaged "occupancy", showing denser annular region and leaner spout region) (see Plate 8 in Color Plate Section at the end of this book).

"occupancy", the proportion of the total run time which the tracer particle spends in each volume element of the object under scrutiny, which is equivalent to time averaged density. All time averaging — of velocities and occupancies — relies on the principle of "ergodicity", that is, that following one particle for a long time is equivalent to following all particles for a short time. Obviously this condition is not always satisfied, particularly when the process in question is time varying, but for many processes in laboratory-scale equipment this condition is satisfied for run times of order 1 h. As the scale increases, so must the run time.

More complex measures of particle behaviour can be obtained by further analysis of PEPT data, as described later.

2.2 Positron-emitting tracers

In most PEPT studies, it is desirable that the tracer particle should have properties indistinguishable from the bulk material. A major emphasis in PEPT-related research has been extending the range of tracer particles available to users, by developing a set of labelling techniques which are appropriate to a wide range of applications (Fan et al., 2006a, b, c). The original approach of direct irradiation in the ^{3}He beam from a cyclotron to generate ^{18}F (half-life 110 min) from oxygen *in situ* was limited to larger particles of high melting point oxides. Using alternative techniques of attaching ^{18}F from solution (produced when pure water is irradiated in the cyclotron beam), a wide range of tracers can now be made. As a result, the minimum tracer size has been reduced in stages from 600 μm to 60 μm for a variety of candidate materials. In most cases, this requires surface chemistry development tailored to the application, introducing bridging ions such as Fe^{3+}.

^{18}F is convenient to produce and process, but has the disadvantage that when tracers are exposed to aqueous environments the attached fluoride rapidly leaches back into solution. For early PEPT studies in water this problem was crudely overcome by painting the surface of the tracer particle with cellulose paint to seal in the activity, but inevitably this can alter the behaviour of the tracer. Recently, techniques have been developed for labelling small particles with ^{61}Cu (half-life 3.4 h) and ^{66}Ga (9.3 h), which survive much better in an aqueous environment. These radioisotopes are produced by irradiating metal foils (^{61}Cu from natural Ni in the reaction $^{60}Ni(d,n)^{61}Cu$ and ^{66}Ga from natural Zn in the reaction $^{66}Zn(p,n)^{66}Ga$) and are then chemically separated using a cation-exchange process, which is significantly more selective than the standard anion-exchange process. In the future, it may become practical to produce and separate ^{64}Cu (12.7 h). The longer-lived nuclides like ^{66}Ga and ^{64}Cu have obvious advantages if PEPT is to be used *in situ*, as discussed in Section 4.

Tracers labelled with ^{61}Cu and ^{66}Ga are suitable for use in aqueous environments, but like most indirectly labelled particles lose their activity rapidly in abrasive systems. For such environments, and also for high temperature applications, directly activated particles are currently preferred, since the radioactivity is incorporated in the matrix rather than attached to the surface. The Positron Imaging Centre at Birmingham has continued to develop direct activation of high melting point oxides (silica, alumina, etc.) and also of small copper particles (producing ^{64}Cu *in situ* through the reaction $^{63}Cu(d,p) \rightarrow ^{64}Cu$). It is difficult to activate very small particles in this way, partly because of practical issues in mounting the particles and focussing the cyclotron beam onto them, but more fundamentally because the heating effect of a highly focussed beam destroys the particles unless they are extremely well cooled.

2.3 Detectors

The original Birmingham positron camera consisted of a pair of rectangular gas-filled multi-wire chambers with rather low sensitivity for detecting the 511 keV γ-rays from positron annihilation. Its successor, an ADAC Forte, has similar geometry (two NaI(Tl) gamma camera heads, each $50 \times 40\,cm^2$) but significantly higher sensitivity and count rate capability. Under optimum conditions this camera records up to 100k coincidence events per second, enabling a tracer to be located to within 1 mm once per millisecond. The characteristics of this camera have been fully described by Parker et al. (2002).

Scanners used for medical PET generally adopt an entirely different approach, using rings of many individual small γ-ray detectors, giving even greater sensitivity and count rate. The geometry of a standard medical scanner is unsuited for many engineering studies (this is one reason why few engineering PET studies have been reported using hospital scanners), but since the architecture of these scanners is inherently modular it is in principle possible to reconfigure them into alternative geometries. This has been done using a donated ex-hospital PET scanner, which has been completely rebuilt as a modular positron camera for performing PEPT studies over a range of length scales: it is transportable and has been used for a series of PEPT studies on a large scale fluidised bed on an industrial site, as described in Section 4. When configured in a compact geometry it achieves a count rate of over 300k events per second, allowing very accurate high-speed tracking, and has been used in this way to study granular motion in high-speed rotating equipment.

2.4 Technique development

Although PEPT is extremely powerful, there are limits to the information that can be obtained by tracking a single particle. In a well-mixed steady-state system, the single particle, if followed for long enough, can demonstrate all features of the flow. However, to observe transients or particle–particle correlations requires simultaneous tracking of more than one particle. The original PEPT algorithm relies on *a priori* knowledge that only a single labelled particle is present in the field of view, and proceeds iteratively to discard outlying events until only a tight cluster of events remains. This algorithm has recently been extended to enable several particles to be tracked simultaneously, the first time that this has been done (Yang et al., 2006, 2007a, b). The standard PEPT algorithm is used to locate one of the tracers (discarding events corresponding to all other tracers), then the discarded data is re-examined and used to find a second tracer, and so on. This approach works reasonably well provided the tracers do not approach very close to each other. The particles can be distinguished if their activities are significantly different.

Multi-particle tracking is obviously of use in studies of segregation, where particles are separated in processing by virtue of differences in their size or density or both. These include fluidisation and gravity-driven flows. A special case of the technique is where three tracers are attached to different points on a larger body — multi-particle tracking can then be used to follow the rotational as well as translational motion of this body. This approach has been applied, for example, to observe the motion of lumps of foodstuff inside rotating cans during heat sterilisation.

3. APPLICATIONS

Applications of positron imaging, particularly PEPT, have been extensive and very diverse. In general, they vary from quick diagnostic studies for industrial users to fundamental work on multi-phase systems, such as "granular gases" (see, e.g., Wildman et al., 2005). Here, we concentrate on work aimed at industrial understanding.

3.1 Fluidised beds

PEPT has been used extensively in fluidised beds. Applications have included: solids circulation and measurement of the characteristic "turnover time" (Stein et al., 2000); scaling studies for solids motion (Stein et al., 2002); motion around in-bed obstacles such as heat-exchange surfaces (Ding et al., 2001b); the effect of cohesion in fluidisation due to surface liquid layers (Seville et al., 2000) and to sintering at high temperature (Seville et al., 1998, 2004); heat transfer by particle convection to in-bed tubes (Wong and Seville, 2006); segregation (Leaper et al., 2004); solids motion in circulating fluidised beds (CFB) (van de Velden et al., 2008); coating in top-spray fluidised beds (Depypere et al., 2009), bottom-spray Wurster coaters (Palmer et al., 2007; Palmer, 2008) and spouted beds (Seiler et al., 2008); and validation of computational approaches to the modelling of fluidised beds (Hoomans et al., 2001; Link et al., 2008).

From a practical point of view, one of the most challenging applications to date has been in the experimental fluidised bed shown in Figure 3, which is a scaled-down version of a polyethylene production unit, of overall height approximately 2 m, with a conical expansion section 1 m above the distributor; the diameter of the lower section is 0.154 m. The bed is constructed of 316 stainless steel and is placed within a gas circulation loop allowing operation at pressures up to 20 bars.

3.1.1 Solids motion studies

Figure 4 shows a typical projection of a (3D) particle trajectory. The essential features of all such trajectories, which we have investigated in previous publications (Stein et al., 2000), are as follows:

- Particles move upwards in a series of discrete fast movements ("jumps"), punctuated by periods of relative inactivity ("idle time" or "quiescent time"). It is during these idle times that the particles are able to sinter in polymer reactors, and control of the length of these periods is necessary to prevent defluidisation by sintering (Seville et al., 1998).
- Jumps are associated with bubble motion.

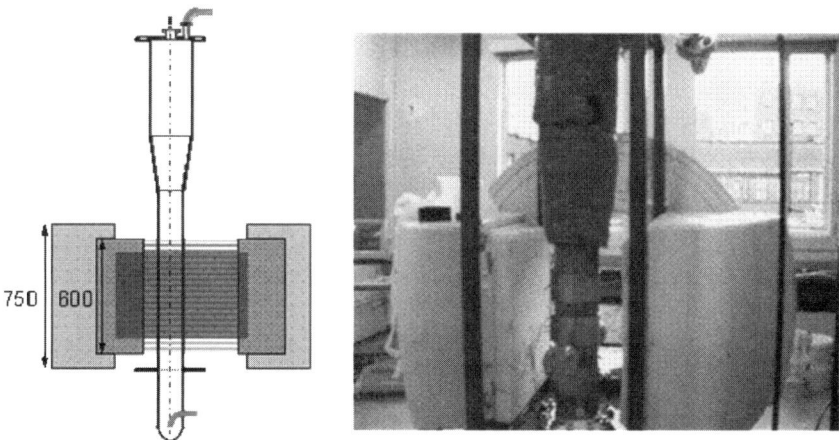

Figure 3 Pressurised fluidised bed-imaging geometry: schematic (left; dimensions in mm), and photograph (right).

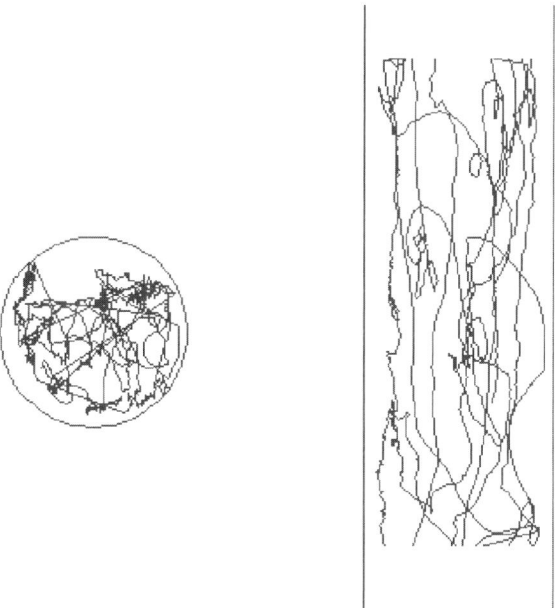

Figure 4 Typical fluidised bed trajectories (Left: plan; Right: elevation).

- Although the average jump velocity and average distance jumped increase with increasing gas velocity, the jump duration stays relatively constant; a theoretical explanation for this is given by Stein et al. (2000).
- Particles move down relatively slowly and (usually) near to the wall.
- The lengths of the trajectories form a wide distribution — not all upward movements terminate with ejection into the freeboard and not all downward movements terminate at the distributor.

Having obtained a trajectory as in Figure 4, it is possible to apply two virtual surfaces within the bed, for example, at heights of one quarter and three quarters of the settled bed height, and to record the times of crossing these surfaces. This enables cycle durations and frequencies to be obtained, which can be useful in diagnosing performance issues.

3.1.2 Motion close to surfaces

Fluidised beds are particularly favoured as chemical reactors, because of their ability to exchange heat through immersed heat exchange surfaces. However, little is known about how the heat exchange process works on a single particle level. The most commonly applied theory of fluidised bed heat exchange is that developed by Mickley and Fairbanks (1955) — the so-called "packet model". Wong and Seville (2006) used PEPT to follow the trajectory of a single tracer particle in a fluidised bed containing heat exchanger tubes. The residence time of particles in the vicinity of the heat exchange surface was determined directly for the first time, allowing the observed heat transfer variations to be interpreted mechanistically

Locational inaccuracy means that it is not possible to determine with certainty whether the tracer makes contact with the tube surface. Therefore, the method shown schematically in Figure 5 was adopted: a notional cylindrical surface is drawn around the heat transfer tube at a certain distance from it, and the time for which the tracer remains within this outer surface is determined. This is termed the residence time.

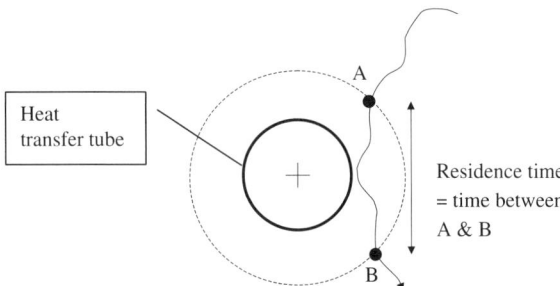

Figure 5 Residence time determination using particle trajectory (Wong and Seville, 2006).

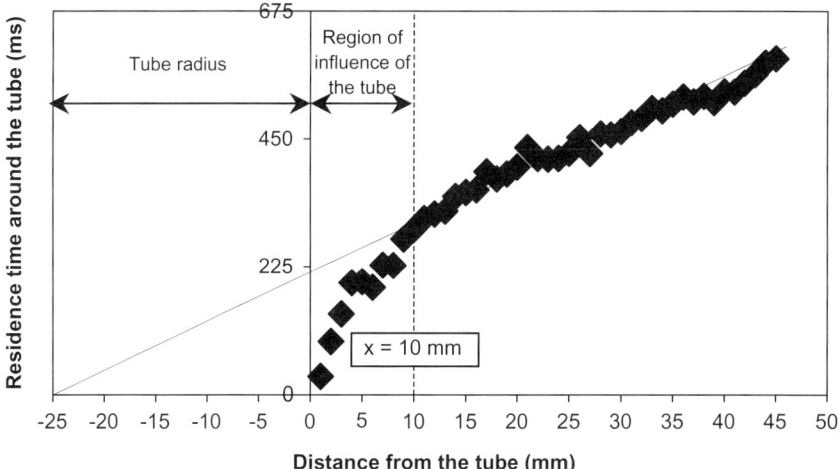

Figure 6 Variation in residence time around tube with increasing distance from the tube: rectangular bed (0.6 m × 0.06 m × 1.5 m), 0.6 mm sand, bed static height 400 mm and tube positioned at a bed level of 280 mm above the distributor (Wong and Seville, 2006).

Figure 6 shows the residence time around a tube, as a function of distance from the tube. Particles were taken randomly from the bulk to be irradiated as tracers. The tracking time for each experiment was approximately 1.5 h.

With increase in the thickness of the annulus inside which the residence time is obtained, it can be seen that there is a change in the slope of the residence time plot. As the thickness of the annulus becomes large by comparison with the radius of the tube, the presence of the tube becomes less significant and a linear relationship would be expected between residence time and annulus dimension (distance from the tube), as observed here for larger distances. The slope of the plot after 10 mm can be extended back to the tube centreline. The point at which the measurements deviate from this linear plot defines the "region of influence" of the tube. The thickness of the region of influence of the tube was estimated at approximately 10 mm, independent of gas velocity. Wong (2003) shows that the mean residence time within this region is inversely proportional to the square root of the excess gas velocity $(U-U_{mf})$, as suggested by Bock (1983).

3.1.3 Circulating fluidised beds

Circulating fluidised beds (CFBs) are attracting increasing interest for both gas–solid and gas-catalytic reactions; however, the operating modes in these two cases are completely different. In modelling and designing

CFBs as reactors, the solids residence time is an important parameter. Previous studies mostly assess operations at moderate values of the solids circulation rates ($\leq 100\,\text{kg}/\text{m}^2\,\text{s}$), whereas gas-catalytic reactions and, e.g., biomass pyrolysis require completely different operating conditions.

Van de Velden et al. (2008) used PEPT to study the movement and population density of particles in the CFB-riser. The PEPT results were used to obtain (i) the vertical particle movement and population density in a cross-sectional area of the riser; (ii) the transport gas velocity required to operate in a fully established circulation mode; (iii) the overall particle movement mode (core flow *versus* core/annulus flow); and (iv) the particle slip velocity. Figure 7 shows an example of PEPT data for the two principal flow regimes.

Using these results Van de Velden et al. (2008) were able to recommend design rules for operation of such reactors in terms of the gas velocity/solids loading parameters. For example, to ensure a narrow residence time distribution (operating, in effect, in plug flow), the superficial gas velocity should exceed the transport velocity by approximately 1 m/s and the solids circulation rate should exceed $200\,\text{kg}/\text{m}^2\,\text{s}$.

3.1.4 Validation of computational models

Hoomans et al. (2001) and Link et al. (2008) have used PEPT to validate computational approaches to the modelling of fluidised beds. In both cases, the computational method used was a 3D Euler–Lagrangian hard-sphere discrete particle model (DPM). Hoomans et al. (2001) simulated a conventional bubbling bed with a uniform gas distributor. Link et al. (2008) simulated the behaviour of a spout-fluid bed. In the hard-sphere collision model used in these studies, rigid particles are assumed to interact through binary, instantaneous collisions. Particle collision dynamics are described by collision laws, which account for energy dissipation due to non-ideal particle–particle and particle–wall interaction by means of the empirical coefficients of normal and tangential restitution, and the coefficient of friction. For systems that are not too dense, the hard-sphere model is considerably faster than soft-sphere models, which is why it was adopted for this work. The particle collision parameters play an important role in the overall bed behaviour, as reported by Goldschmidt et al. (2001), and were determined using impact experiments.

Results are reported by Link et al. (2008) of a combined experimental and simulation study on the flow regimes which can be encountered during operation of a spout-fluid bed. A regime map was composed, employing spectral analysis of pressure drop fluctuations and fast video recordings, and compared with the results of 3D computations. For most investigated regimes, the model is able to predict the appropriate regime.

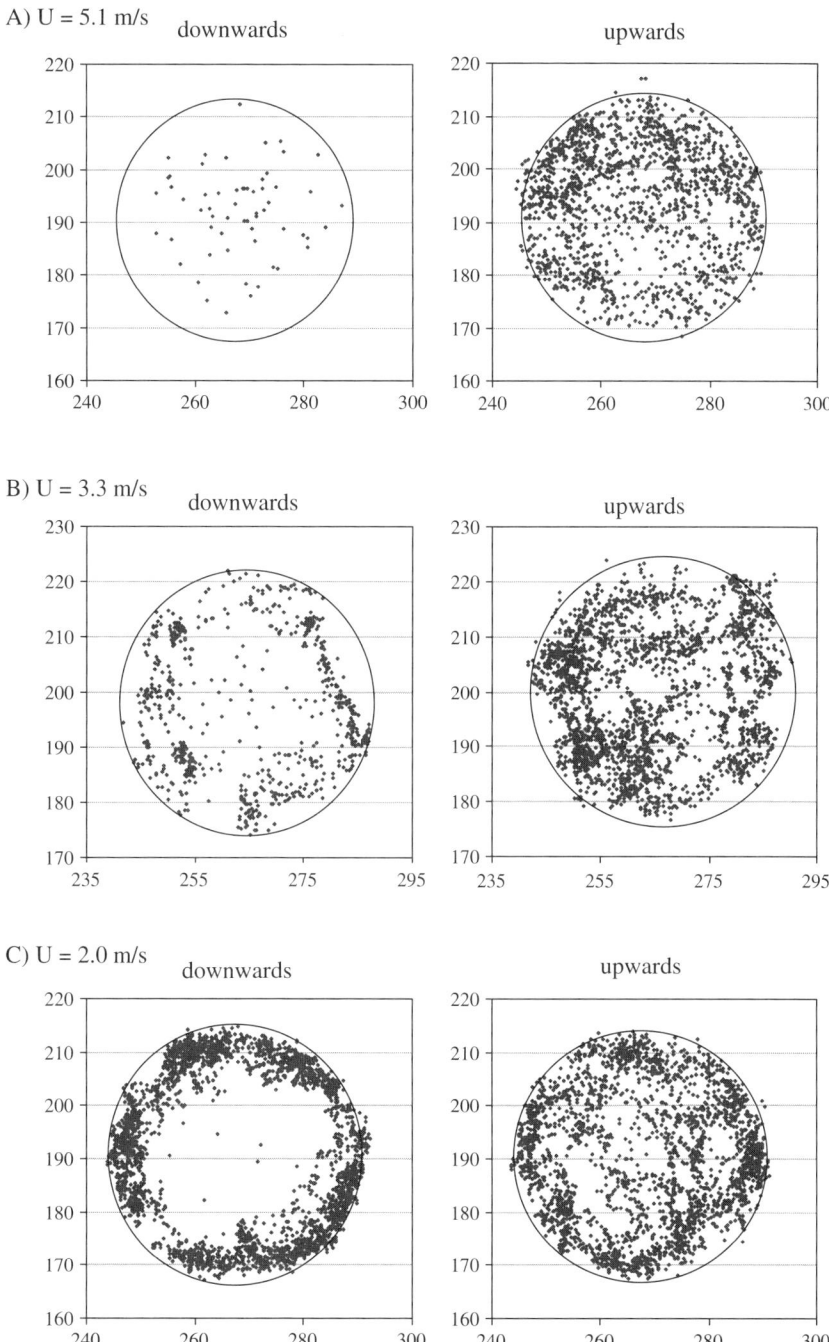

Figure 7 Cross-sectional view of the riser with left: downwards moving particles, and right: upwards moving particles; at $G = 260\,kg/m^2 s$ and indicated superficial gas velocity (U). The plots show all the particle locations over the height of the viewed section, integrated over the time of the run (Van de Velden et al., 2008).

The model is also able to reproduce the measured particle velocity profiles in the bed, including their distribution, for all investigated conditions.

3.2 Rotating drums and kilns

Industrial-scale rotating drums are usually operated in the rolling or slumping mode, according to whether the flow down the inclined surface is continuous or discontinuous, that is, in a series of avalanches. In either case, the bed can be divided into two regions (Figure 8): the "passive" region, in which the particles are carried up as a rigid body by the drum wall, and the "active" region in which they move down. Ding et al. (2001a, 2002) have shown that solids mixing and segregation occur largely in the active region, and will strongly influence bed-freeboard heat and mass transfer and therefore chemical reactions.

Solids exchange between the passive and active regions is important and an expression for the exchange coefficient between the regions has been developed, which is equivalent to a bed "turnover" frequency. An explicit relationship between turnover time and the drum operating parameters has been obtained and tested (Ding et al., 2002). The derivation of this relationship has led to an expression for the transition between the bed slumping and rolling conditions.

In continuous operation, the residence time distribution in a rotating drum depends on the degree of both radial and axial dispersion. In previous work (des Boscs et al., 1997), it was proposed that a particle in a distribution will have a preferred radial position, about which its actual position will be distributed, and will then experience an axial

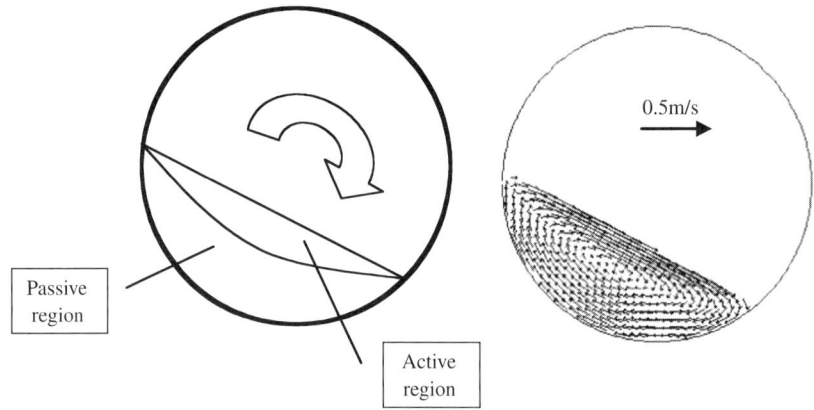

Figure 8 Particle motion in the transverse direction of a rotating drum: (a) schematic and (b) average velocity vectors from PEPT.

advancement per cycle which is statistically distributed around a most probable value corresponding to the angle of maximum slope. The distributions of particle position in successive cycles have been investigated, the most striking result being that even in the rolling mode, and with a narrow size distribution so that segregation is not expected, there is a strong correlation between the radii in the passive region before and after passage through the active layer (Ingram et al., 2005), that is, dispersion between layers is limited. This effect is more marked with wider size distributions, which rapidly segregate into distinct radial bands (Figure 9).

The charge in a rotating drum is also very prone to segregation in an axial direction. PEPT allows tracers of different sizes to be followed (Parker et al., 2005), so that the concentration of fines in the centre and coarse at the wall can be observed. In some cases (perhaps all), axial segregation occurs when the central core of fines running the entire length of the drum breaks out to the surface at regular intervals. PET has been used to image this effect (Figure 10), the first time this has been demonstrated.

3.3 Solids mixing

Both PEPT and PET have been used extensively in studies of solids mixing, in devices ranging from rotating vessels, such as kilns and V-mixers, to bladed mixers with both horizontal and vertical axes.

The V-mixer is a simple device, widely used in the pharmaceutical industry, which consists of two cylinders, joined at 90°, as shown in Figure 11 (Kuo et al., 2002, 2005). It rotates around the axis shown, thus, repeatedly dividing (∧) and recombining (∨) the charge. At each rotation a small amount of material is transferred across the plane of symmetry between the two arms. The reason for studying this device is not because it is of great inherent interest in itself but because it is

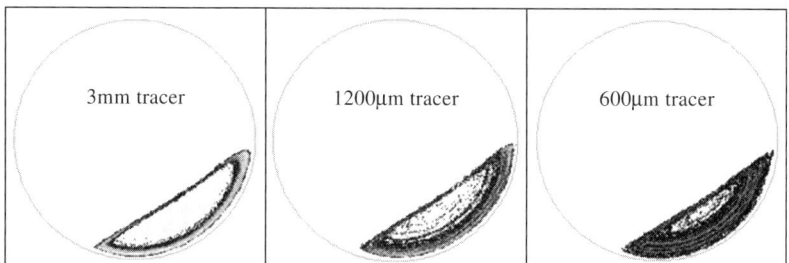

Figure 9 Transaxial PEPT occupancy plots (fraction of run time spent at each location) for three different sizes of tracer particle in a bed containing a wide range of particle sizes; dark indicates high occupancy.

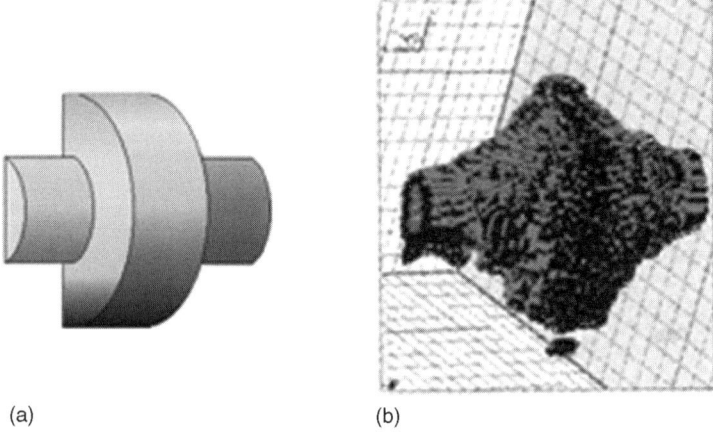

(a) (b)

Figure 10 Axial banding of a binary mixture in a continuous-flow rotating drum: (a) schematic drawing showing the form of axial banding merging with a continuous core and (b) 3D PET image, showing the same features.

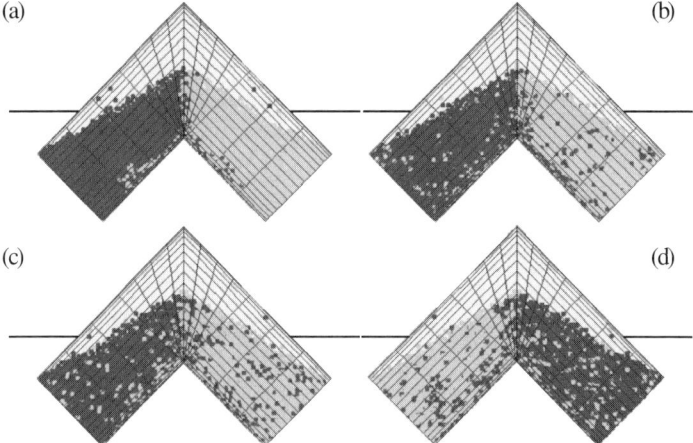

Figure 11 DEM simulations for 1 (a), 4 (b), 8 (c) revolutions in the front views and 8 revolutions in the back view (d) (Kuo et al., 2002).

representative of that generic class of mixers in which the solids are repeatedly divided and recombined. Typical results of discrete element modelling (DEM) studies of this device are shown in Figure 11.

One of the chief challenges in modelling is experimental validation, and single particle trajectory following using PEPT has proved useful in this. Figure 12 shows the close agreement that can be achieved between the average velocity distributions from both model and experiment. A

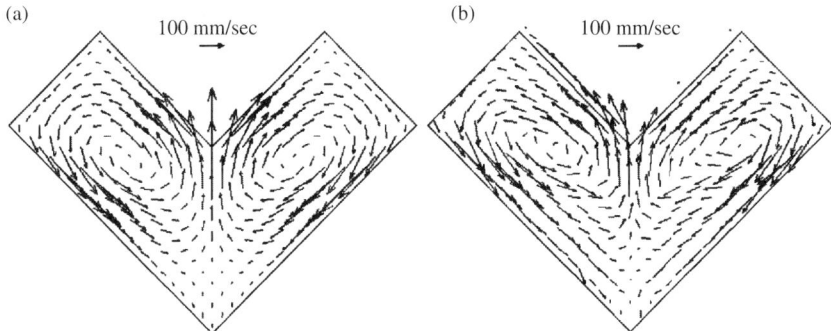

Figure 12 Velocity distributions at 20% fill, 60 rpm: (a) DEM simulations and (b) PEPT experiment (Kuo et al., 2002).

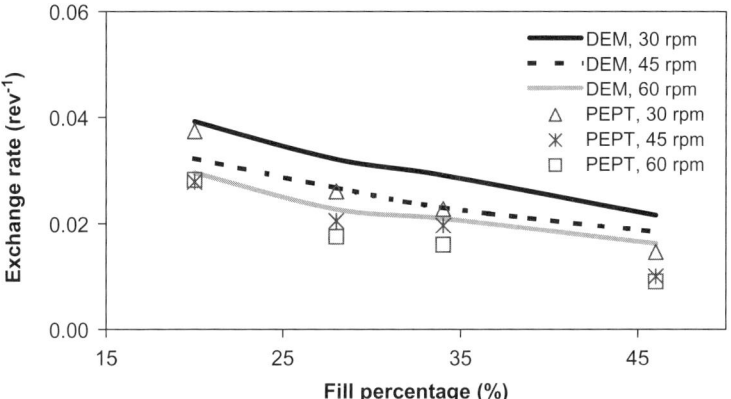

Figure 13 The exchange rate as a function of the fill percentage at different rotational speeds (Kuo et al., 2002).

more quantitative comparison between the two is provided by the exchange rate of solids across the dividing plane, which is plotted as a function of fill level in Figure 13 (the exchange rate is here defined as the probability that a single particle will swop sides per revolution). Again, the agreement is good. It is well-known that the exchange rate reduces as the fill level is increased, but this work also revealed the unexpected result that particle exchange during the division step is more important than during the combination step.

In all types of mixers it is useful to be able to quantify dispersion or particle mobility. Parker et al. (1997) showed that the axial dispersion coefficient, D, could be measured in rotating drums, for example, by plotting the mean value of (axial displacement, $x_d)^2$ versus time:

$$D = <x_d^2>/2t \qquad (2)$$

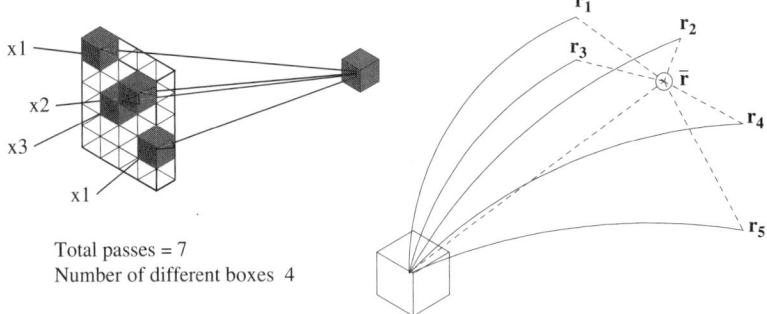

Figure 14 Analysis of dispersion (Martin et al., 2007).

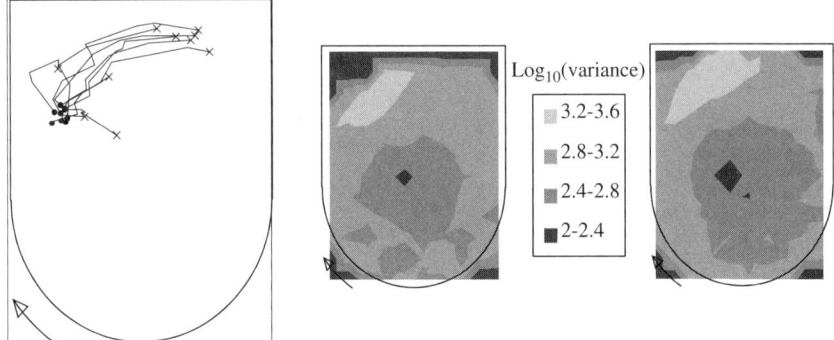

Figure 15 Analysis of dispersion in a high-speed bladed mixer (from Left to Right: multiple trajectories starting from the same volume element; two examples of distributions of variance (Martin et al., 2007).

More interesting, however, is to follow the dispersion of a group of particles originating at the same point (Martin et al., 2007) (Figure 14). The difference between the trajectories is a measure of dispersion related to processes occurring at that point. In single particle tracer studies, each time the tracer passes through a given volume element of the system, its subsequent location (some time or distance later) can be found. When the tracer has passed through each element many times, many traces are obtained with their origins at approximately the same point. Provided that the motion of the solids is time-independent, this is equivalent to tracing the path of many particles from the same location simultaneously. Figure 15 shows an example for a mixer with a high-speed rotating blade.

The variance between the locations is then given by

$$\sigma^2 = \frac{1}{n}\sum_{i=1}^{n}(x_i - \bar{x})^2 + (y_i - \bar{y})^2 + (z_i - \bar{z})^2 \qquad (3)$$

where σ is the standard deviation; $\bar{x}, \bar{y}, \bar{z}$ are the mean final location; and n is the number of traces.

The question arises as to how far (or for how long) the tracer particle should be allowed to move before its location variance is found. In the first place, time is a more appropriate measure for the interval between the start point and the variance measurement, and is consistent with the approach taken in diffusion. Secondly, the elapsed time should be enough to allow measurable movement to take place, but not so much that that movement is no longer related to the conditions which are local to the point of interest. In this particular example, the mixer used was of diameter 150 mm, and the maximum blade tip speed approximately 0.8 m/s. A "dispersion time" of 100 ms thus gives a maximum displacement of approximately 80 mm, which is about the same magnitude as the radius and consistent with the lengths of trajectories shown in Figure 15. An example of the distribution of the resulting variance is also shown in Figure 15.

To obtain a measure of the overall mixer effectiveness, account needs to be taken of the number of times each volume element is visited by the tracer; dividing by the total number of passes through all elements normalises these values. A measure of "mixer effectiveness" (ME) is then given by

$$\mathrm{ME} = \frac{1}{N_p} \sum_{i=1}^{n} \sigma_i^2 n_{pi} \qquad (4)$$

where σ_i^2 is the variance for element i, n_{pi} the number of passes through element i, n the number of elements in the system and N_p the total number of passes.

In the example considered above, the maximum value of ME was approximately 1,050 mm^2.

Martin et al. (2007) showed how the choice of the time-for-dispersion discussed above affects the ME values. ME cannot increase without limit, because it is bounded by the physical size of the mixer. As expected, the ME versus time relationship is of the form:

$$\mathrm{ME} = \mathrm{ME}_{\infty}(1 - \mathrm{e}^{-\lambda t}) \qquad (5)$$

where ME_{∞} is the steady-state ME value at infinite time and λ is a (first-order) rate constant, although this relationship is affected by the existence of parallel mixing processes going on at different scales, for example, local to a blade and on the scale of the whole vessel.

PET is also of use in following mixing processes, although the time necessary to provide a complete tomographic scan of a piece of process equipment is still long — many minutes — so that it is normally necessary in such experiments to stop the process intermittently in order for periodic scans to take place. Armstrong et al. (2005) demonstrate the

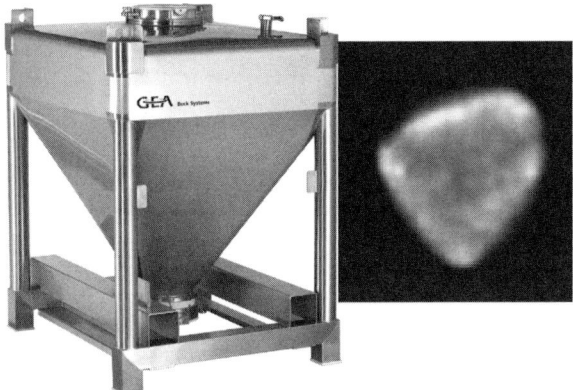

Figure 16 PET image (right) of dispersion of a pseudo drug substance within an excipient contained in a mixing vessel (left) (Armstrong et al., 2005).

use of PET in following the mixing of a pharmaceutical active ingredient in a much larger bulk of excipient material, mixed within a polygonal rotating bin of the type used in the pharmaceutical industry for both intermediate storage and mixing (Figure 16).

3.4 Other applications

The scope of applications of PEPT is given in Table 1, which lists recent projects and includes brief notes on the significance of each as far as use of positron emission techniques are concerned. It will be noted that most applications to date are in gas-phase-continuous systems. PEPT can also be used in liquid-phase-continuous systems, where the challenge is to produce a truly flow-following tracer which does not lose its activity due to leaching into the liquid. Progress in liquid mixing studies is summarised by Barigou (2004), whereas Bakalis et al. (2004, 2006) demonstrate flow-following in viscous liquids. Fishwick et al. (2005a, b) show how PEPT has been used in helping to understand flow in multi-phase reactors.

4. PORTABLE PEPT

PEPT has become an established technique for studying the motion of particles in granular and fluid systems. Until recently, this technique was confined to use with medically derived detectors, which places constraints on the geometry and scale of process equipment that can be viewed. Demand for greater flexibility in the use of the PEPT

Table 1 Applications of PEPT

Project	Investigators[a]	Major outcomes in relation to PEPT[b]
Experimental and computational (DEM) study of fluidisation of cohesive particles	J.P.K. Seville and C. Thornton	First tracking of sub 100 μm particles in gas fluidisation, including under pressure (20 bars). Comparison with DEM predictions
Feasibility of tracking multiple tracers	P.J. Fryer, D.J. Parker and J.P.K. Seville	Multiple tracking algorithm development and experimental demonstration in fluidisation and solid/liquid flows
Application of particulate flavours to foodstuffs	P.J. Fryer and J.P.K. Seville	Tracking of difficult-to-label foodstuffs in rotating food processing machinery
Mixing of dense suspensions and rheologically complex opaque fluids	M. Barigou and A.W. Nienow	Tracking in large liquid-filled vessels (increased scatter) and at high speed with rapidly changing direction. Multimode comparisons with electrical resistance tomography and with CFD.
Development of multi-scale models for powder compaction processing	C.Y. Wu	Tracking of particles in die filling operations
Particle motion in 3D vibro-fluidised granular "gases"	R.D. Wildman (University of Loughborough)	High velocity, high accuracy motion required to extract granular temperature
Product–process relationships in high-shear granulation	M. Ghadiri, R.A. Williams, S. Anthony and Y. Ding (Leeds)	Motion inside metal-walled vessels of complex shape. Scale-up studies

Table 1 *(Continued)*

Project	Investigators[a]	Major outcomes in relation to PEPT[b]
Particle separation by vibrated liquid fluidised beds	M. Biggs, D. Glass and J.Y. Ooi (University of Edinburgh)	Vibrationary solids motion on a small scale in a liquid fluidised bed.
Powder processing in the pharmaceutical industry	J.P.K. Seville, D.J. Parker and A. Ingram	Both PEPT and PET used to characterise dispersion of drug mimic in mixing
Solids motion in circulating fluidised beds	J. Baeyens and J.P.K. Seville	First tracking measurements in circulating beds and cyclones
Coating in a pharmaceutical Wurster coater	J.P.K. Seville and A. Ingram	Provides fundamental input data to coating models
Solids motion in polymer extrusion	A. Ingram and J.P.K. Seville	Tracking molten polymer in thick-walled extruder+comparison with computational model from TU Eindhoven
Novel continuous granulation process	J.P.K. Seville and A. Ingram	Three-phase study in thick metal-walled vessel
Solids motion of polyethylene in fluidised bed reactor	J.P.K. Seville and A. Ingram	Comparison with CFD study by University of Toulouse
Solids motion in partitioned pilot scale fluidised bed	D.J. Parker, J.P.K. Seville and A. Ingram	First *in situ* study using mobile camera. Large-scale fluidisation
Solids motion in rotating waste pyrolyser	A. Ingram	Tracking large eccentric items
Bioseparation in expanded bed absorption columns	O.R.T. Thomas	Tracking 100 μm particles in fluidised chromatography columns

Table 1 *(Continued)*

Project	Investigators[a]	Major outcomes in relation to PEPT[b]
Minerals comminution	I. Govender (University of Cape Town)	Tracking mineral fragments and slurry in ball mills
Metals casting	W.D. Griffiths	First application of PEPT in high temperature liquid flows (molten aluminium) and comparison with computational models

[a]University of Birmingham except where stated.
[b]See publication list.

technique — in imaging larger process equipment and, more importantly, industrial equipment *in situ* — has led to the development of a modular PEPT camera (Leadbeater and Parker, 2009; Parker et al., 2008, 2009). This comprises a set of individual detectors, which can be arranged around the equipment in whatever configuration is appropriate to enable particle tracking. In an initial trial, the modular camera has been used to track particle motion in a 750 mm diameter pressurised fluidised bed reactor under industrially relevant conditions. The results show how the technique can be used reliably on large-scale equipment to measure quantities such as circulation time.

As discussed in Section 2, the present Birmingham positron camera, which has been used in almost all reported PEPT studies, consists of a single pair of large area detectors. In contrast, most medical PET scanners consist of rings of hundreds of small detectors. By distributing the events over many detection elements the problems of dead time and random coincidences are reduced so that higher overall count rates can be achieved. Since count rate is critical for tracking at high speed, it is of considerable interest to investigate extending this approach to PEPT. An important additional benefit of constructing a positron camera from a number of detector modules is that it becomes possible to consider adapting the detection geometry to suit the individual system under study.

A redundant medical PET scanner, a CTI ECAT931/08, was acquired. This comprises 128 detector blocks (Figure 2a), each consisting of four photomultiplier tubes viewing a 30 mm thick crystal of bismuth germanate scintillator approximately $49 \times 56 \, \text{mm}^2$ in area, which is cut into an array of 8×4 elements (each approximately $5.6 \times 12.9 \, \text{mm}^2$,

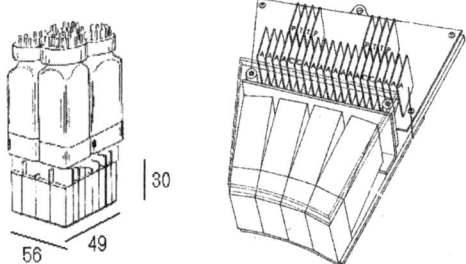

Figure 17 Components of the ECAT931 scanner: (left) detector block (dimensions in mm) and (right) 4 blocks mounted on a bucket (Parker et al., 2008a).

separated by slots 0.6 mm wide). By comparing the light intensities measured in the four photomultipliers, a γ-ray interaction can be unambiguously assigned to a particular element out of 32.

The blocks are grouped in sets of four into "buckets" (Figure 17) on which are mounted the appropriate electronics (preamplifiers and discriminators under microprocessor control). The buckets can thus be considered as detector modules (each with an active area of $200 \times 56 \, mm^2$), and the scanner consists of 32 of these buckets. As originally configured they were mounted in two adjacent rings (16 buckets in each ring), so that in terms of the individual detection elements this corresponded to 8 rings each containing 512 elements.

The scanner is designed to recognise coincidences where events occur in two opposing buckets within a resolving time of 12 ns. In normal operation, only coincidences between two elements in the same ring or adjacent rings (of the 8) were accepted, but for PEPT use this restriction was removed. The data acquisition system of the scanner was also modified so that coincidence data is recorded in list mode with time stamps at 1 ms intervals.

The buckets were removed from the original gantry and reconfigured for trials as two rectangular arrays to mimic the geometry of the conventional positron camera. In considering the optimum layout of buckets, there is a trade-off between sensitivity and field of view. Because it is necessary to detect pairs of back-to-back γ-rays, tracking is only possible when the tracer lies directly between a pair of buckets. Taking into account the overlapping cones of detectable rays, the sensitivity is highest when the tracer lies on the centre line between the two buckets and drops to zero at the edge of the volume between (Figure 18). The same variation is found regardless of whether the buckets are directly opposite to each other or inclined at some angle, though the absolute sensitivity will depend on the orientation and separation. Maximum sensitivity will occur in regions which are in line between several pairs of buckets. On the other hand, to cover an extended field of view it is necessary to spread out the buckets.

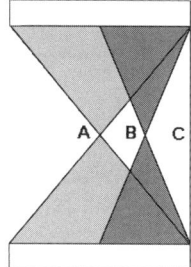

Figure 18 Schematic showing geometrical variation in sensitivity along the centre line between two detectors: at A, a wide cone of back-to-back gammas can be detected; at B, a narrower cone; and at C, the sensitivity drops to zero (Ingram et al., 2007).

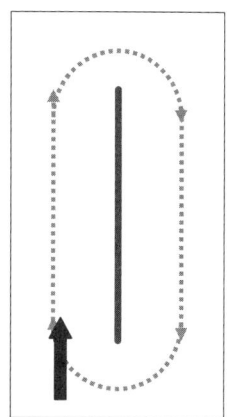

Figure 19 (Left) Detector arrangement *in situ* next to lagged bed and (right) circulation arrangement, with central baffle; arrow shows additional gas injection (Ingram et al., 2007).

The first *in situ* industrial test was carried out at BP's Hull Research and Technology Centre in spring/summer of 2006 (Ingram et al., 2007) on a 750 mm diameter pilot scale fluidised bed with a central baffle plate and asymmetric gas injection to promote particle circulation, as shown in Figure 19. The fluid bed was operated under industrially relevant conditions of elevated temperature and pressure. Owing to the thickness of the lagging, the minimum detector separation was 1150 mm.

In this application, the portable camera is capable of imaging motion over the full width of the fluidised bed but not the full height. Because the main aim of the work was to quantify circulation, it was decided to form the camera elements up into lower and upper banks, intended to follow the flows under and over the baffle, respectively.

Figure 20 shows the tracer particle coordinates over a few seconds for this bed. The y coordinate in this case indicates height and the three shaded strips indicate the upper, middle and lower parts of the bed that are in the field of view. The x coordinate indicates which semi-cylindrical half of the bed the tracer is in. In general, particles were found to move up and down on each side of the central baffle, with occasional movement between semi-cylindrical sections, either under the baffle within the dense bed or over the baffle within the freeboard space. This trial demonstrated (a) the potential for PEPT to be used *in situ* in an industrial environment and (b) the practicality of the configurable layout.

5. SUMMARY AND FUTURE PLANS

In summary, among the most significant achievements of positron emission techniques in recent years are

- provision of realistic tracer particles for a wide range of applications,
- algorithm development and experimental demonstration of multi-particle PEPT,
- repeated demonstrations of the utility of the PEPT technique in validation of computational codes (CFD, DEM, etc.),
- parallel (and recently simultaneous) use of PEPT and electrical impedance tomography,
- parallel use of PEPT & PET in a dispersion/mixing problem,
- application of PEPT in novel fields: mineral engineering, metal casting, bioseparation, etc.,
- *in situ* large scale use of mobile PEPT on an industrial site.

PEPT and PET have great strengths but also certain limitations. Other measurement techniques (x-rays, magnetic resonance imaging, electrical impedance tomography, etc.) have complementary capabilities. Undoubtedly over the next few years there will be an increasing use of multi-modal studies, combining complementary measurement techniques either successively or simultaneously. For some applications, it will become important to be able to integrate PEPT data with computational simulations and data obtained from other modalities.

As in other imaging fields, developments in positron imaging occur in a symbiotic way: new capabilities enable new applications and new demands drive enhancement in capability. In capability development, there is a continuing demand to make tracers which are

- smaller — to extend the capability to its limits at the bottom end of the size range ($<50\,\mu$m),
- more active — to improve performance generally,

Positron Emission Imaging in Chemical Engineering 175

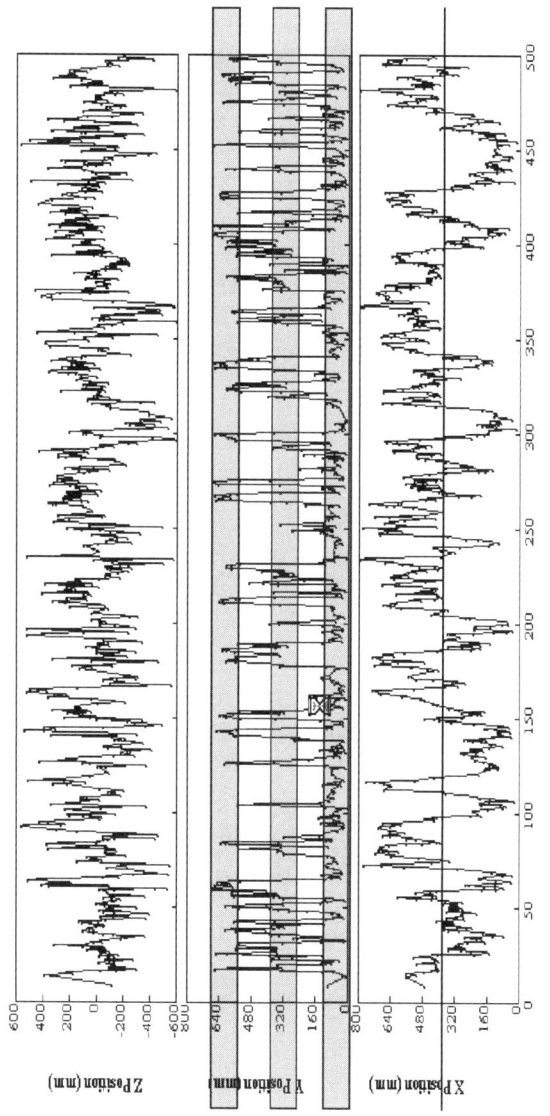

Figure 20 Tracer particle coordinates in 750 mm diameter bed. The y coordinate is vertical (Ingram et al., 2007).

- more robust — to withstand extremes of temperature, abrasion and chemical environment,
- more functional — with specific surface chemistry, so that they will follow one phase rather than another, for example.

Although PEPT is likely to remain the more appropriate choice for most industrially oriented positron-imaging work, PET can be useful in some circumstances, particularly for following slow processes such as diffusion and some classes of diffusion/reaction problems, as well as multi-phase pipeline flow.

In applications, increased capability in PEPT will enable more work with liquids and with multi-phase systems, and product-oriented studies aimed at probing the formation of microstructure. At the other extreme lies large-scale pilot- or process-level study which must be carried out *in situ*, using the new mobile PEPT camera. In large-scale applications, it will not in general be possible to provide trajectory information for the entire process volume. The challenge here is to develop "minimal" PEPT, in which location information is provided at as coarse a level as possible, consistent with the objectives of the study.

As mentioned earlier, work on validation of computational approaches to simulating multi-phase systems under realistic momentum and heat-transfer conditions is growing; PEPT represents one of very few ways of doing this convincingly.

SYMBOLS AND ABBREVIATIONS

D	axial dispersion coefficient
f	fraction of detected events actually used for PEPT location
ME	mixer effectiveness (Equation (4))
ME_∞	steady-state ME value at infinite time (Equation (5))
n	number of trajectory traces passing through a voxel
N	number of events detected during the location interval
N_p	total number of passes through all system voxels
t	time
w	intrinsic spatial resolution of the positron camera
x, y, z	location coordinates
$\bar{x}, \bar{y}, \bar{z}$	mean final location
x_d	axial displacement precision of a PEPT location (Equation (1))
λ	mixing rate constant
σ	standard deviation
PET	positron emission tomography
PEPT	positron emission particle tracking

ACKNOWLEDGEMENTS

This chapter expands on an earlier publication by Seville et al. (2005). The authors gratefully acknowledge the work of a number of colleagues and industrial collaborators, particularly Yulong Ding, Marc Hausard, Peter Knight, Hsui-Po Kuo, Tom Leadbeater, Tom Martin, David Newton, Vince Reiling, Matthias Stein and Charley Wu. Acknowledgement is also made to a succession of generous funding organisations, including the UK Engineering and Physical Sciences Research Council (EPSRC), the Biology and Biological Sciences Research Council (BBSRC), Unilever Research, BP, Innovene, GEA Pharma Systems and Yorkshire Forward.

REFERENCES

Armstrong, B., Fan, X., Ingram, A., and Parker, D. J., *in* "Particulate Systems Analysis 2005" RSC, London (2005).
Bakalis, S., Cox, P. W., Russell, A. B., Parker, D. J., and Fryer, P. J. *Chem. Eng. Sci.* **61**, 1864 (2006).
Bakalis, S., Fryer, P. J., and Parker, D. J. *AIChE J.* **50**, 1606 (2004).
Barigou, M. *Chem. Eng. Res. Des.* **82**, 1258 (2004).
Bock, H. J., 4th International Conference on Fluidization, paper 5.2.1, Kashikojima, Japan, preprints (1983).
Chaouki, J., Larachi, F., and Duduković, M. P., (Eds.), "Non-Invasive Monitoring of Multiphase Flows". Elsevier, Amsterdam (1997).
Depypere, F., Dewettinck, K., Pieters, J., Fan, X., Ingram, A., Parker, D. J., and Seville, J. P. K. (2009), unpublished work.
des Boscs, J. M., Seville, J. P. K., Parker, D. J., Ferlin, P., and Bourlier, C., Proceedings of IChemE Research Event, IChemE, Rugby, UK, p. 533 (1997).
Ding, Y. L., Forster, R. N. G., Seville, J. P. K., and Parker, D. J. *Chem. Eng. Sci.* **56**, 1769 (2001a).
Ding, Y. L., Forster, R. N. G., Seville, J. P. K., and Parker, D. J. *Powder Technol.* **124**, 18 (2002).
Ding, Y. L., Stein, M., Wong, Y. S., and Seville, J. P. K., *in* "Fluidization X" (M. Kwauk, J. Li, and W. C. Yang Eds.), p. 61. Engineering Foundation, New York (2001).
Fan, X., Parker, D. J., and Smith, M. D. *Nucl. Instrum. Methods* **A558**, 542 (2006a).
Fan, X., Parker, D. J., and Smith, M. D. *Nucl. Instrum. Methods A* **562**, 345 (2006b).
Fan, X., Parker, D. J., Smith, M. D., Ingram, A., Yang, Z. F., and Seville, J. P. K. *Nucl. Med. Biol.* **33**, 939 (2006c).
Fishwick, R., Winterbottom, J. M., Parker, D. J., Fan, X., and Stitt, H. *Can. J. Chem. Eng.* **83**, 97 (2005a).
Fishwick, R., Winterbottom, J. M., Parker, D. J., Fan, X., and Stitt, H. *Ind. Eng. Chem. Res.* **44**, 6371 (2005b).
Goldschmidt, M. J. V, Kuipers, J. A. M., and Swaaij, W. P. M. *Chem. Eng. Sci.* **56**, 571 (2001).
Hoomans, B. P. B., Kuipers, J. A. M., Mohd Salleh, M. A., Stein, M., and Seville, J. P. K. *Powder Technol.* **116**, 166 (2001).
Ingram, A., Hausard, M., Fan, X., Parker, D. J., Seville, J. P. K., Finn, N., Kilvington, R., and Evans, M, *in* "Fluidization XII" (F. Berruti, X. Bi, and T. Pugsley Eds.), p. 497. Engineering Conferences International, New York (2007).
Ingram, A., Seville, J. P. K., Parker, D. J., Fan, X., and Forster, R. N. G. *Powder Technol.* **158**, 76 (2005).
Kuo, H.-P., Burbidge, A. S., Knight, P. C., Parker, D. J., Tsuji, Y., and Seville, J. P. K. *Chem. Eng. Sci.* **57**, 3621 (2002).
Kuo, H. P., Knight, P. C., Parker, D. J., and Seville, J. P. K. *Powder Technol.* **152**, 133 (2005).

Leadbeater, T. W., and Parker, D. J., Nuclear Instruments and Methods in Physics Research Section A, in press (2009).
Leaper, M. C., Seville, J. P. K., Hilal, N., Kingman, S. W., and Burbidge, A. S. *Chem. Eng. Proc.* **43**(2), 187 (2004).
Link, J. M., Deen, N. G., Kuipers, J. A. M., Fan, X., Ingram, A., Parker, D. J., Wood, J., and Seville, J. P. K. *AIChE J.* **54**, 1189 (2008).
Martin, T. W., Seville, J. P. K., and Parker, D. J. *Chem. Eng. Sci.* **62**, 3419 (2007).
Mickley, H. S., and Fairbanks, D. F. *AIChE J.* **1**, 374 (1955).
Palmer, S. P., PhD Dissertation, University of Birmingham (2008).
Palmer, S. P., Ingram, A., Fan, X., Fitzpatrick, S., and Seville, J. P. K., *in* "Fluidization XII" (F. Berruti, X. Bi, and T. Pugsley Eds.), p. 433. Engineering Conferences International, New York (2007).
Parker, D. J., Dijkstra, A. E., Martin, T. W., and Seville, J. P. K. *Chem. Eng. Sci.* **52**, 2011 (1997).
Parker, D. J., Fan, X., Forster, R. N. G., Fowles, P., Ding, Y. L., and Seville, J. P. K. *Can. J. Chem. Eng.* **83**, 83 (2005).
Parker, D. J., Forster, R. N, Fowles, P., and Takhar, P. S. *Nucl. Instrum. Methods* **A477**, 540 (2002).
Parker, D. J., Leadbeater, T. W., Fan, X., Hausard, M. N., Ingram, A., and Yang, Z. *Meas. Sci. Technol.* **19**, 094004 (2008).
Parker, D. J., Leadbeater, T. W., Fan, X., Hausard, M. N., Ingram, A., Yang, Z., Nuclear Instruments and Methods in Physics Research Section A, in press (2009).
Seiler, C., Fryer, P. J., and Seville, J. P. K. *Can. J. Chem. Eng.* **86**, 571 (2008).
Seville, J. P. K., Ingram, A., and Parker, D. J. *Trans. Inst. Chem. Eng.* **83**, 788 (2005).
Seville, J. P. K, Salleh, A. M., Ingram, A., McCormack, A., Greenwood, R.W., and Reiling, V., *in* "Fluidization XI" (Arena, U., Chirone, R., Miccio, M., and Salatino, P. Eds.), p. 211. Engineering Conferences International, New York (2004).
Seville, J. P. K., Silomon-Pflug, H., and Knight, P. C. *Powder Technol.* **97**, 160 (1998).
Seville, J. P. K., Willett, C. D., and Knight, P. C. *Powder Technol.* **113**, 261 (2000).
Stein, M., Ding, Y. L, and Seville, J. P. K. *Chem. Eng. Sci.* **57**, 3649 (2002).
Stein, M., Ding, Y. L., Seville, J. P. K, and Parker, D. J. *Chem. Eng. Sci.* **55**, 5291 (2000).
Van de Velden, M., Baeyens, J., Seville, J. P. K, and Fan, X. *Powder Technol.* **183**, 290 (2008).
Wildman, R. D., Martin, T. W., Krouskop, P. E., Talbot, J., Huntley, J. M., and Parker, D. J. *Phys. Rev. E* **71**, 1539 (2005).
Wong Y. S., PhD Dissertation, University of Birmingham (2003).
Wong, Y. S., and Seville, J. P. K. *AIChE J.* **52**(12), 4099 (2006).
Yang, Z., Fan, X., Fryer, P. J., Parker, D. J., and Bakalis, S. *AIChE J.* **53**, 1941 (2007a).
Yang, Z., Fryer, P. J., Bakalis, S., Fan, X., Parker, D. J., and Seville, J. P. K. *Nucl. Instrum. Methods A* **577**, 585 (2007b).
Yang, Z., Parker, D. J., Fryer, P. J., Bakalis, S., and Fan, X. *Nucl. Instrum. Methods* **A564**, 332 (2006).

CHAPTER 5

Electrical Capacitance, Electrical Resistance, and Positron Emission Tomography Techniques and Their Applications in Multi-Phase Flow Systems

Fei Wang[1], **Qussai Marashdeh**[1],
Liang-Shih Fan[1,*] and **Richard A Williams**[2,*]

Contents			
	1.	Introduction	180
	2.	Electrical Capacitance Tomography	182
		2.1 Introduction	182
		2.2 Principle of ECT	183
		2.3 ECT applications	186
	3.	Electrical Resistance Tomography	196
		3.1 Introduction	196
		3.2 Principles of ERT	198
		3.3 ERT applications	204
	4.	Positron Emission Tomography	209
		4.1 Introduction	209
		4.2 Principle of PET	209
		4.3 PET applications	211
	5.	Concluding Remarks	216
		Nomenclature	216
		References	217

1 Department of Chemical and Biomolecular Engineering, The Ohio State University, Columbus, OH 43210, USA
2 Institute of Particle Science and Engineering, School of Process, Environmental and Materials Engineering, University of Leeds LS2 9JT, UK

*Corresponding author
E-mail address: fan@chbmeng.ohio-state.edu
E-mail address: r.a.williams@leeds.ac.uk

Abstract This article describes the recent progress in research and development on electrical capacitance tomography (ECT), electrical resistance tomography (ERT), and positron emission tomography (PET). Specifically, the article highlights several aspects of the three technologies and illustrates their application and performance through selected demonstration cases studies. The principles and results from the methods provide quantitative and/or qualitative assessment of the significance of each technique. The measurement techniques lend themselves for widespread application in multi-phase flow imaging research and some for industrial-scale measurements due to their non-invasive nature.

1. INTRODUCTION

Multi-phase flows are widely used in industrial operations such as fluidized beds, bubble columns, slurry bubble columns, and solid–liquid and solid pneumatic conveying. They have been utilized extensively in the chemical, petrochemical, metallurgical, food, and pharmaceutical industries. To investigate multi-phase flow behavior, measurement techniques are vital for understanding and quantifying multi-phase system parameters. These techniques are classified generally into intrusive and non-intrusive based on the mechanism of acquiring measurement signals. In intrusive techniques, the measurement probe penetrates the wall of the process to acquire the desired signal from direct contact with the flow, which may perturb the physical flow, whereas non-intrusive techniques are based on remotely acquiring the measurement signal from sensors mounted away from the flow. In most case, these sensors are mounted at the periphery of the process walls internally or externally. In order to fully investigate different multi-phase flow behavior, a number of measurement techniques, both intrusive and non-intrusive, have been developed.

Intrusive probes that have been used to measure multi-phase flow systems include capacitance probe (e.g., Geldart and Kelsey, 1972; Gunn and Al Doori, 1985; Ho et al., 1983; Lanneau, 1960; Sharma et al., 2000; Shi et al., 1991; Werther and Molerus, 1973), optical fiber probe (e.g., Bayle et al., 2001; Cui and Chaouki, 2004; Du et al., 2003; Liu et al., 2003a, 2003b; Nakajima et al., 1991; Okhi and Shirai, 1976; Smith et al., 1999; Yasui and Johanson, 1958), endoscopic probe (e.g., Du et al., 2004a; Peters et al., 1983), and pressure transducer probe (e.g., Fan et al., 1981; Geldart and Xie, 1992; Gibilaro et al., 1988; Kang et al., 1967; Lirag and Littman, 1971; Sitnai, 1982). These techniques have been used intensively to measure gas–solid fluidized beds. However, intrusive probes pose unique challenges to their operation based on the modality of probe used.

For example, the capacitance probe is sensitive to electrostatic charges influenced by temperature and relative humidity, making it difficult to define the measuring volume. As for the optical probe, the measuring volume is not well defined and varies with the solid concentrations in the bed, introducing an error into the measured signal (Du et al., 2003). A reliable and precise calibration is essential in this case. Additionally, pressure probes can only provide local, averaged information inside a fluidized bed, making it difficult to trace the cause of pressure changes. For example, the bubble flow, bubble coalescence, and bubble breakup, bubble burst at the top surface, bubble formation at the distributor can all generate the pressure variations in the same bed. Moreover, it is evident that the pressure probe measurement interferes with the bubble flow behavior in fluidized beds. In this case, bubbles are vulnerable to break, accelerate, or elongate by the immersed probe (Rowe and Masson, 1981). Small probes are often used to minimize the interference with the flow. However, such probes can be easily damaged. Comprehensive reviews of principles and applications of intrusive techniques in gas–solid fluidized beds are provided by Yates and Simons (1994), Werther (1999), and Du et al. (2003).

Non-intrusive techniques, on the other hand, have the advantage of avoiding interference with the internal flow of a multi-phase flow system. The information provided by non-intrusive techniques is usually in the form of two-dimensional (2D) cross-sectional concentration profiles or 3D volume images. Non-intrusive techniques used to measure multi-phase flow systems include X-ray (e.g., Gilbertson et al., 1998; Hulme and Kantzas, 2004; Hubers et al., 2005; Kai et al., 2005; Kantzas and Kalogerakis, 1996; Rowe and Matsuno, 1971; Rowe and Partridge, 1965; Rowe and Yacono, 1976; Yates et al., 1994), γ-ray tomography (GRT) (e.g., Baumgarten and Pigford, 1960; Boyer and Fanget, 2002; Clough and Weimer, 1985; Orcutt and Carpenter, 1971; Patel et al., 2008; Schubert et al., 2008; Wang et al., 2001; Weimer et al., 1985; Wu et al., 2007), positron emission tomography (PET) (e.g., Dechsiri et al., 2005a, b; Hoffmann et al., 2005), radioactive particle tracking (RPT) (e.g., Larachi et al., 1997; Larachi and Chaouki, 2000), magnetic resonance imaging (MRI) (e.g., Fennell et al., 2005; Holland et al., 2009; Savelsberg et al., 2002), electrical resistance tomography (ERT) (e.g., Bennett and Williams, 2004; Dickin and Wang, 1996; Vilar et al., 2008; Williams et al., 1993), electrical capacitance tomography (ECT) (e.g., Du et al., 2003, 2005; Halow and Nicoletti, 1992; Ormiston et al., 1965; Wang et al., 1995) and electrical magnetic tomography (EMT) (e.g., Binns et al., 2001; Williams and Beck, 1995). Although such methods (ECT, ERT, EMT) tend to non-intrusive due consideration may need to be given to the influence secondary effects, such as any influence on the effect of electromagnetic fields on the materials and their response within the process. Other sensing principle might also include dielectric spectroscopy, microwave, ultrasound, and various conventional and diffuse optical methods (Cullivan and Williams, 2005).

Although non-intrusive imaging techniques span a wide range of sensing modalities, this chapter will focus on ECT, ERT, and PET techniques in multi-phase flow measurements. The reconstruction techniques applied to ECT, ERT, and PET are similar to algorithms developed for computed tomography (CT) in the medical field. However, in processes applications, these techniques can be studied after being categorized as algebraic reconstruction techniques (ART) or optimization reconstruction techniques (ORT). In the first category, a set of algebraic equations are formed to model the response of the system to density variations in the imaging domain. The measured signal is then used to obtain a density map based on matrix manipulation to solve the set of independent equations. Whereas in the optimization techniques, a set of objective functions are optimized together to obtain the most likely image associated with the measured signal. The performance of each reconstruction approach is different when applied to different tomography modalities and is usually related to the level of non-linearity between the signal detected and the density distribution. For example, ECT and ERT are two electric-based imaging modalities where the density distribution is non-linearly related to the electric field distribution inside the imaging domain. Optimization techniques in these two examples provide better imaging results when compared to algebraic techniques. Further details on reconstruction techniques are provided in (Benton and Parker, 1997; Warsito and Fan, 2001a, b; Williams and Beck, 1995; Yang and Peng, 2003). Investigating different reconstruction algorithms is beyond the scope of this chapter, we focus here on the applications of ECT, ERT, and PET techniques in fluidized beds, bubble columns, slurry bubble columns, and gas–solid pneumatic conveying.

2. ELECTRICAL CAPACITANCE TOMOGRAPHY

2.1 Introduction

Imaging technologies provide an invaluable insight into the dynamics of a multi-phase flow system operation. Such systems include fluidized beds, bubble columns, slurry bubble columns, and gas–solid pneumatic conveyers. The real-time rendering potential of electrical-based imaging technologies, along with low construction cost, high safety, and suitability with different vessel sizes, led to an increased interest in capacitance-based imaging systems. These systems based on electrical capacitance sensors are considered prominent among other imaging modalities and have been widely used to study the hydrodynamics of multi-phase flow systems (Warsito and Fan, 2001a, b, 2003).

2.2 Principle of ECT

An ECT system is composed of three basic components: (1) a capacitance sensor, (2) a data acquisition system, and (3) a computer system for reconstruction and viewing. Figure 1 is a sketch of the ECT system with all three components (Warsito and Fan, 2003). The capacitance sensor is made of n_e capacitance electrodes distributed around the wall of the process vessel. The n_e capacitance electrodes provide up to $n_e(n_e-1)/2$ combinations of independent capacitance measurements between the electrode pairs. The capacitance measurements are related to the local dielectric constant (permittivity) filling the process vessel between electrode pairs (Figure 2) (Warsito and Fan, 2001b). The relation between the electric potential and the permittivity distributions follows Poisson equation shown in Equation (1).

$$\varepsilon(x,y)\nabla^2\phi(x,y) + \nabla\varepsilon(x,y)\nabla\phi(x,y) = 0 \qquad (1)$$

where $\varepsilon(x,y)$ is the dielectric constant (permittivity) distribution and $\phi(x,y)$ is the potential distribution for an electrode pair of the $n_e(n_e-1)/2$

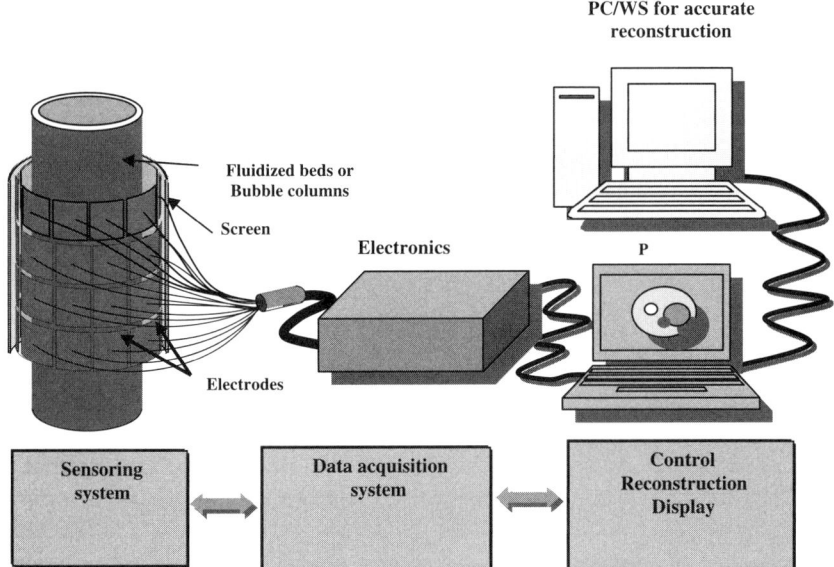

Figure 1 Sketch of the ECT system including sensor, data acquisition system, and computer for reconstruction (Warsito and Fan, 2003).

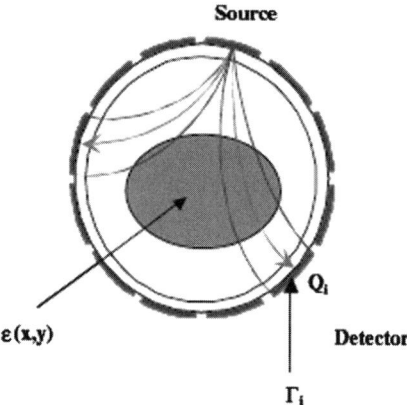

Figure 2 Principle of ECT capacitance measuring system (Warsito and Fan, 2001b).

combinations. The accumulated charge on the detector electrode (Q_i) (Figure 2) is expressed in Equation (2)

$$Q_i = \oint_{\Gamma_i} \varepsilon(x,y) \nabla \phi \cdot \hat{n} dl \qquad (2)$$

where Γ_i is a surface enclosing the detector electrode and $\hat{n}$ is the normal vector to Γ_i. The capacitance between the source and the detector electrodes for the ith electrode pair, C_i, is expressed in Equation (3).

$$C_i = \frac{Q_i}{\Delta V_i} = \frac{1}{\Delta V_i} \oint_{\Gamma_i} \varepsilon(x,y) \nabla \phi \cdot \hat{n} dl \qquad (3)$$

From the above equation, the measured capacitance between the source and the detector electrodes of the electrode pair i is determined from the given dielectric constant (permittivity) distribution of the medium under investigation. The processes of finding the capacitance for a given permittivity distribution is referred to as the *forward problem*. On the other hand, the process of finding the permittivity distribution from a set of capacitance measurements is referred to as the *inverse problem*.

Various schemes have been applied to solve the forward problem of ECT for the visualization of multi-phase flow components. Forward solutions based on the sensitivity model (Huang et al., 1989; Xie et al., 1992) are the most widely used due to the simplicity and speed of applying this model. The simplest form of the sensitivity model is the single iteration *linear forward projection* (LFP) which is explained in Equation (4).

$$C = SG \qquad (4)$$

where G represents the $M \times 1$ image vector (permittivity distribution in the imaging domain), C is the $N \times 1$ measured vector of capacitances between the n_e electrodes pairs, and S is an $N \times M$ so-called sensitivity matrix. This matrix is constructed by solving Equation (3) usually by using computational methods or by experimental measurements. The basic idea of the sensitivity model is to assume that each permittivity pixel contributes linearly to the overall measured capacitance, and the forward problem can be reduced to a simple matrix multiplication.

Solving the inverse problem, on the other hand, is a higher level of complexity as the problem is often ill-posed and ill-conditioned. Various reconstruction techniques have been developed to solve the problem, and the most widely used is based on a single iteration *linear back projection* (LBP). Despite the simplicity of the LBP technique and its capability of providing real-time imaging, it suffers from severely blurred imaging, which led researchers to develop an iterative scheme of reconstructive algorithms (Yang and Peng, 2003). On the basis of iterative reconstruction, the error between the forward solution and the measured capacitance is reduced iteratively until reaching a preset error level. Iterative image reconstruction techniques (Yang and Peng, 2003) are mainly classified into ART and optimization techniques. Iterative linear back projection (ILBP) (Yang et al., 1999) is one of the most used ART. Iterative reconstruction techniques are generally described by Equation (5).

$$G^{k+1} = G^k + \alpha^k S^T \left(C - S G^k \right) \qquad (5)$$

where G^k is the estimated permittivity vector in the kth iteration, α is a relaxation factor. The structures of α and the procedure to correct the estimated image vector (permittivity distribution) varies based on different ART.

However, due to the severity of the ill-posedness of the ECT inverse problem, iterative schemes often diverge if a limit on the number of iterations is not preset. In an attempt to provide better reconstruction results for quantitative imaging, researchers developed multi-objective optimization schemes for electrical tomography image reconstruction. One of the most prominent reconstruction techniques in this regard is the *neural network multi-criterion image reconstruction technique* (NN-MOIRT). The NN-MOIRT has been shown to provide better reconstructed image resolution, guaranteed convergence to a local minimum in the optimization process, and immunity to measurement noise (Warsito and Fan, 2001b). This reconstruction technique is even more valued when considering 3D volume reconstruction, as the inverse problem is severely ill-posed (Marashdeh et al., 2008; Warsito et al., 2007). NN-MOIRT is designed to optimize a set of objective functions that include the mean square error (MSE), entropy objective, and the smoothness functions as depicted in Equations (6–8), respectively. The entropy objective function is a measure of the information coded in the image. The smoothness function is equivalent to a filter for eliminating measurement noise.

$$f_{mse}(G) = \frac{1}{2}\gamma_1 \|SG - C\|^2 \tag{6}$$

$$f_i(G) = \gamma_2 \sum_{j=1}^{N} G_j \ln(G_j) \tag{7}$$

$$f_s(G) = \frac{1}{2}\gamma_3 (G^T X G + G^T G) \tag{8}$$

In the above equations X is an $M \times M$ non-uniformity matrix and γ_1, γ_2, γ_3 are normalized constants between 0 and 1. Figure 3 shows the comparison of reconstruction results for a two-phase flow system from (a) noise-free data and (b) noise data using LBP, ILBP, SIRT, and NN-MOIRT techniques (Warsito and Fan 2001b). When there is no noise added, the ILBP, SIRT, and NN-MOIRT techniques provide relatively accurate reconstructed images, and the images consist the original permittivity distribution. LBP only provides a rough estimation of the original image. When noise is added, reconstructed images from LBP are worse, and images from ILBP and SIRT are not accurate enough compared to the original image. It is clear from this study that the NN-MOIRT technique is more effective in reconstructing an image and eliminating noise when compared to other widely used reconstruction techniques.

2.3 ECT applications

2.3.1 Circulating fluidized beds

ECT is used to image flow behaviors in circulating fluidized beds (CFB) models. The ECT imaging of choking in CFB is presented here. Choking is a complex phenomenon in gas–solid fluidization, where a sudden small change in gas or solid flow rate introduces a large change in the hydrodynamic behavior such as the pressure drop or solids holdup in the gas–solid flow (Du et al., 2004a, b). To fundamentally understand choking, it is important to observe the CFB operating variables during the choking transition from dilute to dense fluidization and the underlying mechanism. The initiation of choking, or the collapse of dilute suspension, introduces a sharp increase in the pressure drop in the vertical CFB riser. It has been deduced that only when the gas flow rate is lower than the transport velocity, U_{tr}, or the solids circulation rate is smaller than the transport solid circulation rate, $G_{s,tr}$, that the initiation of choking occurs with increasing solids circulation rate at a given gas velocity or with decreasing gas velocity at a given solids circulation rate in a vertical CFB riser. The qualitative and quantitative information

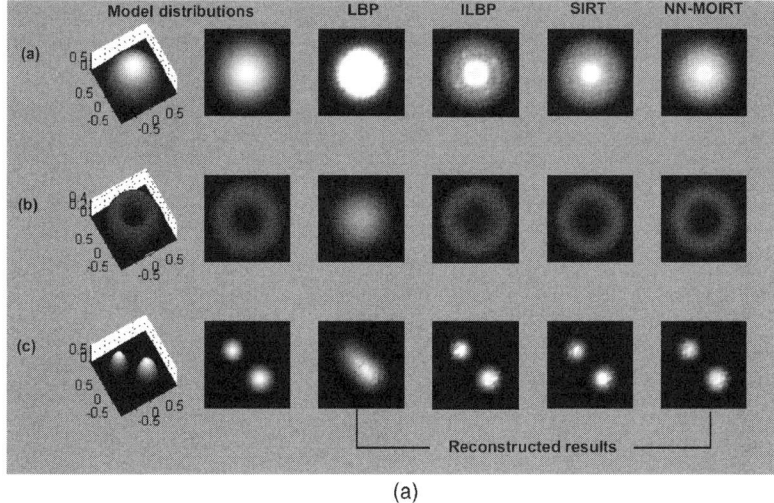

(a)

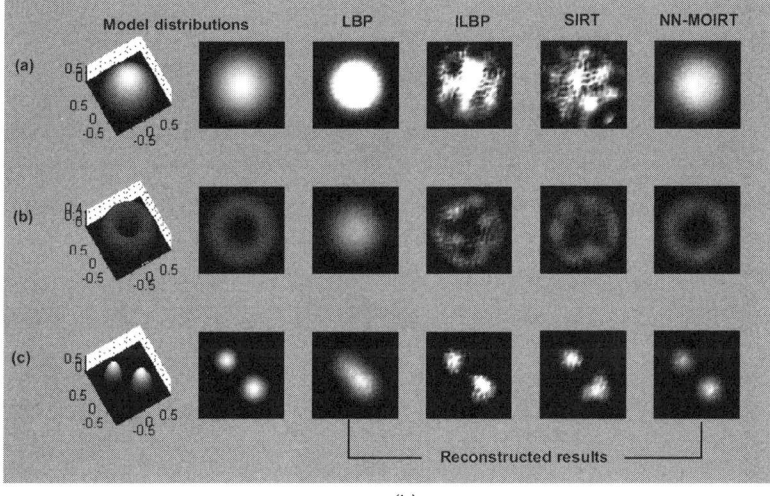

(b)

Figure 3 Comparison of reconstruction results for a two-phase flow system from (a) noise-free data and (b) noise data using LBP, ILBP, SIRT, and NN-MOIST techniques (Warsito and Fan, 2001b).

regarding microscopic and macroscopic flow structure of the gas–solid fluidization is needed to fully understand the choking mechanism. Pressure drop sensors and intrusive probes have been used to provide general information of the flow behavior. However, a measurement system capable of providing detailed information of the flow behavior during choking is required to fully understand the phenomena. In this

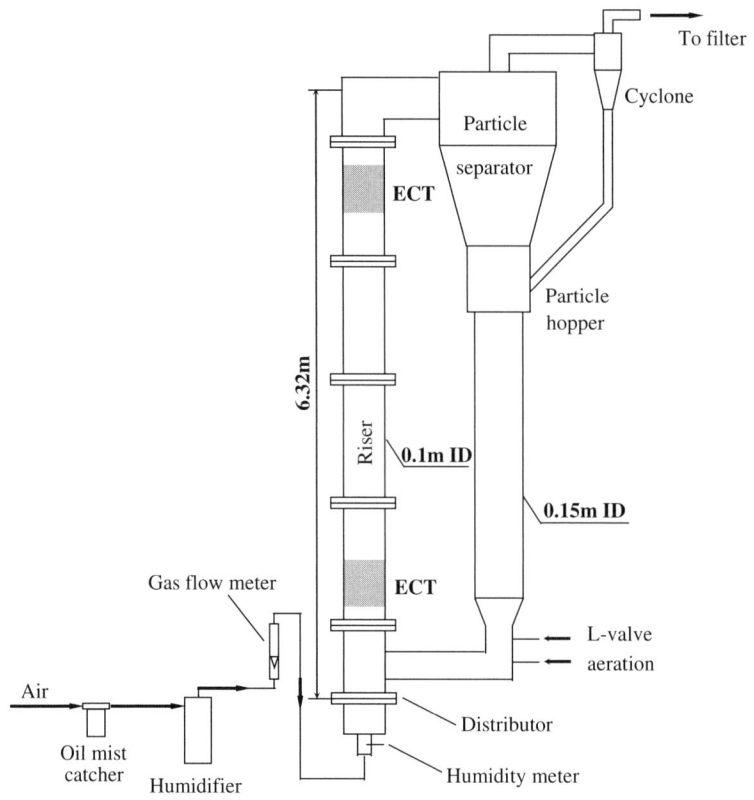

Figure 4 Sketch of a circulating fluidized bed (Du et al., 2004a).

regard, ECT is a reliable instrument for accurate, real-time imaging of multi-phase flow and can be applied for acquiring in-depth information of the flow during choking in CFB risers.

Figure 4 is the sketch of a circulating test fluidized bed (Du et al., 2004a). The CFB riser includes a 0.1-m ID riser with a height of 6.32 m, a separator and a secondary cyclone system, a large volume solids storage hopper, and an L-valve. FCC catalysts with a mean diameter of 60 μm and particle density of 1400 kg/m^3 are used as the fluidized particles in the CFB riser. ECT sensors are installed at both the lower and the upper position of the CFB riser. Quasi-3D flow structure of the CFB is provided after reconstruction of the capacitance signal from the ECT sensors using NN-MOIRT reconstruction algorithm developed by Warsito and Fan (2001b). Figure 5 is the Quasi-3D flow structures at lower part of a CFB riser (Du et al., 2004a). At that lower part of the CFB riser, a persistent dilute gas ring is clearly observed between the solid core and the solid ring near the wall with a low solid circulation rate $G_{s,tr}$ of 2.3 kg/m^{2}s and

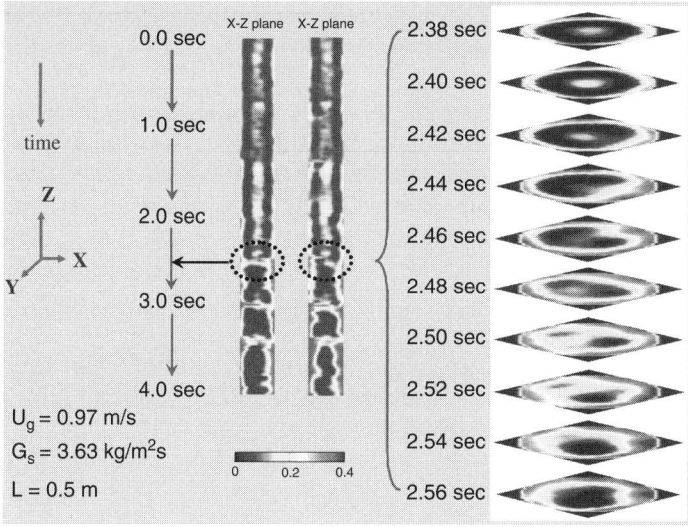

Figure 5 Quasi-3D flow structures for the choking transition at lower part of a CFB riser (Du et al., 2004a) (see Plate 9 in Color Plate Section at the end of this book).

a gas velocity U_g ($U_g < U_{tr}$) of 0.97 m/s, which is shown in the X-Z and Y-Z plan images between 0.0 sec and 2.3 sec in the left side of Figure 5. A large dilute gas core occurs in the central region and the size of the solid ring near the wall increases solid circulation rate $G_{s,tr}$ of 3.63 kg/m²s at same gas velocity U_g of 0.97 m/s. This occurrence is shown in the left side of Figure 5 in the X-Z and Y-Z plane images between 2.6 sec and 4.0 sec. An abrupt flow structure change, or choking transition, to dense-phase fluidization occurs when the solid circulation rate undergoes a step increase from 2.3 to 3.63 kg/m²s, which is also shown in the X-Z and Y-Z plane images around time 2.4 sec in Figure 5. The detailed flow structure variation and choking transition in the CFB riser is provided in the X-Y plane images in the right side of Figure 5. Before the choking transition, the solid concentration in the core region and the solid ring near the wall measured by ECT is about 0.15 and 0.25, respectively. During the choking transition, solids in the core region move toward the wall. Subsequently, solids keep moving between the core region and the wall region in the riser, which is described as a dynamic process, or choking transition, until the new structure forms. After the choking transition, a dilute gas core region in the center of the riser and a large dense solids ring near the wall of the riser are formed. The concentration in the dense solid ring near the wall measured by the ECT is about 0.4. The ECT provided a real-time quasi-3D cross-sectional flow structure of the circulating fluidized bed. It also helped reveal flow conditions before and after the choking as well as the underlying mechanism of the choking formation.

2.3.2 Pneumatic solid conveying

Transportation of solids is commonly required in many solid processing systems. Among numerous solid transportation techniques, pneumatic conveying is widely used because of its advantages in requiring low routine operation, maintenance, and low labor costs. Additionally, this technique is appealing because of its characteristics of being dust free and flexible in routing and spacing (Fan and Zhu, 1998; Ostrowski et al., 1999). However, the pneumatic solid conveying system consumes more power than other bulk solid transportation systems. It also introduces high wearing and abrasive effects on the solids due to their high collision velocity. The pneumatic solid conveying system is classified based on the following criteria: (1) the angle of the inclination of the pipelines; (2) operational conditions; (3) flow characteristics. A fundamental understanding of the transport phenomena associated with the pneumatic solid conveying system is of great importance to design and optimize the transportation systems. The ECT is considered here a robust, reliable, and non-intrusive technique for accurate, real-time imaging of gas–solid flows in the pneumatic solid conveying system. The qualitative and quantitative information of the flow structures in this system are obtained by using the ECT. For example, ECT was used to study the horizontal pneumatic solids conveying test plant behaviors for the industrial-scale pneumatic conveyors (McKee et al., 1995a). Eight electrodes in this case were used to obtain a total of 28 independent capacitances for each measurement. The 2D reconstruction images calculated from the dielectric constant (permittivity) distribution show the solid distribution of the test cross-sections of the pipe. Figure 6 depicts the solid distribution of the cross-sections at upstream (0.88 m away from the solids feeder) and downstream (4.08m away from the feeder) of a horizontal pipe with different solids loading in a pneumatic solids conveying system. Acetal resin pellets with a diameter of 2.85mm, a bulk density of $879\,kg/m^3$, a particle density of $1350\,kg/m^3$, and different solids loading factors (10:1; 7:1; 4:1; 1:1) are used for the test. At high solid loading rate (10:1, 7:1) and an air conveying velocity of 28 cm/s, solid deposition is observed from ECT results both at the upstream and downstream of the horizontal pipe. At low solid loading rate (4:1, 1:1) with the same air conveying velocity, the ECT at the downstream of the horizontal pipe shows the particle conveying flow fully suspended with no solid deposition observed (McKee et al., 1995a). These results from a solid conveying system show that the ECT is capable of providing online cross-sections of solid concentration in pneumatic solid conveying systems. The methods were further refined to

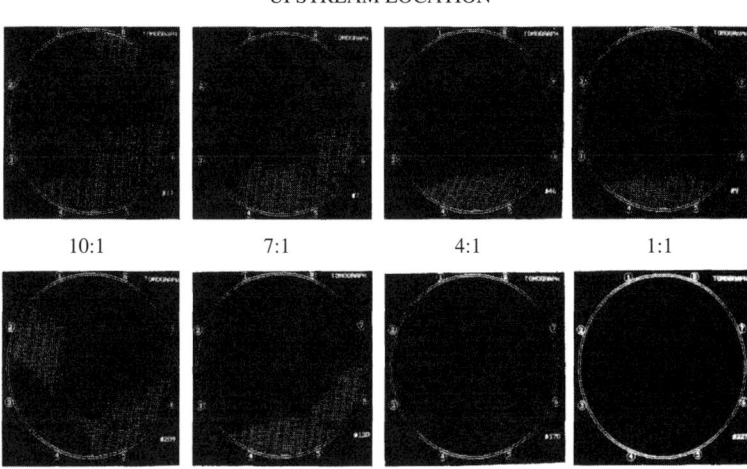

UPSTREAM LOCATION

10:1   7:1   4:1   1:1

DOWNSTREAM LOCATION

Figure 6 Solid distributions at upstream and downstream of a horizontal pipe with different solids loading (McKee et al., 1995a).

demonstrate applications for real-time control of pneumatic conveying systems based on flow regime characterization and recognition (Ostrowski et al., 2000). This also provided significant detail that could be utilized for direct visualization of the shape of powder slugs and the estimation of velocity profiles within the conveyor (Williams, 2005). Some of the data handling and modeling methods derived are also applicable to other electrical tomographic methods.

2.3.3 Hydrodynamic behaviors in bubble and slurry bubble columns

Gas–liquid bubble columns and gas–liquid–solid slurry bubble columns are widely used in the chemical and petrochemical industries for processes such as methanol synthesis, coal liquefaction, Fischer–Tropsch synthesis and separation methods such as solvent extraction and particle/gas flotation. The hydrodynamic behavior of gas–liquid bubble columns and gas–liquid–solid slurry bubble columns are of great importance for the design and scale-up of reactors. Although the hydrodynamics of the bubble and slurry bubble columns has been a subject of intensive research through experiments and computations, the flow structure quantification of complex multi-phase flows are still not well understood, especially in the three-dimensional region. In bubble and slurry bubble columns, the presence of gas bubbles plays an important role to induce appreciable liquid/solids mixing as well as mass transfer. The flows within these systems are divided into two regimes, homogeneous and heterogeneous. In the homogenous regime, the variation of the bubble size is small and no coalescence of bubbles occurs. In the heterogeneous regime, substantial coalescence and breakup of bubbles make the flow very complex and the variation of the bubble size extremely large. The qualitative and quantitative information regarding flow structures of the bubble and slurry bubble columns is required for understanding the heat and mass transfer behaviors within these systems. Among available tomography technologies, ECT is one of the most promising techniques for dynamic flow imaging measurements. ECT provides real-time qualitative and quantitative cross-sectional imaging data of the multi-phase flow behaviors in bubble and slurry bubble columns. Some examples of measurements mapping flow regimes and detailed microstructure development were achieved using ECT (Bennett et al., 1999). For aqueous-continuous fluidized systems, it may be possible to perform measurements that can be probed using ERT (e.g., Jin et al., 2006). Dielectric methods based on ECT are normally better for systems at high gas voidage. This is because the presence of gas bubbles (infinite resistance) can cause spurious outcomes using current injection methods, although some remedies have been proposed which will be discussed later.

An improvement to the ECT system utilizes capacitance sensors to provide direct 3D imaging in what is known as electrical capacitance volume tomography (ECVT). This innovative approach was used to investigate the hydrodynamic behaviors in gas–liquid bubble columns and gas–liquid–solid slurry bubble columns (Warsito and Fan, 2005). The 3D real-time gas holdup distribution, the transient phenomena in the entrance region of gas–liquid bubble columns and gas–liquid–solid slurry bubble columns, and the 3D bubble plume spiral motion and liquid vortex dynamics were investigated using the 3D ECVT sensor and the 3D-NN-MOIRT reconstruction technique. Figure 7 depicts a design of an ECVT sensor that reveals its 3D features (Warsito and Fan, 2005). The electrical potential distribution between the electrode pairs is also depicted in the figure. Twelve capacitance electrodes were arranged in two planes. The capacitances between the electrode pairs of different layers (top-right) provided the feasibility for measuring the capacitance in the 3D volume of the test region. A related 3D sensitivity model was established to solve the volume image reconstruction problem which provided a $20 \times 20 \times 20$ voxels image (bottom-left). Figure 8 is a 3D real-time volume tomography image of gas–liquid flow in a bubble column (Warsito and Fan, 2005). The spatial and time resolution is $5 \times 5 \times 8\,\mathrm{mm}^3$ and 12.5 ms, respectively. The X-Z, Y-Z, and X-Y planes to render the images are defined by the coordinate system in the bottom-right of the figure. The phase concentration (holdup) of the multi-phase system is calculated from the permittivity distribution of the flow volume

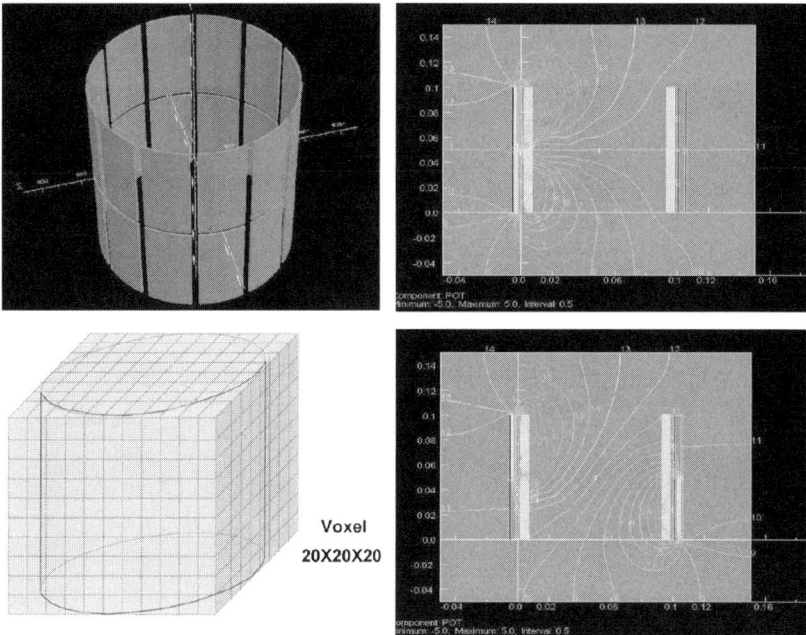

Figure 7 Three-dimensional ECVT sensor configuration and electrical potential distribution (Warsito and Fan, 2005) (see Plate 10 in Color Plate Section at the end of this book).

reconstructed by ECVT. The blue and red colors represent low and high gas holdup, respectively. The top-left two pictures of Figure 8 show a snapshot of the bubble gas holdup distributions in the X-Z and Y-Z planes, respectively. The bottom-left two pictures of the figure show the 3D perspective volume image and 3D isosurface image of the bubble swarm within the bubble column. The 3D isosurface image displays the bubble swarm boundary with a cut-off boundary value of 10%. The cut-off boundary has no strict criterion. It is arbitrarily used to distinguish boundary between the fast moving bubble swarm and the surrounding liquid flow as well as to show the motion of the bubble swarm clearly. A photograph of the gas–liquid flow in the bubble column with the same condition taken by a high-speed digital video camera system is displayed on the top-right image of the figure. It is compared to the X-Z plane, Y-Z plane, 3D volume, and 3D isosurface images. Two main flow regimes, a fast bubble flow regime with a fast motion and high gas hold up, and a vertical-spiral liquid flow regime with a slow motion and low gas holdup, are observed in the gas–liquid bubble columns. The ECVT images clearly show that the fast bubble flow regime, or the spiral bubble plume flow regime, has circular motion when it is rising. This observation is consistent with those observed by Chen et al. (1994).

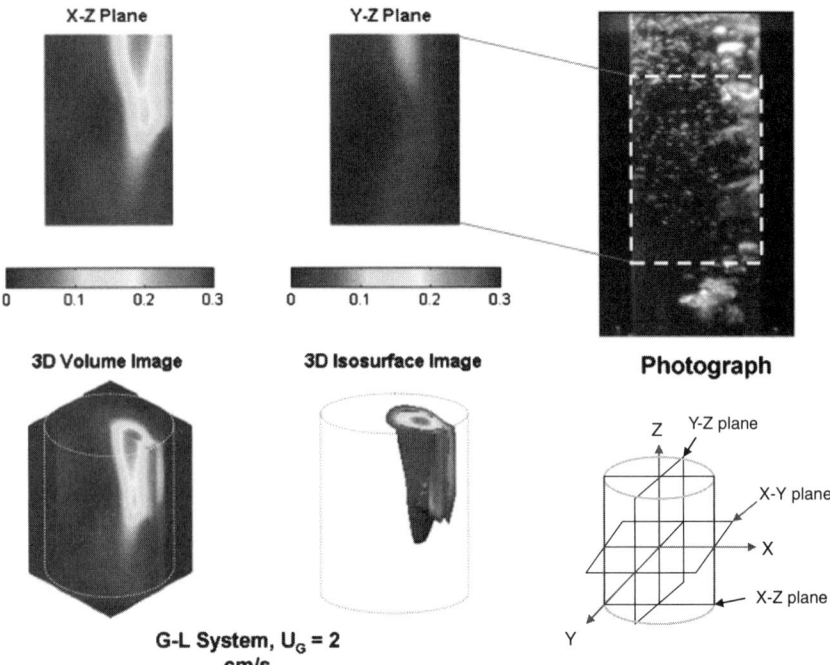

Figure 8 Three-dimensional ECVT images of gas–liquid flow in a bubble column (Warsito and Fan, 2005) (see Plate 11 in Color Plate Section at the end of this book).

Although ECVT is a promising non-intrusive technique, this system is limited to measuring bubbles greater than the voxel resolution. A bubble smaller than the voxel resolution cannot be measured by the ECVT. A potential solution to this issue is to increase the number of sensor electrodes and modify the related sensitivity which could increase the resolution to track the smaller, single bubble.

The unique capability of ECVT in viewing 3D real-time images of multi-phase flow systems is best demonstrated in the example of horizontal gas penetration in a gas–solid fluidized bed reported in Wang et al. (2008). A gas–solid fluidized bed with a 0.3 m ID and a total bed height of 2.4 m and FCC particles with a mean diameter of 60 μm and a particle density of $1400\,kg/m^3$ are used in this example. A horizontal tube is mounted on the wall of the testing section of the fluidized bed, 0.3 m above the distributor to provide horizontal jets in the fluidized bed. The testing region of the fluidized bed with the horizontal side injection is covered with the ECVT sensor. The sensor consists of three layers with four electrodes in each layer. Each electrode has a rectangular plate shape. The edges of the two side-by-side electrodes in the middle layer are modified to accommodate the penetration tube. Figure 9a shows a 3D

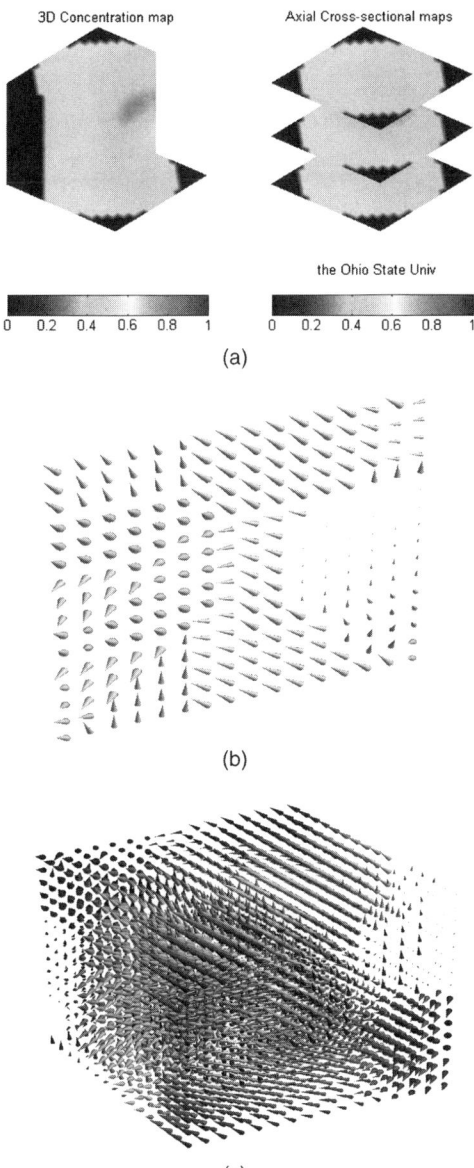

Figure 9 (a) Three-dimensional view of 3D solid concentration with a horizontal gas jet in the fluidized bed; (b) 3D voxel-volume-averaged solid phase velocity vector map in the Y-Z plane of the fluidized bed; (c) 3D voxel-volume-averaged solid phase vector map (Wang et al., 2008) (see Plate 12 in Color Plate Section at the end of this book).

view of 3D solid concentration with a horizontal gas jet in the fluidized bed. Blue and red colors represent low and high solid concentrations in the fluidized bed, respectively. The 3D shape of the horizontal jet and its development in the gas–solid fluidized bed are viewed by ECVT. The voxel-volume-averaged phase velocity vector field of the gas–solid fluidized bed with the horizontal jet is obtained from quantitative ECVT images. Figure 9(b) shows the 3D voxel-volume-averaged solid phase velocity vector in the Y-Z plane (Y: the direction of the horizontal jet; Z: vertical direction relative to the jet) of the fluidized bed. Figure 9(c) shows the 3D voxel-volume-averaged solid phase velocity vector in the whole ECVT test region of the gas–solid fluidized bed. In this example, quantitative ECVT provides the maximum horizontal jet penetration length, maximum horizontal jet width, and the voxel–volume-averaged phase velocity vector. It is generally perceived that increasing the electrode sensor numbers and acquisition channel numbers for the ECVT system would yield images of higher resolution. A 32-channel ECVT system providing real-time 3D images has been demonstrated recently for such multi-phase flow systems for a high resolution effect (Wang et al., 2008). Figure 10 depicts such 32-channel system for this demonstration. It is seen that the sensor used in this figure is of four layers of shifted planes, with eight electrodes in each layer. All the 32 channels in all 4 planes are used to obtain each image frame. Figure 10b shows a multi-phase flow column with 32-channel ECVT sensors mounted on the wall along with acquisition hardware. The column is of a 0.3 m ID and a total height of 0.6 m. The ECVT sensor is connected to a 32-channel acquisition box. A laptop is used to control the acquisition hardware and perform image reconstruction. In the images shown in Figure 10c, imaging is conducted with liquid used in the column.

3. ELECTRICAL RESISTANCE TOMOGRAPHY

3.1 Introduction

Electrical resistance tomography is a technique used for medical and industrial applications. It is based on injecting current into the reactor domain of interest and measuring conductivity changes that can be used to reconstruct a conductivity map of the electrode plane region. The variation in electrical conductivity of different materials can be used to interpret the ERT images. Current injection measurements are often made at a single frequency using a sinusoidal or pulsed waveform. But multiple frequencies can also be utilized through spectroscopic methods. This enables both "real" and "imaginary" components of the electrical signals to be analyzed and in effect giving an impedance map. ERT is the

(a)

(b)

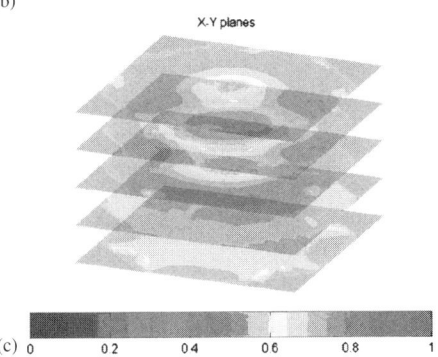

(c)

Figure 10 (a) Configuration of the 32-channel ECVT sensor; (b) fluidization system mounted with the 32-channel ECVT acquisition system; (c) liquid holdup image obtained by the 32-channel ECVT system (Wang et al., 2008) (see Plate 13 in Color Plate Section at the end of this book).

real component of this map and is most widely used. However, full electrical impedance tomography (EIT) is also possible from which real (resistivity) and imaginary (dielectric) data can be abstracted (West et al., 2002). ERT exhibits the same advantages of other electrical imaging modalities except that it generally requires an invasive insertion of probes to inject current. In industrial applications, ERT offers complementary and relatively low-cost measurements in multi-phase flow applications such as solid/liquid and miscible liquid/liquid mixing, hydrocyclone flow visualization, high-speed flow imaging in slurry conveying system, and the spatial concentration of the dispersed phases in pipelines, separators, and reactors (Dickin and Wang, 1996; Fangary et al., 1998; George et al., 2000; Lucas et al., 1999; Primrose, 2008; Vilar et al., 2008; Williams et al., 1993; Williams et al., 1999). ERT uses conductive sensors on the wall of the process vessel to inject current and then sense voltage differences from which the conductivity of the electrolyte inside the process vessel can be measured. Like other electrical methods, various approaches can be used to analyze the data, which are discussed below. Although there is much attention given to reconstructions using algorithms to interpret the internal distribution of resistivity in the process vessel, other methods are also extremely valuable and will be considered below. Since ERT is a soft field tomography modality, similar to other electrical tomography modalities, many of the reconstruction techniques presented in the previous section for ECT can be used for ERT as well. The distribution of the resistivity provides quantitative imaging of the phase distribution in the process vessel. ERT has been widely used to study the hydrodynamics of conductive multi-phase flows in the multi-phase systems.

3.2 Principles of ERT

An ERT system is built on a sensor, a data acquisition system, and some form of interpretation facility, often an image reconstruction algorithm, which are three key elements of a tomography system (Wang et al., 1994). Figure 11a is a sketch of the ERT system with all its components. The sensor electrodes are made of conductive materials, usually with a relative conductivity much higher than the electrolyte, penetrating the process vessel where one electrode injects current in the electrolyte whereas the rest act as receivers. The sensor shown in Figure 11a has four layers of electrodes with equal intervals in height on the axis direction of the process vessel. Each layer has an equal number of electrodes with equal intervals around the process vessel. The sensor arrays provide resistivity measurement between the electrodes both in the layer and between different layers. The data acquisition system is designed to provide resistivity measurements from measured currents at the

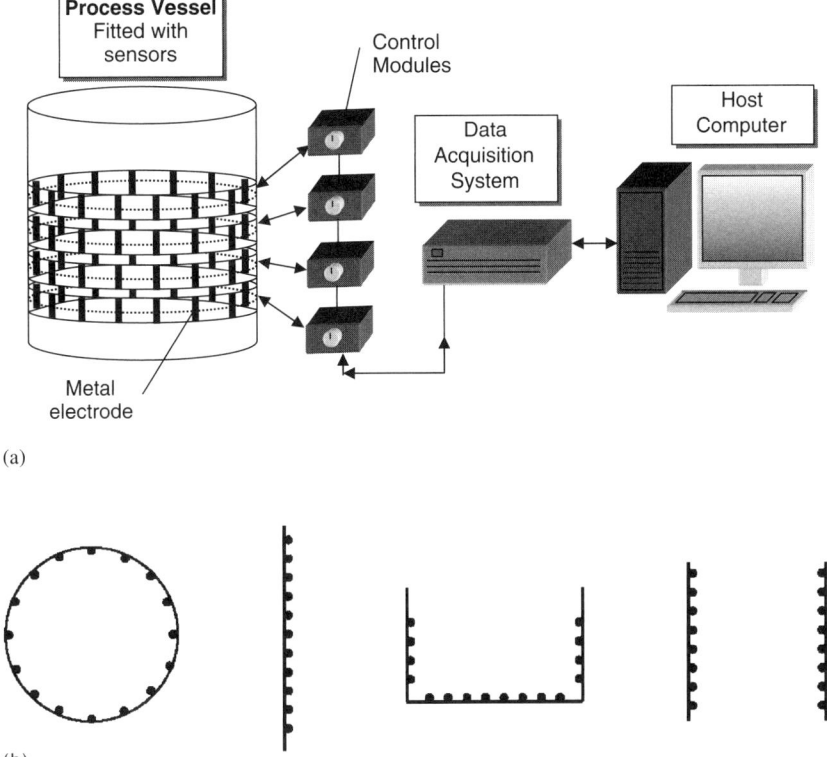

Figure 11 (a) Sketch of the ERT system including a multi-layer circular sensor array, data acquisition system, and computer for reconstruction and (b) various electrode arrangements in form of circular, probe, channel, and dual probe arrays.

receiving electrodes. Different designs are usually implemented in distributing the electrodes for proper 3D resistivity measurements (Dickin and Wang, 1996).

The objective of the sensor is the detection of differences of concentration in the multi-phase flow that is sensed through detecting changes in local electrical properties of the different phases. In ERT, the applied source is electrical current 0–75 mA between 75–500 kHz, the voltage is sensed, and the electrical property obtained is the conductivity (mS/cm). The change of the electrical properties in the multi-phase flow is sensed by the sensor arrays (Figure 11a). In addition a spare electrode, referred to as the ground electrode, positioned away from the measurement electrodes but in electrical contact with the internal fluid is required to ensure all voltage measurements are fixed against a common ground source. The different configurations of the electrodes depend on the process vessel to be measured. The electrodes are normally located equi-distantly around inner cross-sectional circumference of pipe or process system (Figure 11b). The electrodes are made of metals (platinum, silver palladium, etc.); stainless steel is preferred for its resistance to chemical attack. There are five main strategies employed in EIT depending on the injection of current and voltage measurement, shown in Figure 12:

1. Adjacent: In this strategy current is applied through two neighboring electrodes and the voltages measured from successive pairs of neighboring electrodes (Figure 12a). Current is then applied through the next pair of electrodes and the voltage measurements repeated. The adjacent measurement strategy yields N^2 measurements, where N is the number of electrodes. Furthermore, to avoid the effects from electrode/electrolyte contact impedance problems, the voltage is not measured at a current-injecting electrode so the total number of independent measurements is $N(N-3)/2$. The main shortcoming of the adjacent strategy is that it has a non-uniform current distribution since most of the current travels near the peripheral electrodes. Therefore, the current density at the centre of the vessel is relatively low which makes the strategy very sensitive to measurement error and noise (Hua et al., 1993). However, the strategy requires minimal hardware to implement and image reconstruction which is the main advantage.
2. Opposite: In this strategy current is applied through diametrically opposite electrodes (Figure 12b. The voltage reference is the electrode adjacent to the current-injecting electrode. Thus, for a particular pair of current-injecting electrodes, the voltages are measured with respect to the reference at all the electrodes except the current-injecting ones. The next data set is obtained by switching the current to the next pair of opposite electrodes in the clockwise direction and the voltage reference electrode is changed accordingly. The voltages are again measured using the same procedure and the whole procedure is repeated. Compared with the adjacent strategy, the

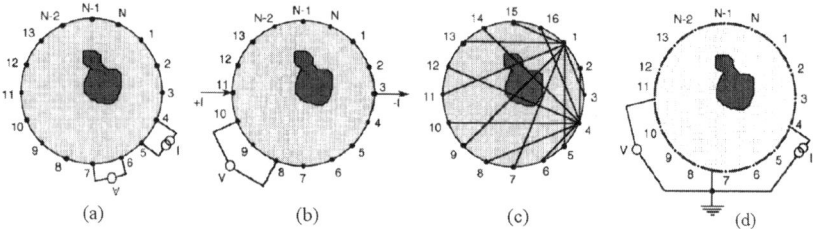

Figure 12 Common electrical excitation strategies for ERT measurement (adapted from Wang, 2002).

opposite strategy is less sensitive to conductivity changes at the boundary since most of the current flows through the central part of the region. However, it suffers from a serious disadvantage: for the same number of electrodes N, the number of independent measurements is less than for the adjacent strategy. Breckon and Piddcock (1987) showed that, for the opposite strategy, the number of independent measurements is given by $N/4N/2–1$).

3. Diagonal: In this strategy, currents are injected between electrodes separated by large dimensions (Figure 12c. The outcome, with respect to the adjacent strategy, is a more uniform current distribution in the region being imaged. For a 16-electrode tomography measurement, the data collection procedure is as follows: electrode 1 is fixed as a current reference and electrode 2 as the voltage reference; current is then applied successively to electrodes 3, 5, till 15. For each of these current pairs, the voltages from all electrodes except the current electrodes are measured with respect to electrode 2. The current reference is then changed to electrode 4 and the voltage reference to electrode 3 and current applied through electrodes 6, 8, till 16 and 2. As before, the voltage is measured on all other electrodes except the current-injecting ones. Thus, for each pair of current electrodes, 13 voltage measurements are obtained and seven different current electrode pairs are selected to give $7 \times 13 = 91$ data points. A further 91 points can be obtained by changing the voltage and current reference. Therefore, a 16-electrode system yields a maximum of 182 data points out of which only 104 are independent. The diagonal strategy does not yield a high sensitivity in the periphery compared with the adjacent method. However, it has better sensitivity over the entire region and is not as sensitive to measurement error and thus produces a better quality image (Hua et al., 1993).

4. Conducting boundary: The conducting boundary strategy was devised by Wang (1994) as a measurement strategy for use on process vessels and pipelines with electrically conducting boundaries (Figure 12d). As opposed to the preceding four-electrode measurement strategies, the conducting boundary strategy employs only two electrodes. The relatively large surface area of the conducting boundary is employed as the current sink to reduce the common-mode voltage across the measurement electrodes. Consequently, neither common-mode feedback nor earthed (load) floating measurement techniques are necessary in this strategy. The earthed conducting boundary also acts as a shield, reducing effects of electromagnetic interference. The conducting boundary strategy has a significantly lower common-mode voltage component than does the adjacent electrodes strategy. Wang (1994) reported it to be 800

times smaller for the conducting boundary strategy. Conversely, the amplitude of the measured voltages for identically shaped process vessels was approximately factor of 7 lower for the conducting boundary strategy in comparison with the adjacent strategy sensitivity is similar to that of the adjacent strategy.

5. Simultaneous current injection: Pioneering work based on medical electrical imaging used simultaneous current injection and voltage measurements, for example, the adaptive current tomographic system (ACT4) at Renslaaer group (Ross, 2003; Saulnier et al., 2006). It is a high-speed, high-precision, multi-frequency, multi-channel instrument, able to support up to 72 channels. The instrument is able to apply either voltage or currents to all the electrodes simultaneously and respectively measure the resulting currents or voltages. It can also control both the phase and the amplitude of the voltage or current excitation. Such a system can be configured to support 60 electrodes in two 5×6 radiolucent array and each electrode is driven by a 16-bit precision voltage source and has a circuit for measuring the resulting current and voltage. These circuits are digitally controlled to produce and measure signals at 5–1000 kHz. The magnitude and phase of each source are controlled independently. Each source, and each voltage and current measurement, is calibrated to a common standard reference (Saulnier et al., 2006). During each measurement, 59 orthonormal excitation patterns are used to maximize distinguishability and applications have focused on different medical flows, for example, pulmonary function (Kim et al., 2007).

The mathematical model of the ERT sensor electrodes is described in detail by Somersalo et al., (1992) and Kaipio et al. (2005), and principles of description are similar in many respects to those described earlier for ECT. The data measured using the EIT electrodes is transformed and analyzed by a data acquisition system. The analysis of the raw data enables to convert it into variables that describe the process to be measured. Different alternatives can be considered in the analysis of the data and some specifically for multi-phase systems. First, the use of raw measurements for statistical analysis to observe trends or for obtaining different statistical correlations. Second, the application of a process-based model using raw data as input. The distribution of concentration of the multi-phase flow is obtained by a model based on the characteristic parameters of the process (West et al., 2000). Also use of statistical methods such as Markov chain Monte Carlo (MCMC) based on prior information about the process have been successfully applied to reconstruct the images (West et al., 2004). Third, conversion of the raw data into images using a reconstruction algorithm. For ERT, for example, the application of different algorithms such as LBP, Newton–Raphson method (NRM), sensitivity conjugate gradients (SCG) enables to obtain a group of images that show an estimation of the distribution of concentration of the multi-phase flow in the process (Wang, 2002; Xie, 1995). The mapping of the distribution of concentration enables the calculation of the parameters of the fluid such as the concentration and the velocity profiles. The reconstruction processes have their own limitations and require careful use to avoid misinterpretation.

The reconstruction of the images is independent of the parameters of a specific process; it is based entirely on the results obtained from the measurement and the image reconstruction algorithm. The choice of image reconstruction algorithm is a trade-off between accuracy of image and time required for reconstruction. Several image reconstruction algorithms have been developed to suit different tomography applications (Xie, 1995). There has been a demand for fast image reconstruction algorithms that can be used for the real-time imaging of fast-moving processes. Therefore, much effort has been focused on the development of image reconstruction algorithms, both non-iterative and iterative for ERT. A comprehensive comparison between image reconstruction algorithms was reported by Yorkey et al. (1987a). In particular the most employed techniques of reconstruction in ERT are filtered back-projection between equipotent lines (Barber, 1990); back projection using sensitivity coefficient (Kotre, 1989); perturbation method (Yorkey et al., 1987b); double-constraint method (Wexler et al., 1985); Newton–Raphson method (Abdullah 1993) and the SCG method (Wang, 2002).

In summary, a linearization of the forward problem follows the same concept discussed in ECT as both systems rely on the electric property distribution. The forward problem can be written in a matrix expression as in Equation (9).

$$Y \cdot v = c \qquad (9)$$

where Y is a global admittance matrix (George and Liu, 1995), v is a vector representing the potentials of the mesh-node point, c is a vector representing the boundary current. Several back-projection algorithms and sensitivity coefficient methods (Barber et al., 1983; Breckon and Pidcock, 1987; Kotre, 1994) are developed for the qualitative reconstruction showing the changes of resistivity to the original reference distribution. Different quantitative algorithms (Abdullah, 1993; Loh, 1994; Yorkey, 1986) are also developed to provide the real values of the resistivity distribution on the pixels within the pixel vessel. ERT reconstruction algorithms are very similar to ECT reconstruction as the field distribution in both systems is governed by similar equations. A major practical difference however is the necessity to use corrections for the conducting boundary effects (Wang et al., 2005).

3.3 ERT applications

3.3.1 Hydrocyclone flow visualization and comparison with computational fluid dynamics

A hydrocyclone is a device to separate solids/particles from liquid from a suspension or to separate liquids of different density in fluid mixtures by means of difference in size or density (Cullivan and Williams, 2003). The hydrocyclone is widely used in industry such as separation of oil from water; separation of sand, staples, plastic particles from pulp, and paper mills; and separation of metal particles from cooling liquid in metal working. Figure 13 is a sketch of a hydrocyclone configuration and the flow structure in the unit. It consists of a cylindrical section with a tangential inlet at the top to feed the slurry suspension or liquid mixtures, a conical base for the separation, and two exits. One exit is at the bottom for underflow with the denser fraction, whereas the other is at the top for overflow with the lighter liquid fraction. The tangential inlet generates high swirl of the slurry suspension or liquid mixtures. Because of centrifugal force, the larger particles are thrown against the wall and move to the underflow. On the other hand, small particles, or lighter liquid, are dragged inwards to the overflow. The mechanisms of separation and models of fluid flow in hydrocyclone separators have been studied (Chocha et al., 1998; Dyakowski and Williams, 1998) but are still not well understood because of their complex high swirl and varying turbulence conditions, the presence of an interface inside the separator, and the requirement for online measurement data for the model development (Williams et al., 1999). Non-intrusive measurements such as ERT and computational fluid dynamics (CFD) are effective techniques to understand the internal flow of the hydrocyclone separator. Figure 14 is the ERT/EIT reconstructed image of the hydrocyclone cross-section compared to ultrasound tomography (UST) (Cullivan and Williams, 2003). Eight parametrically reconstructed sections are obtained in the conical base of the hydrocyclone. The light and dark colors represent

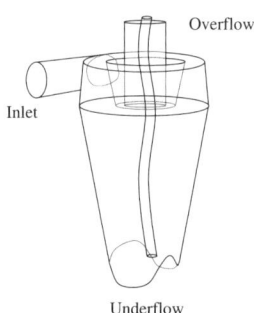

Figure 13 Sketch of the hydrocyclone flow structure.

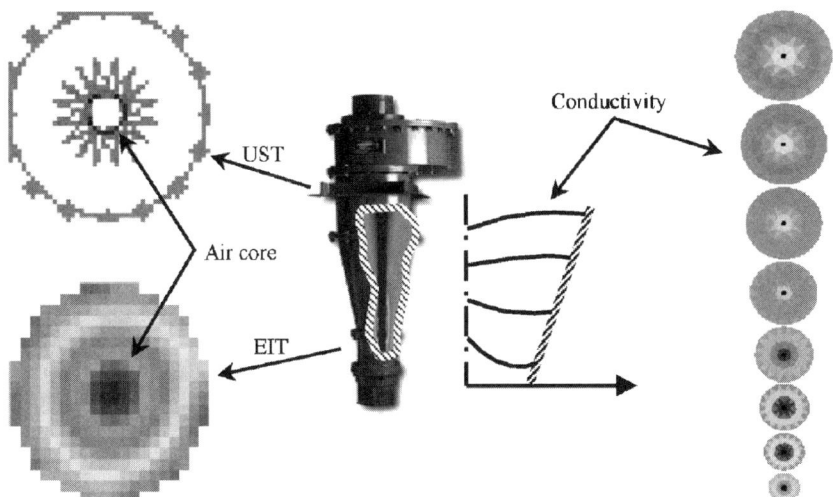

Figure 14 ERT tomographic image of the hydrocyclone cross section and the comparison with UST (Cullivan and Williams, 2003).

high and low conductivity material in the images, respectively. An air cone shape which has low conductivity is clearly shown as a dark region in the conical chamber of the hydrocyclone. The air cone shape obtained from ERT and that from the UST (top-left of Figure 14) are consistent with each other. ERT hydrocyclone measurements are solid validation for the CFD results on hydrocyclone simulations. Figure 15 is the positive axial-velocity contours in the hydrocyclone (Cullivan et al., 2004). The central axial-velocity is positive toward the underflows and an air-core is developed. ERT and CFD results show good agreement for the flow structure in the hydrocyclone.

3.3.2 High-speed flow imaging in slurry conveying

Hydraulic conveying systems use fluid to transport solid particles in pipes over long distance and is widely applied in chemical processes such as coal and mineral processing, oil drilling, pharmaceutical, and food industries (Laskovski et al., 2007; Lim, 2007; Yin and Wang, 2003). The solids concentration, size, density, distribution, velocity, and destabilization of the suspended solids are yet to be understood in the opaque and flowing slurry conveying system. Understanding the suspended solid flow behaviors and the solid distribution in hydraulic conveying systems is of great importance for the optimization and design of the conveyor, pressure drop estimation in the conveying pipe, and operations at maximum solids concentration (Fangary et al., 1998). ERT is being used to obtain conductivity maps across a slice of the conveying

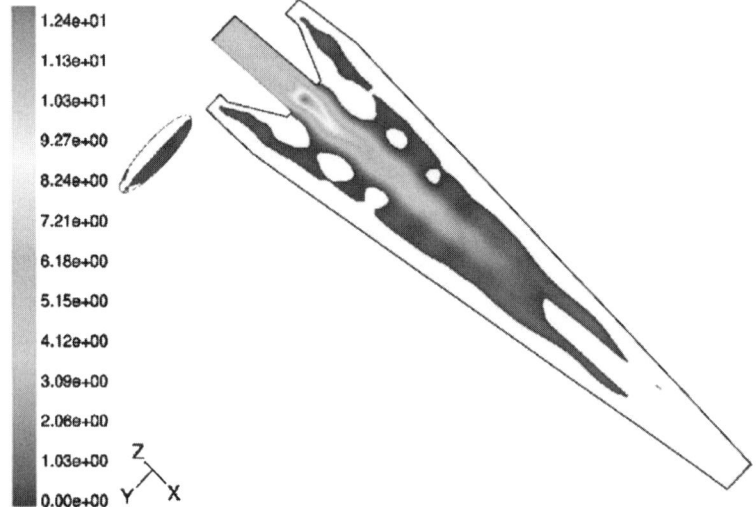

Figure 15 Axial velocity contours in hydrocyclone (Cullivan et al., 2004) (see Plate 14 in Color Plate Section at the end of this book).

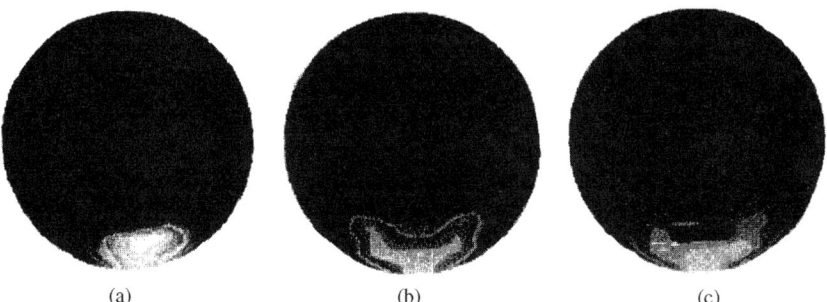

Figure 16 Tomographic images of three different deposit thicknesses in slurry conveying system (Fangary et al., 1998).

pipe to investigate the solid concentration and transport behaviors in the opaque and flowing slurry conveying system.

The images in Figure 16 represent the locations of the deposit-slurry interface for three different deposit thicknesses (13 mm, 31 mm, and 38 mm) in a hydraulic conveying system using ERT (Fangary et al., 1998). The fluid media is an electrolyte of 2.1 mS/cm and the working solids are spherical glass beads with a diameter of 5 mm. The surface of the deposit estimated from ERT images is higher than the actual sediment height. Nevertheless, they are consistent with each other. The results show that ERT is a useful tool in capturing images of hydraulic conveying flow with a small amount of solid sediment. ERT can also be used to capture the solid distribution and online solid deposition in hydraulic slurry conveying systems. This has been further developed to enable real-time imaging of solid deposition in pipeline systems, and through use of multiple electrode planes disposed along the pipe, it is possible to cross-correlate signals between adjacent planes in order to elucidate information on velocity (Lucas et al., 1999; Williams, 2005). For example, the detailed flow structure of swirling suspensions in pipelines can be sensed and used to optimize operational conditions (Wang et al., 2003) and to develop strategies for unblocking plugged pipelines. Such methods are in use industrially (Primrose, 2008).

3.3.3 Visualization of dispersions in an oscillatory baffled reactor

Formulation of multi-component emulsions and mixtures are of interest in chemical and industrial processes (Vilar, 2008; Vilar et al., 2008). Standard stirred tank reactors (STR) and oscillatory baffled reactors (OBR) are traditional methods for the formulation of liquid–liquid mixtures and liquid–solid emulsions. Compared with STR, oscillatory baffled reactors provide more homogeneous conditions and uniform mixing with a relatively lower shear rate (Gaidhani et al., 2005; Harrison and Mackley, 1992; Ni et al., 2000). Figure 17 is a sketch of a typical oscillatory baffled reactor. It consists of the reactor vessel, orifice plate baffles, and an oscillatory movement part. The orifice plate baffles play an important role in the OBR for the vertex generation in the flow vessels as well as the radial velocities of the emulsions and mixtures. They are equally spaced in the vessel with a free area in the center of each baffle

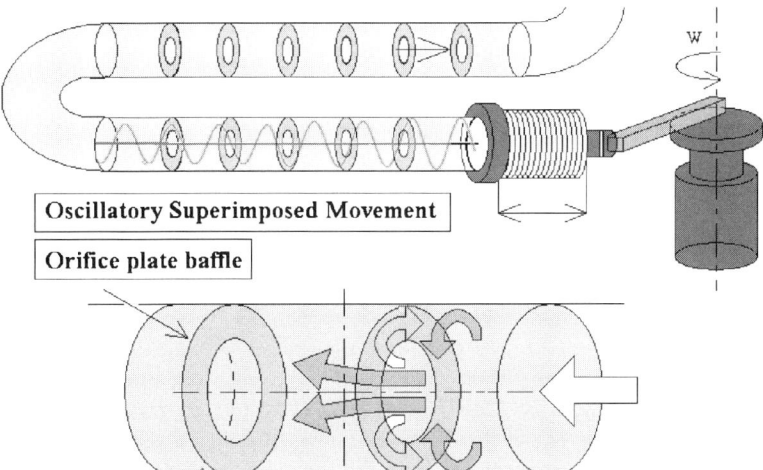

Figure 17 Sketch of the oscillatory baffled reactor (top) and flow within baffled sections (bottom).

for the fluid to flow. The part of oscillatory movement places an oscillatory motion on the fluid and, with the specific baffles, provides good axial/radial mixings for the working fluids and solids with highly efficient heat and mass transfer. A mineral oil, with a light density of 850 kg/m^3 working as a dispersed phase, is introduced in droplet form into the continuous phase water flow with a high density of 997 kg/m^3 at the top of the flow vessel. Figure 18 shows images of local oil fraction distributions in oscillatory baffled reactor obtained by ERT (Vilar et al., 2008). The green and blue colors represent low and high conductivity materials which are oil and water, respectively. The oil volume fraction could be calculated based on the conductivity map of the OBR. The oil phase is introduced at the top of the reactor and can be seen at the top of the images on the upper left section of Figure 18. The vortex formed from the oscillatory movement and the baffles enhance the mixing of oil and water. By using adjacent electrode rings (lower right, Figure 18), characteristics of the velocities of the phases can be seen (upper right and lower left). Here we see that ERT is used to visualize dispersions in the oscillatory baffled reactors and can give clear and quantitative mapping of the emulsions and mixtures of liquid–solid and liquid–liquid flows. Use of high-speed ERT systems operating at more than 1000

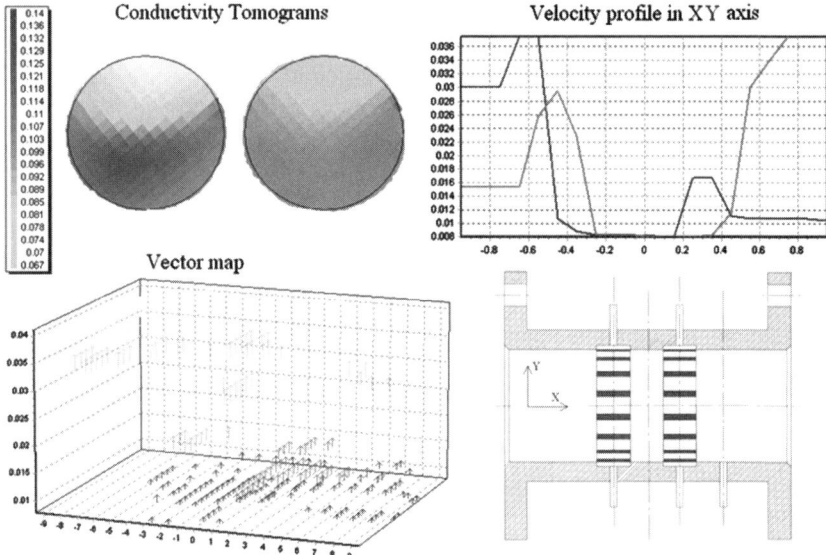

Figure 18 Tomographic images of local oil fraction distributions in oscillatory baffled reactor (top left), with estimated velocity profiles (top right) from cross-correlation signals between adjacent electrode rings (bottom right) and vector map (bottom left) (Vilar et al., 2008; Vilar, 2008) (see Plate 15 in Color Plate Section at the end of this book).

frames per second have enabled abstraction of concentration and velocity data in the reactor and synchronization of images with the stroke of the pulsed reactor to elucidate detailed information on inter-stage mixing (Issa et al., 2009).

4. POSITRON EMISSION TOMOGRAPHY

4.1 Introduction

Positron emission tomography is a radioisotope emission technique widely used in diagnostic medicine and has found new applications outside the medical field (Hoffmann, et al. 2005; Parker and McNeil, 1996). PET is a specific type of GRT techniques. For the PET technique, a radioactive source or tracers is introduced into the selected component of a multi-component system. The decay of two back-to-back γ-rays generated by radioactive tracers is monitored by external detectors. Back-projection techniques are used to obtain the distribution of the tracers in the monitoring region, from which the multi-component flow structure in the monitoring region is obtained. PET is a widely used technique for tracking multi-phase flows in engineering studies.

4.2 Principle of PET

In PET, radioactive tracers are introduced into a selected component of a multi-component system. Emission of a positron, which is the anti-particle of the electron having the same mass and equal but opposite charge, as one of its decay products is involved in the form of β-decay from the radioisotope. Two back-to-back 511 keV γ-rays are formed by the annihilation of the positron and the electron. External position-sensitive detectors are used to monitor the straight line formed by the two back-to-back γ-rays or the line along which the annihilation has taken place. The back-to-back γ-rays pair detected by the two separate external position sensitive detectors should be within a coincidence resolving time of approximately 20 ns. The number of γ-rays along each line is accurately and quantitatively measured by the two separate external position-sensitive detectors to reflect the tracer density integrated along the line, or called a "projection." The entire tracer distribution in the monitoring region is provided by the back-projection techniques imposed on measurements of all possible projections (Benton and Parker, 1997). The flow structure of the selected component in the multi-component system is then obtained. Medical PET scanners or cameras have been extensively developed in the past 30 years. A typical medical PET scanner consists of a ring of scintillators to provide a 2D image of the tracer distribution in the slice of the ring. More coaxial rings of scintillators provide extended 2D images or even fully 3D images of the tracer distribution. The most famous non-medical PET camera is the Brimingham positron camera (Hawkesworth et al., 1986; Parker and McNeil, 1996), which is shown in Figure 19. The camera consists of two external position-sensitive detectors, each having an area of $600 \times 300\,\text{mm}^2$. The camera takes a 2D picture of the plane parallel to the detector faces. Three-dimensional high-resolution images are obtained by rotating the detectors around the monitored object. Methods based on single particle tracking and multiple particle tracking are under development with the aim to design and use labels that are well matched to the measurement system under investigation. Work on the hardware is also seeking to develop versions of the equipment that are higher resolution and more portable for on-site use at suitable industrial plants.

The absorption and scattering of the γ-rays in the presence of matter between the tracers and the detectors, such as human bones and body tissues in medical application as well as vessel walls and internals of a multi-phase flow system in non-medical application, will introduce

Figure 19 Photography of the first Birmingham University positron camera for industrial imaging (Parker and McNeil, 1996).

signal loss of the γ-rays. In medical applications, reduction of signal intensity is a key problem since the amount of radioactive tracers needs to be at a minimum in order to not threaten patient's health. However, in industrial applications, reduction of signal intensity is not a problem since the amount of radioactive tracers is of little concern; PET is thus a more easy to use technique for tracking multi-phase flows in engineering studies.

4.3 PET applications

4.3.1 Slurry mixtures in stirred tanks

Slurry mixing of opaque multi-phase flow systems is intensively developed in pharmaceuticals, polymers, and food industries. Fundamental understanding of the hydrodynamics of slurry mixing behavior in opaque multi-phase flow systems is of great importance for the design and optimization of the STR. The operating conditions for the STRs, such as the minimum stirrer-spinning speed of the impeller to suspend all the solids/liquid in the reactor without any dead zone and the stirrer speed to provide homogeneous solid–liquid/liquid–liquid emulsions and mixtures, need to be determined to obtain high-quality products (Barigou, 2004). PET has the capability to track the movements and concentrations of one component, such as sand particles or selected fluids, in a multi-component system. Figure 20 shows a PET image of the sand concentration distribution of the sand–water slurry in a STR with a cylindrical vessel shape (McKee et al., 1995b). Pure coarse sands with a diameter of 600–700 µm were used as the suspended solids in the left image, whereas a mixture consisting of 80% coarse sands and 20% fine sands with a diameter of 150–210 µm were used in the right image. The two figures show only the coarse sand concentration distribution since only the coarse sands were traced with radioactive tracers. The gray scale

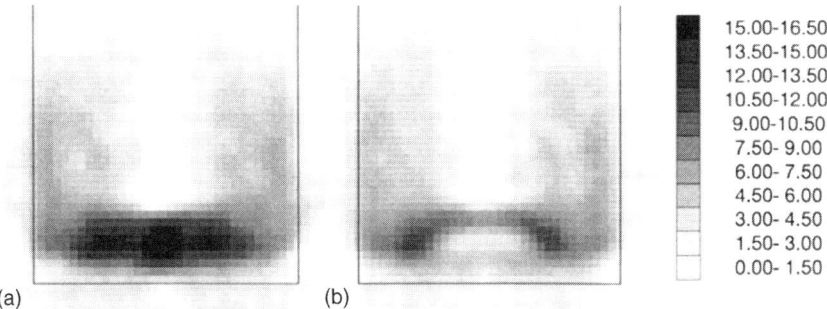

Figure 20 Tomographic sand distribution in slurry mixtures (McKee et al., 1995b).

shows the volume percentage of coarse sands in the reactor. Noise exists outside the reactor and has a value less than 3%. The single projected view of the STR provides a 3D solid distribution, since the axial stirrer in the cylindrical vessel provides a solids distribution with rotational symmetry (McKee et al., 1995b). Images of the sand concentration distribution in a sand–water STR demonstrate the capability of PET in tracking multi-phase flow in engineering applications. A number of studies have been performed on multi-phase systems in process equipment (Conway-Baker et al., 2002; West et al., 1999) and also detailed comparisons of the dynamic granular flow models in which predictions from discrete elements models are compared with real behaviors (e.g., Link et al., 2008).

4.3.2 Dispersion of particle pulse in gas–solid fluidized beds

Gas–solid fluidized bed technology provides high mass and heat-transfer coefficients between gas and solids in the reactor. It has been applied widely in industries such as the oil, metallurgical, and chemical industries (Fan and Zhu, 1998; Kunni and Levenspiel, 1991). Although gas–solid fluidization has been developed for several decades and considerable research results on hydrodynamics of the fluidized beds have been achieved, accurate solid movement and dispersion in gas-solid fluidized beds are still required for the industrial design, operation, and fundamental understanding of gas–solid fluidized beds. PET is an effective technique to accurately record the movement and dispersion of solids in these reactors. For dispersion of solid measurements in the fluidized beds, the tracer solids should have the same physical properties and fluidization characteristic as the bulk solids. Dechsiri et al. (2005b) studied the dispersion of particle pulse in a gas–solid fluidized bed. The FCC catalyst particles (Geldart group A) with an average of 79.5 μm and a density of $1464\,\text{kg}/\text{m}^3$ are used for the bulk and tracer solids in the bed. Three grams of FCC solids are labeled with radioactive material and uniformly placed at the surface of the bed as the tracers. The fluidized bed is made of a cylindrical glass column with a height of 35 cm and diameter of 15 cm. A cylindrical PET camera is located outside the fluidized bed vessel to capture the movement and dispersion of the tracer. Figure 21 depicts a PET image of the dispersion of pulses initially placed at the surface of a gas–solid fluidized bed (Dechsiri et al., 2005b). The numbers at the top of Figure 21 are the particle pulse positions and movements over the range of 12 s. The two numbers at the bottom of Figure 21 are the 3D and contour plots of relative averaged intensity of the tracer solid concentration in the horizontal plane of the fluidized bed vs. height in the bed and time, respectively. The dispersion of the solids shows asymmetry on the two sides of the axis of the fluidized bed. The images of the tracer solids movement and dispersion show that the

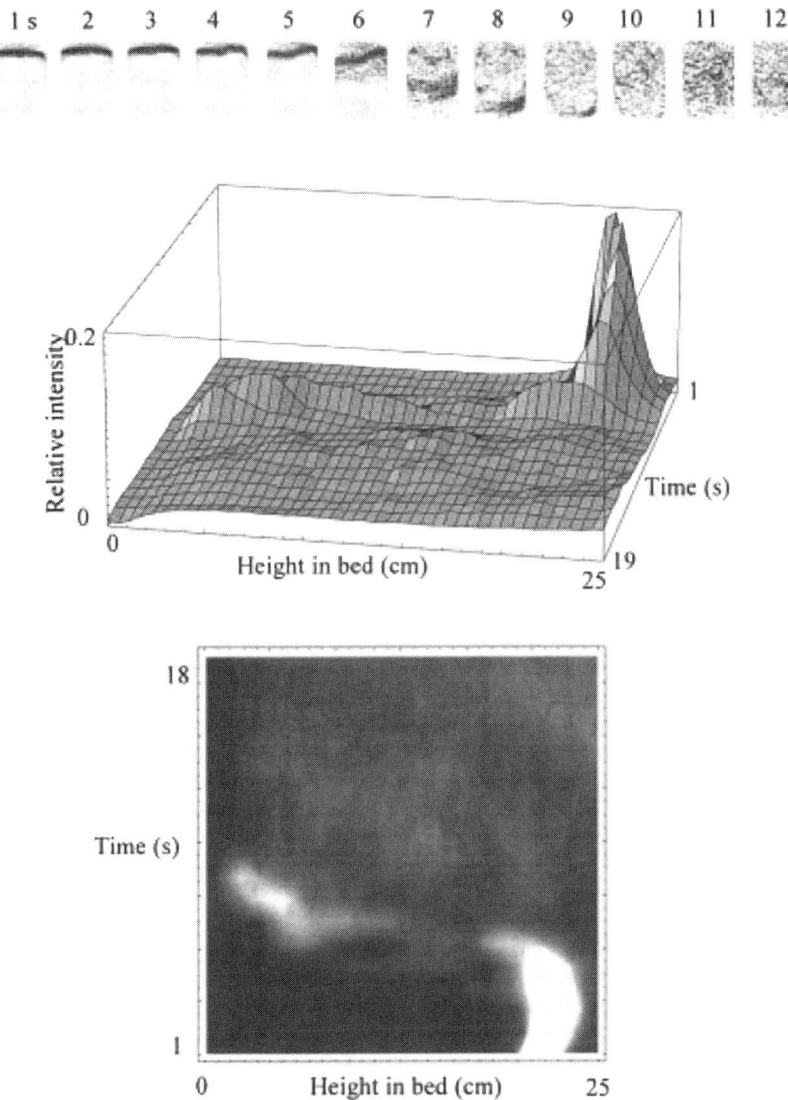

Figure 21 Dispersion of particle pulse initially placed at the surface of a gas–solid fluidized bed (Dechsiri et al., 2005b).

downward velocity of the pulse solids in the bulk phase has a value of 0.05 m/s.

4.3.3 Visualization of multi-phase fluids through sudden expansions

Sudden expansions of beds are very common in many industrial facilities such as heat exchange, mixing, and feeding systems (Arola et al., 1998).

Owing to limited measurement techniques, understanding the mechanism of the particle dispersion through sudden expansions is still a fundamental research topic. Particle collisions effect the particle concentration and introduce inhomogeneity in the suspensions. Accurate and quantitative measurements of the opaque flow behaviors through sudden expansions are needed. PET has been successfully applied for non-intrusive measurements of pipe flows of solids suspensions such as papermaking and pulp suspensions. The behavior of papermaking fiber suspension passing through a sudden expansion has been studied using PET (Heath et al., 2007). The papermaking fibers are labeled using Fluorine-18 (^{18}F). The concentration profiles are reflected by the radioactivity distribution obtained by a PET camera. Figures 22 and 23 contain PET images (a. top view; b. cross-section view; c. side view; d. 3D

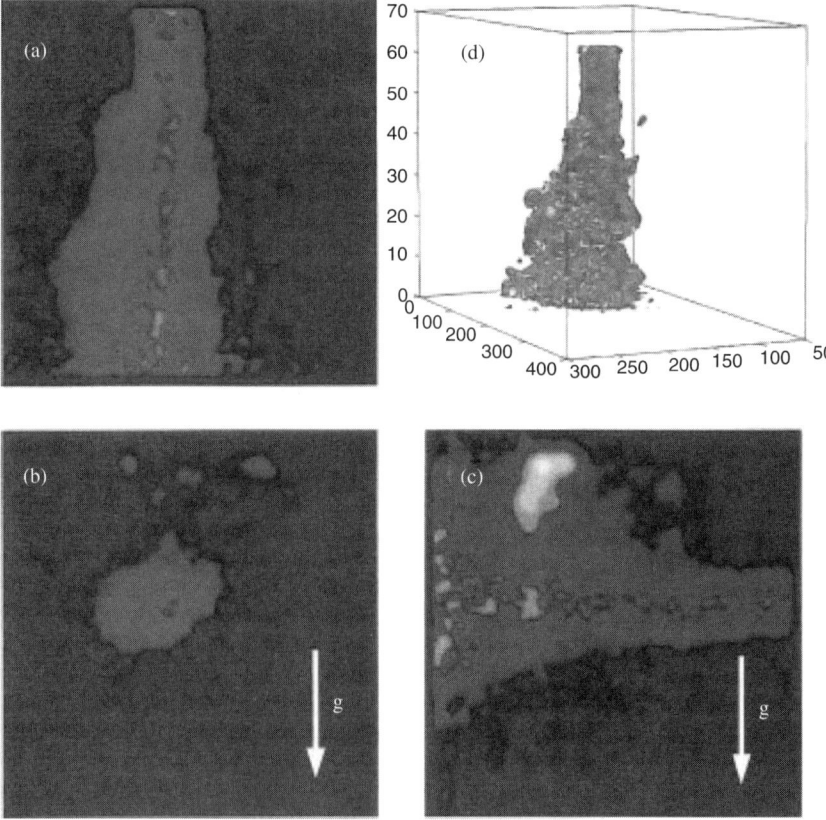

Figure 22 Flow of a fiber suspension through a sudden expansion with an upstream velocity of 0.5 m/s (Heath et al., 2007) (see Plate 16 in Color Plate Section at the end of this book).

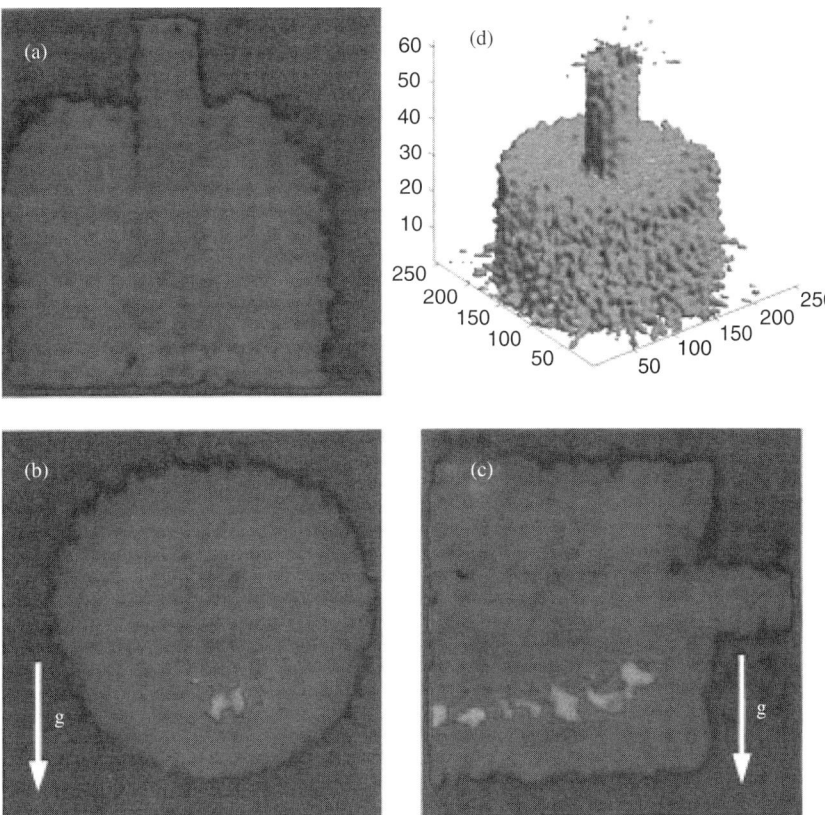

Figure 23 Flow of a fiber suspension through a sudden expansion with an upstream velocity of 0.7 m/s (Heath et al., 2007) (see Plate 17 in Color Plate Section at the end of this book).

reconstruction of the tracer distribution) of the concentration profiles of the papermaking fiber suspension through the sudden expansion with the upstream velocity of 0.5 m/s and 0.7 m/s, respectively (Heath et al., 2007). At a low upstream velocity of 0.5 m/s, the papermaking fibers are not well-mixed after passing through the expansion part due to low shear from the fluid, which is characterized as "plug flow." At a high upstream velocity of 0.7 m/s, the papermaking fibers are well-mixed after passing through the expansion part due to high shear from the fluid, which is defined as "fluidized" (Heath et al., 2007). PET measurements provide the quantitative 3D concentration of the papermaking fiber suspension passing through the sudden expansion. PET shows its solid capability for visualization of solids suspension multi-phase flow behaviors through sudden expansions.

5. CONCLUDING REMARKS

In this review, tomography systems of capacitance (ECT), resistance (ERT), and positron emission (PET) are presented. Their applications in imaging industrial processes highlight the importance and advantages of each technique. However, the common favored advantage among the three techniques is their ability to provide non-invasive measurements of various processes.

Principles of ECT and the latest developments in the technology are highlighted with emphasis on the volume ECT (ECVT) for 3D imaging. The significance of ECT technique in process engineering is presented in the framework of industrial application. Fluidized beds, pneumatic solids conveying, and slurry bubble columns are examples of ECT's capability to provide quantitative and qualitative understanding of the internal process dynamics.

ERT is presented of its similarities to ECT as both techniques are governed by similar equations of electrical field and current distributions. Examples of ERT in hydrocyclone flow visualization, comparison with CFD, high-speed flow imaging in slurry conveying, and visualization of dispersions in an oscillatory baffled reactor are discussed to highlight applicability of ERT in industrial processes. Both ECT and ERT are electrical modalities for process imaging and share the common characteristics of high imaging speed, safety, suitability for various sizes of vessels, and non-linearity of interrogating signals in regard to phase distribution.

PET, on the other hand, is a radioactive modality of tomography. Examples of PET applications in industrial process include imaging of slurry mixtures in STR, dispersion of particle pulse in gas–solid fluidized beds, and visualization of multi-phase fluids through sudden expansions. Although such methods require sophisticated facilities, due to the nature of the radiation source, they provide accurate data for detailed model comparison which is of major benefit for industrial design and operation.

NOMENCLATURE

NOTATION

C $1 \times N$ measured vector of capacitances between the n_c electrode pairs

C_i capacitance between the source and the detector electrodes for the ith electrode

c vector representing the boundary current

f_i	entropy objective function
f_{mse}	MSE function of G
f_s	smoothness function
G	$1 \times M$ image vector (permittivity distribution in the domain)
G^k	estimated permittivity vector in the kth iteration
$G_{s,tr}$	transport solids circulation rate
N	number of ERT electrodes
$\hat{n}$	normal vector to Γ_i
n_e	number of capacitor electrodes
Q_i	charge in the detector electrode for the ith electrode pair
S	$N \times M$ sensitivity matrix
S^T	transposed matrix of S
U_g	gas velocity
U_{tr}	transport velocity
ΔV_i	voltage difference for the ith electrode pair
X	$N \times N$ uniformity matrix
Y	global admittance matrix

GREEK LETTERS

α	relaxation factor
Γ_i	closed curve enclosing the detector electrode
$\gamma_1, \gamma_2, \gamma_3$	normalized constants between 0 and 1
$\varepsilon(x, y)$	dielectric constant (permittivity) distribution
$\phi(x,y)$	potential distribution
v	vector representing the potentials of the mesh-node point

REFERENCES

Abdullah, M. Z., Electrical impedance tomography for imaging conducting mixtures in hydrocyclone separators. PhD Thesis, UMIST (1993).

Arola, D., Powell, R. L., McCarthy, M. J., Li, T. Q., and Odberg, L. *AIChE J.* **44**, 2597–2606 (1998).

Barber, D. C. *Clin. Phys. Physiol. Meas.* **11**, 45–46 (1990).

Barber, D. C., Brown, B. H., and Freeston, I. L. *Electron. Lett.* **19**, 933–935 (1983).

Barigou, M. *Chem. Eng. Res. Des.* **82**(A9), 1258–1267 (2004).

Baumgarten, P. K., and Pigford, R. L. *AIChE J.* **6**, 115–123 (1960).

Bayle, J., Mege, P., and Gauthier, T., Dispersion of bubble flow properties in a turbulent FCC fluidized bed, in "Proceedings of 10th Engineering Foundation Conference, *Fluidization X*" (M. Kwauk, J. Li, and W.-C. Yang Eds.), pp. 125–132. Beijing, China (2001).

Bennett, M. A., and Williams, R. A. *Min. Engn.* **17**, 605–614 (2004).

Bennett, M. A., West, R. M., Luke, S. P., Jia, X., and Williams, R. A. *Chem. Eng. Sci.* **54**(21), 5003–5012 (1999).

Benton, D. M., and Parker, D. J., Non-medical applications of positron emission tomography, *in* "Non-Invasive Monitoring of Multiphase Flows" (J. Chaouki, F. Larachi, and M. P. Dudukovic Eds.), pp. 161–184. Elsevier Science, New York (1997).
Binns, R., Lyons, A. R. A., Peyton, A. J., and Pritchard, W. D. N. *Meas. Sci. Technol.* **12**, 1132–1138 (2001).
Boyer, C., and Fanget, B. *Chem. Eng. Sci.* **57**, 1079–1089 (2002).
Breckon, W. R., and Pidcock, M. K. Mathematical aspects of impedance imaging, *Clin. Phys. Physiol. Meas. A* **8**, 77–84 (1987).
Chen, R. C., Reese, J., and Fan, L.-S. *AIChE J.* **40**, 1093–1104 (1994).
Chocha, F., Castro, B., Ovalle, E., and Romero, J. *Phys. Separ. Technol.* 35–60 (1998).
Clough, D. E., and Weimer, A. W. *Ind. & Eng. Chem. Fund.* **24**(2), 235–241 (1985).
Conway-Baker, J., Barley, R. W., Williams, R. A., Jia, X., Kostuch, J., McLoughlin, B., and Parker, D. J. *Min. Engn.* **15**(1-2), 53–59 (2002).
Cui, H., and Chaouki, J. *Chem. Eng. Sci.* **59**(16), 3413–3422 (2004).
Cullivan, J. C., and Williams, R. A. *Part. Sci. Technol,* **21**, 83–103 (2003).
Cullivan, J. C., and Williams, R. A., Velocity, size and shape and concentration measurements of particulate mixtures. Chapter 14.4, *in* "Multiphase Flow Handbook" (C. Crowe Ed.), pp. 42–87. Taylor and Francis, London CRC Mechanical Engineering Series (2005).
Cullivan, J. C., Williams, R. A., Dyakowski, T., and Cross, C. R. *Min. Engn.* **17**, 651–660 (2004).
Dechsiri, C., Ghione, A., van de Wiel, F., Dehling, H. G., Paans, A. M. J., and Hoffmann, A. C. *Can. J. Chem. Eng.* **83**(1), 88–96 (2005a).
Dechsiri, C., van der Zwan, E. A., Dehling, H. G., and Hoffmann, A. C. *AIChE J.* **51**(3), 791–801 (2005b).
Dickin, F., and Wang, M. *Meas. Sci. Technol.* **7**, 247–260 (1996).
Du, B., and Fan, L.-S. *Ind. & Eng. Chem. Res.* **43**(18), 5507–5520 (2004b).
Du, B., Warsito, W., and Fan, L.-S. *AIChE J.* **49**(5), 1109–1126 (2003).
Du, B., Warsito, W., and Fan, L.-S. *AIChE J.* **50**(7), 1386–1406 (2004a).
Du, B., Warsito, W., and Fan, L.-S. *Ind. & Eng. Chem. Res.* **44**(14), 5020–5030 (2005).
Dyakowski, T., and Williams, R. A., Hydrocyclone flow modeling—a continuous research challenge, Chapter 5, *in* "Innovation in Physical Separation Technologies: Richard Mozley Symposium Volume," The Institution. Mining & Metallurgy, London, pp. 61–73 (1998).
Fan, L.-S., and Zhu, C., "Principles of Gas-Solid Flows". Cambridge University Press, New York (1998).
Fan, L. T., Ho, T. C., Hiraoka, S., and Walawender, W. P. *AIChE J.* **27**, 388 (1981).
Fangary, Y. S., Williams, R. A., Neil, W. A., Bond, J., and Faulks, I. *Powder Technol.* **95**, 61–66 (1998).
Fennell, P. S., Davidson, J. F., Dennis, J. S., Gladden, L. F., Hayhurst, A. N., Mantle, M. D., Mueller, C. R., Rees, A. C., Scott, S. A., and Sederman, A. J. *Chem. Eng. Sci.* **60**(7), 2085–2088 (2005).
Gaidhani, H. K., McNeil, B., and Ni., X. *Trans IChemE, Part A, Chem. Eng. Res. Des.* **83**(A6), 640–645 (2005).
Geldart, D., and Kelsey, J. R. *Powder Technol.* **6**(1), 45–50 (1972).
Geldart, D., and Xie, H. Y. *Fluid. VII, Proc. Eng. Found. Conf. Fluid.* 749–756 (1992).
George, A., and Liu, J. W., "Computer solution of large positive definite sparse matrices". Prentice-Hall, New York (1995).
George, D. L., Torczynski, J. R., Shollenberger, K. A., O'hen, T. J., and Ceceio, S. L. *Int. J. Multiphase Flow* **26**, 549–581 (2000).
Gibilaro, L. G., Di Felice, R., Foscolo, P. U., and Waldram, S. P. *Chem. Eng. J.* **37**, 25–33 (1988).

Gilbertson, M. A., Cheesman, D. J., and Yates, J. G., Observations and measurements of isolated bubbles in a pressurized gas-fluidized bed. Fluidization IX, Proceedings of the Engineering Foundation Conference on Fluidization 9 th, Durango, CO, United States, May 17–22, 61–68 (1998).
Gunn, D. J., and Al-Doori, H. H. *Int. J. Multiphase Flow,* **11**, 535–551 (1985).
Halow, J. S., and Nicoletti, P. *Powder Technol.* **69**(3), 255–277 (1992).
Harrison, S. T. L., and Mackley, M. R. *Chem. Eng. Sci.* **47**(2), 490–493 (1992).
Hawkesworth, M. R., O'Dwyer, M. A., Walker, J., Fowles, P., Heritage, J., Stewart, P. A. E., Witcomb, R. C., Bateman, J. E., Connolly, J. F., and Stephenson, R. *Nucl. Instrum. Methods A* **253**, 145–157 (1986).
Heath, S. J., Olson, J. A., Buckley, K. R., Lapi, S., Ruth, T. J., and Martines, D. M. *AIChE J.* **53**, 327–334 (2007).
Ho, T. C., Yutani, N., Fan, L. T., Walawender, W. P., and Song, J. C. *Chem. Eng. Sci.* **38**(4), 575–582 (1983).
Hoffmann, A. C., Dechsiri, C., Van der Zwan, E. A., and Dehling, H. G. *AIChE J.* **51**(3), 791–801 (2005).
Holland, D. J., Marashdeh, Q., Muller, C. R., Wang, F., Dennis, J. S., Fan, L.-S., and Glandden, L. F. *Ind. Eng. Chem. Res.* **48**, 172–181 (2009).
Hua, P., Webster, J. G., Tompkins, W. J., Effect of the measurement method on noise handling and image quality of EIT imaging, IEEE 9th Annual Conference on Engineering in Medicine and Biological Science, pp. 1429–1430 (1993).
Huang, S. M., Plaskowski, A., Xie, C. G., and Beck, M. S. *J. Physica E* **22**, 173–177 (1989).
Hubers, J. L., Striegel, A. C., Heindel, T. J., Gray, J. N., and Jensen, T. C. *Chem. Eng. Sci.* **60**(22), 6124–6133 (2005).
Hulme, I., and Kantzas, A. *Powder Technol.* **147**(1-3), 20–33 (2004).
Issa, M., Wang, M., Vilar, G. and Williams, R. A., Measurements using high speed EIT on an oscillatory flow reactor 3rd International Tomography Workshop Japan, Tokyo, May (2009).
Jin, H., Wang, M., and Williams, R. A. *Chem. Eng. J.* **130**(2-3), 179–185 (2006).
Kai, T., Misawa, M., Takahashi, T., Tiseanu, I., and Ichikawa, N. *Can. J. Chem. Eng.* **83**(1), 113–118 (2005).
Kaipio, J., Duncan, S., Seppänen, A., Somersalo, E., and Voutilainen, A., *in* "Process imaging for automatic control" (D. Scott, and H. McCann Eds.), Taylor & Francis Group, Boca Raton (2005).
Kang, W. K., Sutherland, J. P., and Osberg, G. L. *Ind. & Eng. Chem. Fund.* **6**(4), 499–504 (1967).
Kantzas, A., and Kalogerakis, N. *Chem. Eng. Sci.* **51**(13), 3555 (1996).
Kim, B., Isaacson, D., Xia, H., Kao, T., Newell, J. C., and Saulnier, G. J. *Physiol. Meas.* **28**, 237–246 (2007).
Kotre, C. J. *Clin. Phys. Physiol. Meas.* **11**, 275–281 (1989).
Kotre, C. J. *Phys. Meas. A* **15**, 125–136 (1994).
Kunni, D., and Levenspiel, O., "Fluidization Engineering". Butterworht-Heinemann, Stoneham (1991).
Lanneau, K. P. *Transactions of the Institution of Chemical Engineers* **38**(3), 125–143 (1960).
Larachi, F., and Chaouki, J. *Recents Progres en Genie des Procedes* **14**(75), 347–353 (2000).
Larachi, F., Chaouki, J., Kennedy, G., and Dudukovic, M. P. Non-Invasive Monitoring of Multiphase Flows. 335–406. Elsevier Science, New York (1997).
Laskovski, D., Stevnson, P., Zhou, J., and Galvin, K. P. *Powder Technol.* **179**, 59–64 (2007).
Lim, E. W. C. *Chem. Eng. Sci.* **62**, 4529–4543 (2007).
Link, J. M., Deen, N. G., Kuipers, J. A. M., Fan, X., Ingam, A., Parker, D. J., Wood, J., and Seville, J. P. K. *AIChE J.* **54**(5), 1189–1202 (2008).
Lirag, R. C., and Littman, H. *Chem. Eng. Prog. Symp. Series* **67**(116), 11–22 (1971).

Liu, J., Grace, J. R., and Bi, X. *AIChE J.* **49**(6), 1405–1420 (2003a).
Liu, J., Grace, J. R., and Bi, X. *AIChE J.* **49**(6), 1421–1432 (2003b).
Loh, W. W., Array processor for use in electrical impedance tomography. MSc Thesis UMIST (1994).
Lucas, G. P., Cory, J., Waterfall, R., Loh, W. W., and Dickin, F. J. *J. Flow Meas. Instrum.* **10**, 249–258 (1999).
Marashdeh, Q., Fan, L.-S., Du, B., and Warsito, W. *Ind. Eng. Chem. Res.* **47**, 3708–3719 (2008).
McKee, S. L., Dyakowski, T., Williams, R. A., Bell, T. A., and Allen, T. *Powder Technol.* **82**(1), 105–114 (1995a).
McKee, S. L., Parker, D. J., and Williams, R. A., Visualization of size-dependent particle segregation in slurry mixers using positron emission tomography, in "Frontiers in Industrial Process Tomography" (D. M. Scott, and R. A. Williams Eds.), pp. 249–259. AIChE and Engineering Foundation, New York (1995).
Nakajima, M., Harada, M., Asai, M., Yamazaki, R., and Jimbo, G., Bubble fraction and voidage in an emulsion phase in the transition to a turbulent fluidized bed, in "Circulating Fluidized Bed III" (P. Basu, M. Horio, and M. Hasatani Eds.), p. 79. Pergamon, Oxford (1991).
Ni, X., Cosgrove, J. A., Arnott, A. D., Greated, C. A., and Cumming, R. H. *Chem. Eng. Sci.* **55**, 3195–3208 (2000).
Okhi, K. and Shirai, T., Particle velocity in a fluidized bed, in *"Fluidization Technology"* (D. L. Keairns Ed.), Hemisphere, New York (1976).
Orcutt, J. C., and Carpenter, B. H. *Chem. Eng. Sci.* **26**(7), 1049–1064 (1971).
Ormiston, R. M., Mitchell, F. R. G., and Davidson, J. F. *Trans. Inst. Chem. Eng.* **43**(7), T209–T216 (1965).
Ostrowski, K. L., Dyakowski, T., Luke, S. P., and Williams, R. A. *Powder Technol.* **104**, 287–295 (1999).
Ostrowski, K. L., Luke, S. P., Bennett, M. A., and Williams, R. A. *Chem. Eng. J.* **77**(1-2), 43–50 (2000).
Parker, D. J., and McNeil, P. A. *Meas. Sci. Technol.* **7**, 287–296 (1996).
Patel, A. K., Waje, S. S., Thorat, B. N., and Mujumdar, A. S. *Powder Technol.* **185**, 239–250 (2008).
Peters, M. H., Fan, L.-S., and Sweeney, T. L. *Chem. Eng. Sci.* **38**(3), 481–485 (1983).
Primrose, K, Industrial applications of electricial tomography: case studies, see www.itoms.com (Manchester) (2008).
Ross, A. S., An adaptive current tomograph for breast cancer detection, PhD Thesis, Rensselaer Polytechnic Institute, Troy, NY, USA (2003).
Rowe, P. N., and Masson, H. *Trans. Inst. Chem. Eng.* **59**(3), 177–185 (1981).
Rowe, P. N., and Matsuno, R. *Chem. Eng. Sci.* **26**(6), 923–935 (1971).
Rowe, P. N., and Partridge, B. A., X-ray study of bubbles in fluidized beds. *Trans. Inst. Chem. Engr.* (London) **43**(5), T157-T175 Pub. in *Chem. Engr. (London)* No. 189 (1965).
Rowe, P. N., and Yacono, C. X. R. *Chem. Eng. Sci.* **31**(12), 1179–1192 (1976).
Saulnier, G. J., Ross, A. S., and Liu, N. *Physiol. Meas.* **27**, 221–236 (2006).
Savelsberg, R., Demco, D. E., Blumich, B., and Stapf, S. *Phys. Rev. E: Statistical, Nonlinear, and Soft Matter Physics* **65**, 020301(R) (2002).
Schubert, M., Hessel, G., Zippe, C., Lange, R., and Hampe, U. *Chem. Eng. J.* **140**, 332–340 (2008).
Sharma, A. K., Tuzla, K., Matsen, J., and Chen, J. C. *Powder Technol.* **111**(1-2), 114–122 (2000).
Shi, T. M., C.G. Xie, C. G., Huang, S. M., Williams, R. A., and Beck, M. S. *Meas. Sci. Technol.* **2**, 923–933 (1991).
Sitnai, O. *Chem. Eng. Sci.* **37**(7), 1059–1066 (1982).

Smith, M., Bayle, J., and Gauthier, T., Bubble flow study in a turbulent FCC fluidized bed. *EFCE, European Congress of Chemical Engineering*, 2nd. Montpellier, October 5–7 47 (1999).

Somersalo, E., Cheney, M., and Isaacson, D. *SIAM J. Appl. Math.* **52**, 1023–1040 (1992).

Vilar, G., On-line measurement of dispersions in an oscillatory baffled reactor using electrical impedance tomography, PhD Thesis, University of Leeds, Leeds UK (2008).

Vilar, G., Williams, R. A., Wang, M., and Tweedie, R. *J. Chem. Eng. J.* **141**, 58–66 (2008).

Wang, M., Electrical impedance tomography on conducting walled vessels, PhD Thesis, UMIST, Manchester UK (1994).

Wang, M. *Meas. Sci. Technol.* **13**, 101–117 (2002).

Wang, M., Dickin, F. J., and Williams, R. A., Electrical resistance tomography, UK and worldwide patent application 9404766.9 (11 March 1994).

Wang, S. J., Dyakowski, T., Xie, C. G., Williams, R. A., and Beck, M. S. *Chem. Eng. J. (Lausanne)* **56**(3), 95–100 (1995).

Wang, Z. C., Afacan, A., Nandakumar, K., and Chuang, K. T. *Chem. Eng. Process.* **40**, 209–219 (2001).

Wang, M., Jones, T. F., and Williams, R. A. *Trans. IChemE* **81**(Part A8), 854–861 (2003).

Wang, M., Ma, Y., Holliday, N., Dai, Y., Williams, R. A., and Lucas, G. *IEEE Sensors Journal* **5**(2), 289–299 (2005).

Wang, F., Marashdeh, Q., Warsito, W., and Fan, L.-S., Imaging gas/solid jet penetration in a gas-solid fluidized bed using electrical capacitance volume tomography, AIChE Annual Meeting, Philadelphia, PA, USA, 16–21 November, Section 03B01 (2008).

Warsito, W., and Fan, L.-S. *Chem. Eng. Sci.* **56**, 6455–6462 (2001a).

Warsito, W., and Fan, L.-S. *Meas. Sci. Technol.* **12**, 2198–2210 (2001b).

Warsito, W., and Fan, L.-S. *Chem. Eng. Sci.* **58**, 823–832 (2003).

Warsito, W., and Fan, L.-S. *Chem. Eng. Sci.* **60**, 6073–6084 (2005).

Warsito, W., Marashdeh, Q., and Fan, L.-S. *IEEE Sens. J.* **7**, 525–535 (2007).

Weimer, A. W., Gyure, D. C., and Clough, D. E. *Powder Technol.* **44**(2), 179–194 (1985).

Werther, J. *Powder Technol.* **102**(1), 15–36 (1999).

Werther, J., and Molerus, O. *Int. J. Multiphase Flow* **1**, 103–122 (1973).

West, R. M., Jia, X., and Williams, R. A. *Chem. Eng. Commun.* **175**, 71–79 (1999).

West, R. M., Jia, X., and Williams, R. A. *Chem. Eng. J.* **77**(1-2), 31–36 (2000).

West, R. M., Scott, D. M., Sunshine, G., Kostuch, J. A., Heikkinen, L., Vauhkonen, M., Hoyle, B. S., Schlaberg, H. I., Hou, R., and Williams, R. A. *Meas. Sci. Technol.* **13**, 1890–1897 (2002).

West, R. M., Aykroyd, R. G., Meng, S., and Williams, R. A. *Physiol. Meas.* **25**, 181–194 (2004).

Wexler, A., Fry, B., and Neuman, M. R. *Appl. Opt.* **24**, 3985–3992 (1985).

Williams, R. A., Mineral and material processing, Chapter 11, *in* "Process Imaging for Automatic Control" (D. M. Scott, and H. McCann Eds.), pp. 359–400. Taylor and Francis (2005).

Williams, R. A., and Beck, M. S. (Eds.), Process tomography-principles, *in* "Techniques and Applications" p. 550. Butterworth-Heinemann, Oxford (1995).

Williams, R. A., Mann, R., Dickin, F. J., Ilyas, O. M., Ying, P., Edwards, R. B., and Rushton, A. *AIChE. Symp. Ser.* **293**, 8–15 (1993).

Williams, R. A., Jia, X., West, R. M., Wang, M., Cullivan, J. C., Bond, J., Faulks, I., Dyakowski, T., Wang, S. J., Climpson, N., Kostuch, J. A., and Payton, D. *Min. Engn.* **12**(10), 1245–1252 (1999).

Wu, C. N., Cheng, Y., Ding, Y. L., Wei, F., and Jin, Y. *Chem. Eng. Sci.* **62**, 4325–4335 (2007).

Xie, C. G., Image reconstruction, *in* "Process Tomography—Principles, Techniques and Applications" (R. A. Williams, and M. S. Beck Eds.), pp. 281–323. Butterworth-Heinemann, Oxford (1995).

Xie, C. G., Huang, S. M., Hoyle, B. S., Thorn, R., Lean, C., Snowden, D., and Beck, M. S. *IEE Proceedings G* **139**, 89–98 (1992).

Yang, W. Q., and Peng, L. *Meas. Sci. Technol.* **14**(1), R1–R13 (2003).

Yang, W. Q., Spink, D. M., York, T. A., and McCann, H. *Meas. Sci. Technol.* **10**, 1065–1069 (1999).

Yasui, G., and Johanson, L. N. *Am. Inst. Chem. Engrs. J.* **4**, 445–452 (1958).

Yates, J. G., and Simons, S. J. R. *Int. J. Multiphase Flow* **20**(Suppl., Annual Reviews in Multiphase Flow 1994), 297–330 (1994).

Yates, J. G., Cheesman, D. J., and Sergeev, Y. A. *Chem. Eng. Sci.* **49**(12), 1885–1895 (1994).

Yin, W. L., and Wang, H. X., Quantification of swirling flow in hydraulic conveying from resistance tomography images, *Instrumentation and Measurement Technology Conference* (Conference Proceedings, Vail, CO, USA), 20–22 May (2003).

Yorkey, T. J., Comparing reconstruction algorithms for electrical impedance tomography," PhD Thesis, University of Wisconsin (1986).

Yorkey, T. J., Webster, J. G., and Tompkins, W. J. *IEEE Trans. Biomed. Eng.* **34**, 843–852 (1987a).

Yorkey, T. J., Webster, J. G., and Tompkins, W. J. *IEEE Trans. Biomed. Eng.* **34**, 898–901 (1987b).

CHAPTER 6

Time-Resolved Laser-Induced Incandescence

Alfred Leipertz and **Roland Sommer**

Contents		
	1. Introduction	224
	2. Measurement Principle	225
	2.1 Energy balance and LII signal	225
	2.2 Determination of primary particle size and its distribution	228
	3. Flame Investigations	236
	4. Technical Applications	237
	4.1 Control of nanoparticle production processes	237
	4.2 Automotive soot investigations	251
	4.3 Particle suspensions	262
	5. Conclusions	265
	Nomenclature	267
	Acknowledgements	267
	References	268

Abstract   Online characterization of nanoscaled particles is an important issue in basic research, e.g., combustion soot formation and oxidation, and in several different technical applications, e.g., in nanoparticle production reactors or in automotive raw exhaust. For the determination of mass concentration and primary particle size, a possible in situ measurement technique is time-resolved laser-induced incandescence (TIRE-LII). The basic principle of this technique is the heating-up of the nanoscaled particles by a high-energetic laser pulse and the subsequent detection and analysis of the spectrally and temporally resolved enhanced thermal

Department of Engineering Thermodynamics and Erlangen Graduate School in Advanced Optical Technologies, University of Erlangen-Nuremberg, Am Weichselgarten 8, 91058 Erlangen, Germany radiation. At later times after the laser pulse, heat conduction to the ambient gas is the dominant heat loss mechanism and so particles with different specific surface area cool down differently. From the temporal signal decay, the size distribution of the primary particles can be derived. Furthermore, the signal maximum is proportional to the mass concentration. Here, besides an introduction to the basic principle of this technique, an overview is given on current technical applications using this measurement technique and on its use in basic combustion research, for nanoparticle characterization, with emphasis on carbonaceous particles. Measurements have been performed at different nanoparticle production reactors on the one hand side and directly in the diesel engine raw exhaust and in ambient air on the other hand. Thereby, the results were compared with established measurement methods such as transmission electron microscopy (TEM), conventional adsorption analysis and gravimetrical techniques. Furthermore, a new approach is shown applying this technique also in liquids for suspended particles.

1. INTRODUCTION

The industrial production of nanoscaled particles has become very important in recent years due to the increased number of application areas of these materials. Exemplarily, carbon blacks are important nanoparticles, which are included as crucial additives in tires, paints and varnishes. One particle property that determines together with other attributes the quality of these products is the primary particle size. To measure and control this quantity during the production process, an appropriate size characterization method is required. For this, online and in situ process monitoring is necessary. Other areas in which the characterization of carbonaceous particles plays an important role are soot measurements in engines, mainly diesel raw exhaust and in ambient air. Unfortunately, current measurement techniques such as transmission electron microscopy (TEM), chemical analysis, dynamic light scattering (DLS), impactors or scanning mobility particle sizers (SMPS) do not allow fast, non-intrusive or selective determination. This is however possible using time-resolved laser-induced incandescence (TIRE-LII).

Laser-induced incandescence (LII) has been proven to be a powerful measurement tool for soot characterization in different flames for many years. First theoretical considerations for soot characterization by laser heating were proposed by Melton (1984), followed by first experimental investigations by Dasch (1984). Basic approaches for the determination of the soot concentration have been carried out by Tait and Greenhalgh (1993); Vander Wal and Wieland (1994) and Vander Wal and Choi (1995).

First quantitative investigations in flames were conducted in the following years by different groups (Shaddix and Smyth, 1996; Mewes and Seitzman, 1997). The determination of the primary particle size was first experimentally conducted by Will et al. (1995, 1998) introducing the TIRE-LII technique and nearly at the same time by Roth and Filippov (1996). In recent years it has been further developed for particle characterization in technically important systems (see, e.g., in Leipertz and Dankers, 2003). Direct accessible parameters are the volume concentration and the primary particle diameter. Meanwhile, it has been extended to evaluate even the size distribution of particles (Lehre et al., 2003; Dankers and Leipertz, 2004). In combination with light scattering measurements, the volume equivalent aggregate particle size and the number of the primary particles per aggregate can also be determined (Will et al., 1996). The potential and features of this technique are presented exemplarily by a few selected measurements, briefly for flame investigations and more extended for the online primary particle size characterization within a carbon black production reactor (Dankers et al., 2003) or the soot analysis in engine exhaust (Schraml et al., 2004). In particular, for engine applications, the method has been implemented to a robust sensor, which allows a reliable soot measurement even for modern low emission vehicles. This sensor has a high temporal resolution of up to 20 Hz and a very low detection limit of just a few $\mu g/m^3$, which qualifies the system also for emission measurement and environmental health control. Moreover, the potential of this measurement technique for particle sizing in suspensions is described. Additional details on the application of this technique can be found in published overview articles (Leipertz et al., 2002; Santoro and Shaddix, 2002; Leipertz and Dankers, 2003). The state of the art in modeling the LII process is described in some detail by Schulz et al., (2006), Michelsen et al. (2007), Michelsen et al. (2008) and in the compilation of several LII-related papers published in a special feature issue in *Applied Physics B*, Vol. 83, 333–485 (2006).

2. MEASUREMENT PRINCIPLE

2.1 Energy balance and LII signal

The basic principle of LII is the rapid heating of nanoparticles often up to their sublimation temperature within a few nanoseconds by means of a short intense laser pulse and the subsequent detection and evaluation of the enhanced thermal radiation. First, the particles are heated up by absorbing the laser radiation, which results in an increased internal energy. Considering carbonaceous particles, their maximum particle temperatures reached are about 4000–4500 K for a short time period of several nanoseconds. The subsequent heat loss is dominated by different mechanisms, mainly sublimation and heat conduction. The contribution of the thermal radiation to the energy loss is small for all times. By setting up the energy balance for one single particle, the resulting differential equation yields

$$C_{abs}E_i - \Lambda (T - T_0) \pi d_p^2 + \frac{\Delta H_S}{M} \frac{dm}{dt}$$
$$- \pi d_p^2 \int \varepsilon(d_p, \lambda) M_\lambda^b(T, \lambda) d\lambda - \frac{\pi d_p^3}{6} \rho c_P \frac{dT}{dt} = 0 \quad (1)$$

The temporally resolved particle temperature can be calculated by solving the differential equation including the laser absorption in the Rayleigh-Regime (absorption efficiency C_{abs}, laser energy E_i), heat conduction (Knudsen number dependent heat transfer coefficient Λ, particle temperature T, temperature T_0 of the surrounding gas), sublimation (sublimation enthalpy ΔH_V, molar mass M) and radiation (emission coefficient ε, spectral energy density M_λ^b, wavelength λ), as well as the change of the internal energy (mass density ρ, specific heat c_P). Thereby, Equation 1 includes the assumption of spherical primary particles (diameter d_p), which have only point contact to other particles inside the aggregates. Applying Planck's radiation law, the detectable signal can be determined (Santoro and Shaddix, 2002).

Accurate modeling is only possible by the consideration of wavelength-dependent optical and temperature-dependent thermodynamic parameters and the correct application of the thermal accommodation coefficient which is dependent on the ambient particle conditions and is described in detail elsewhere (Schulz et al., 2006; Daun et al., 2007). Moreover, Michelsen (2003) suggested the inclusion of a nonthermal photodesorption mechanism for heat and mass loss, the sublimation of multiple cluster species from the surface, and the influence of annealing on absorption, emission, and sublimation. A more general form of the energy equation including in more detail mass transfer processes has been derived recently by Hiers (2008). For practical use, Equation (1) turns out to be of sufficient physical detail.

One attribute of this measurement technique is the non-linear behavior of the maximum peak signal with the excitation laser energy. An initial increase in laser fluence leads to a rapid rise in the peak LII signal, whereby it levels off to a so-called "plateau" region at a laser fluence of about 0.2 J/cm^2 for a laser wavelength of 532 nm, from which no significant signal change is evident (Will et al., 1998). This observation is found in previous studies on aerosols and flames (Snelling et al., 2000; Bladh and Bengtsson, 2004). Noteworthy is the plateau width that is strongly dependent on the laser spatial energy distribution. In the

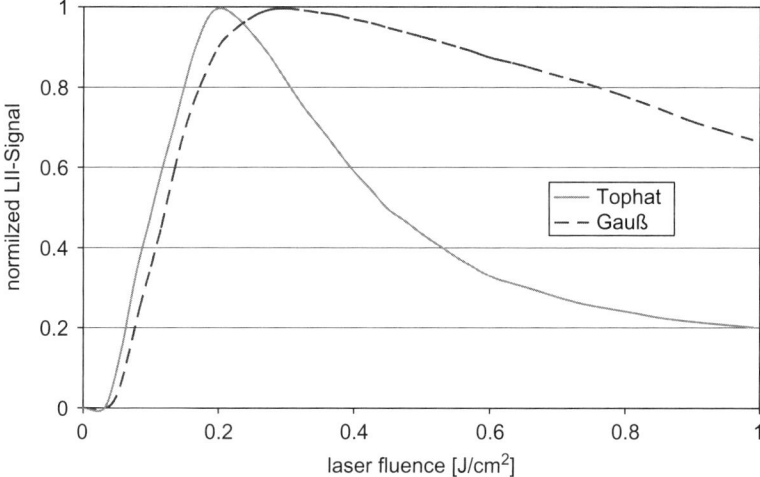

Figure 1 Maximum LII signal dependence on laser fluence.

plateau region, the particles reach their peak temperature, sublimation acting as the limiting factor. At higher laser excitations, the particles shrink due to surface sublimation of carbonaceous matter that, in turn, leads to lower signal intensities. Moreover, due to high laser fluences, morphological changes can also occur possibly modifying the particle properties (Vander Wal et al., 1998; Vander Wal and Choi, 1999). For technical applications, this "plateau" region is of advantage as changes in the laser fluence do not affect the signal behavior. In Figure 1 the plateau shape is shown in dependence of the laser spatial profile type.

Under the assumption of sufficiently high laser excitation energies, the heat loss at peak particle temperature is dominated by sublimation. Hence, the maximal signal is proportional to the particle volume

$$S_{\text{LII}} \propto d_p^{3+(\lambda/0.154)} \tag{2}$$

by using long detection wavelengths λ (Melton, 1984; Will et al., 1995; Mewes and Seitzman, 1997). If quantitative data for the mass concentration are required, it is necessary to perform an appropriate calibration. This can easily be done by a single line-of-sight extinction measurement or comparison to gravimetric methods.

However, recent studies by Reimann et al. (2008) and Bladh et al. (2008) showed that for atmospheric flame conditions the particle size influences the relationship between "prompt" LII signal (maximum LII signal) and the volume concentration to a low degree in the low-fluence regime but has a clear influence in the high-fluence regime. A possible consideration of this is to determine the volume fraction as well as the primary particle size (see Chapter II.B) in parallel.

Even more flexibility of LII can optionally be achieved if it is combined with elastic light scattering, which is discussed in more detail elsewhere (see, e.g., Will et al., 1996). By this also the agglomerate size is accessible. This is, more precisely, an optically equivalent diameter of the agglomerates (radius of gyration), which correlates with the widely determined diffusion diameter of the particles. In this context, it should be pointed out that elastic light scattering in contrast to LII is not selective to the desired fraction of the particles. As also other particle components, e.g., volatiles, contribute to the scattered signal, an appropriate sample conditioning is required for performing combined scattering/LII experiments.

2.2 Determination of primary particle size and its distribution

At later times after the laser pulse heat conduction becomes the dominating heat loss mechanism (Will et al., 1995), and therefore, particles with different specific surface areas cool down differently, which is shown schematically in Figure 2 (Roth and Filippov, 1996; Will et al., 1995, 1998). The local gas temperature adjacent to particles turns out to be the most critical parameter for the accuracy of size determination (Will et al., 1998), which however can be derived from the temperatures of the soot particles themselves (Schraml et al., 2000).

For the determination and evaluation of the LII signal and the decay time, two strategies exist. The complete temporal LII signal decay can be

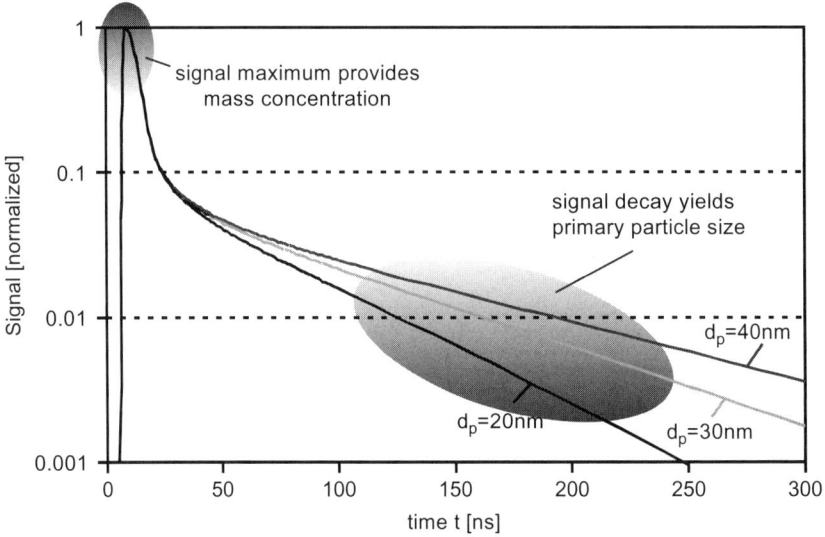

Figure 2 Temporal curve of the LII signal for different primary particle diameters.

detected by using a fast photomultiplier tube (PMT). This procedure, however, allows only the acquisition of pointwise information but this is no drawback in systems with spatially distributed particles owing all the same attributes. If this is not the case, two-dimensionally resolved measurements can be performed providing information over an extended part of the combustion field simultaneously. For this purpose ICCD cameras with short gating times are used. Since there exists a minimum temporal interval between two recordings of a CCD camera, it is only possible to detect the LII signal at certain times after the laser pulse. From the ratio of the signals at different times, the signal decay time can be calculated and processed for each pixel as for the pointwise case. This strategy has been used for the measurements in the flame study section (Chapter III).

For all the other measurements presented in this overview, the temporally resolved LII signals were detected with a fast PMT which guarantees a good signal-to-noise ratio. For times after the laser pulse when heat conduction is the most dominant process for energy release from the particles, the LII signal decay of a monodisperse class of particles is nearly a single exponential. As a first approach, it is reasonable to evaluate the mean primary particle size of this monodisperse particle collective in the following way. From the experimental LII signal curve, a signal decay time τ is determined by an exponential fit in a time interval in which heat conduction dominates particle cooling (Figure 3). This τ is compared to a numerically calculated signal decay time (based on the modeling of the power balance) taking into account the surrounding gas temperature T_0 and assuming a monodisperse particle size distribution with diameter $d_{P,mono}$. Hence, the simultaneous determination of the ambient gas temperature is necessary. Moreover, this means that $d_{P,mono}$ is slightly shifted toward larger particles

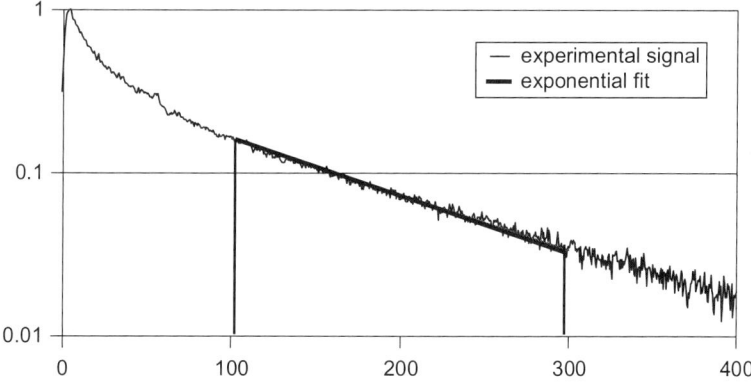

Figure 3 Experimental LII signal with exponential fit.

compared with the medium size $d_{P,med}$ of the real particle size distribution, which is more polydisperse in nature, because the LII signal scales nearly with d_p^3. Thus, TIRE-LII yields this mean primary particle diameter $d_{p,mono}$ without further calibration by comparison of the experimental signal behavior with theoretical predictions since the signal decay can be evaluated independently of the total signal strength.

The accuracy of the technique has first been tested with carbon black particles of the known size as distributed by Degussa AG which also has been used for the investigation of in liquid suspended particles (see Chapter IV.C). In Figure 4, the results are displayed with particles in the size range between 10 and 95 nm (Dankers et al., 2002). Two of the carbon blacks with a primary particle size of 25 nm (Degussa Printex U and Printex 55) had a different aggregate size formed by a different number of primary particles. As the LII measured primary particle size was the same for both aggregates, an influence of the aggregate size on the LII results can be excluded which is true for aggregates with fractal dimensions between 1.5 and 1.9 normally given for combustion-generated soot. This also has been found during carbon black production (see Chapter IV.A.2, Figure 19) and for the investigation of in liquid suspended particles (Chapter IV.C, Figure 39).

A slightly different approach has been introduced by Snelling et al. (2001) and Lehre et al. (2005) which is based on the simultaneous detection of the TIRE-LII signal at two different wavelengths (two-color-LII). By applying Planck's law, the particle ensemble temperature can be calculated from the measured ratio of the TIRE-LII signals at the two wavelengths selected. The major advantage of this technique is that the particle temperatures are experimentally determined independent of the

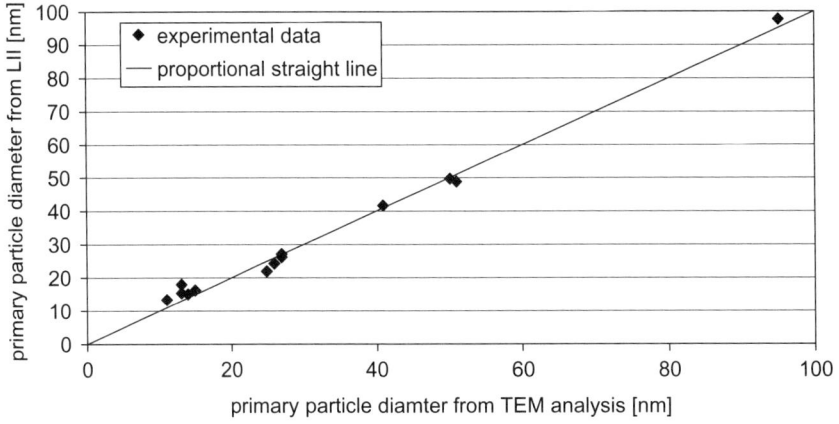

Figure 4 Comparison of LII with TEM.

laser absorption process during the laser pulse which enters most of the models used for simulating the LII signals.

The approach to evaluate the mean primary particle size of a monodisperse particle collective is sufficient if the size distribution is narrow in technical systems, which is typically not always the case. An optional extension of the mean primary particle size determination yields its distribution function. This method relies on the fact that the superposition of several monodisperse signal curves gives, in contrast to monodispersial decay functions, a total signal with a non-exponential shape. In principle, any arbitrary distribution function can be recovered by employing an inverted Laplace algorithm. As this is a very time-consuming issue and is very sensitive to statistical errors and noise on the signal course, a more robust method has to be chosen. One favorable way assumes the presence of any fixed distribution function, which is given by its median diameter and its distribution width. In this case, the shape of the distribution function is assumed to be well characterized by these two parameters. A broadly used and well justified assumption for this distribution is given by a logarithmic normal function

$$P(d_p) = \frac{1}{\sqrt{2\pi}d_p\sigma} \exp\left\{-\frac{[\ln(d_p) - \ln(d_{p,med})]^2}{2\sigma^2}\right\} \quad (3)$$

which is typical for combustion processes (Dankers and Leipertz, 2004; Liu et al., 2006a–c). Here, $d_{p,med}$ denotes the count median diameter and $\sigma = \ln(\sigma_g)$ is the width, σ_g being the geometric standard deviation. On the basis of this assumption, higher moments of the particle size distributions can be determined.

Bockhorn et al. (2002) have introduced an approach, by which a calculated signal course is fitted to the whole experimental signal decay under the variation of different distribution parameters such as the width or the mean primary particle diameter. A major drawback of this method is the relative high computing effort for the evaluation which makes the application for online measurements questionable. Roth and Filippov (1996) have introduced an inversion algorithm that also suffers from a complicated and time-consuming calculation procedure and does not lead to unique solutions.

To overcome these deficiencies, a simple online approach has been developed that relies on the fact that the ratio of the contribution to the LII signal of different size classes changes with time after the induced laser pulse. Smaller particles cool down faster, and therefore, a broad size distribution leads to a deceleration of the signal decay as the long-lasting signal of bigger particles become more important at later times. Smaller particles show a faster signal decay and have therefore a constantly decreasing influence on the total signal of the particle collective. This

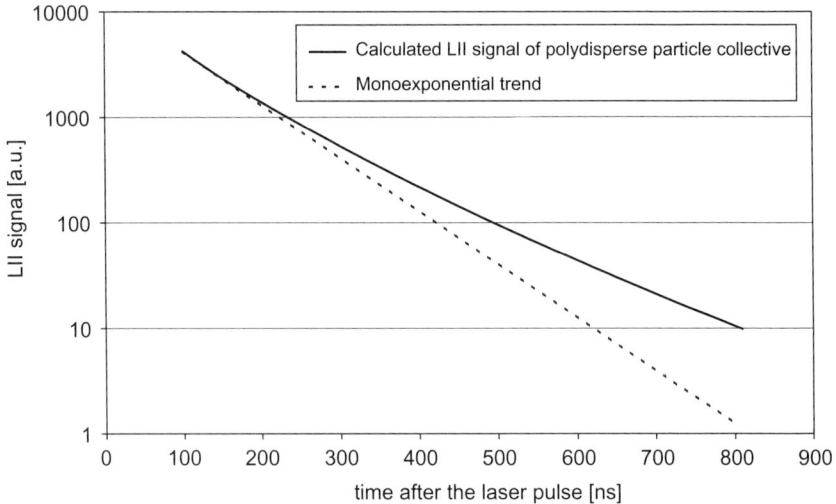

Figure 5 Deviation of the calculated LII signal decay of a polydisperse particle collective (log-normal, $d_{p,med} = 14$ nm, $\sigma = 0.4$) from the monoexponential trend for times after the laser pulse when heat conduction dominates the energy loss (Dankers and Leipertz, 2004).

results in a deviation from an exponential decay, i.e., the signal decay time increases with time (see Figure 5).

For the reconstruction of the size distribution, the theoretical signal course of a particle collective is calculated directly by a weighted summation of LII signals from monodisperse size classes, which are computed from the numerical solution of the power balance. The count median particle diameter $d_{p,med}$, the distribution width σ and the ambient temperature T_0 are input parameter for this calculation. Fitting of this theoretical signal of a size-distributed particle ensemble in two different time intervals after the laser pulse $((\Delta t)_1, (\Delta t)_2)$ provides two characteristic signal decay times $(\tau_1 = \tau((\Delta t)_1), \tau_2 = \tau((\Delta t)_2))$.

The calculation of two signal decay times τ_1 and τ_2 can be done under variation of mean particle size, distribution width and ambient temperature yielding the functions $d_{p,med} = f(T_0, \tau_1, \tau_2)$ and $\sigma = f(T_0, \tau_1, \tau_2)$, which are shown exemplarily for an ambient temperature of 1000 K in Figure 6 (Dankers and Leipertz, 2004). It is clearly observable that τ_1 and τ_2 are correlated with σ and $d_{p,med}$ unambiguously. Thus, with the assumption of a certain form of the size distribution, e.g., the log-normal distribution in Equation (3), $d_{p,med}$ and σ can be determined clearly from the experimental curves. To do this the corresponding signal decay times are derived from exponential fits of the experimental data in the selected time intervals.

In Figure 7 this is exemplarily shown for a curve detected in a carbon black reactor with two fits in the intervals 100–300 ns and 400–800 ns after

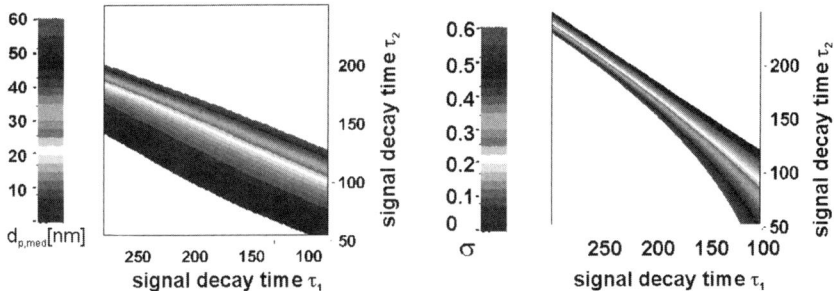

Figure 6 Representation of $d_{p,med} = f(T_U, \tau_1, \tau_2)$ and $\sigma = f(T_U, \tau_1, \tau_2)$ for an ambient temperature of 1000 K (Dankers and Leipertz, 2004).

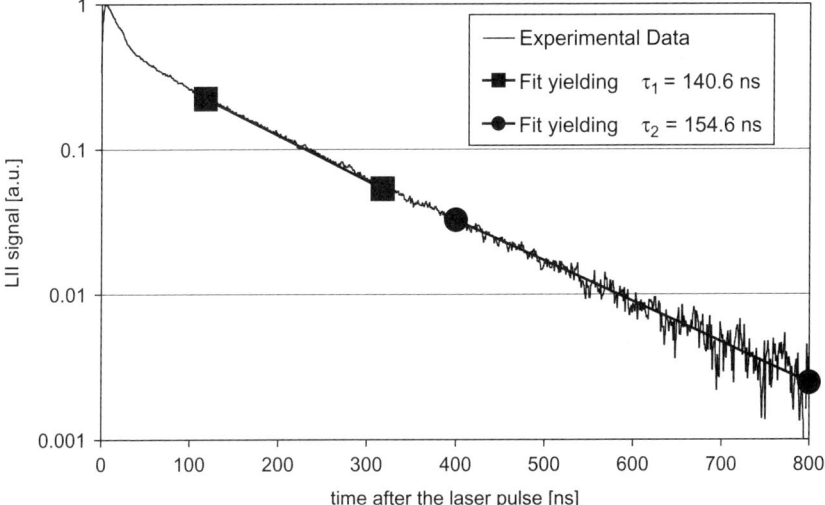

Figure 7 Experimental LII curve with two exponential fits for the reconstruction of primary particle size distribution (Dankers and Leipertz, 2004).

the laser pulse yielding $\tau_1 = 140.6$ ns and $\tau_2 = 154.6$ ns, respectively. The description of the corresponding experimental setup and the performance of the measurements can be found elsewhere (Leipertz and Dankers, 2003). The model signal is not needed in its analytical form; a pre-calculated data base contains only the functions $d_{p,med} = f(\tau_1, \tau_2)$ and $\sigma = f(\tau_1, \tau_2)$ for different ambient temperatures and the fit effort is restricted to two exponential decays.

For the experimental curve shown in Figure 7, under assumption of a long-normal size distribution, the signal decay times yield $d_{p,med} = 32$ nm and $\sigma = 0.11$ which corresponds to a distribution as depicted in Figure 8.

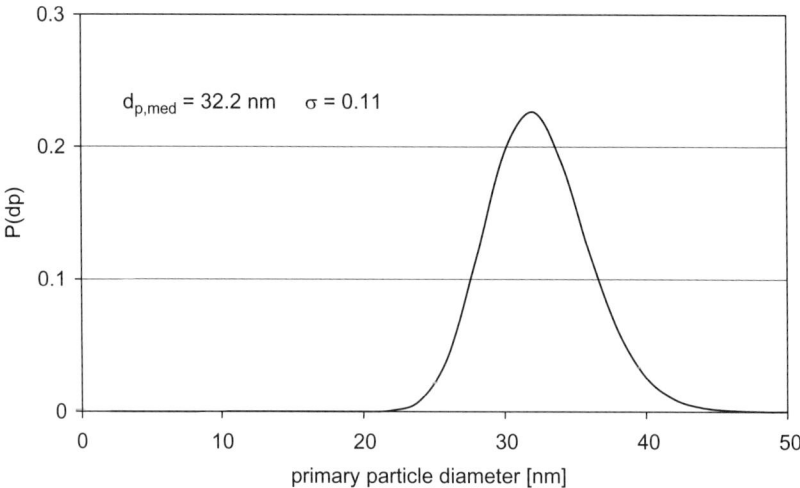

Figure 8 Reconstructed primary particle size distribution by evaluation of the experimental curve shown in Figure 7 (Dankers and Leipertz, 2004).

Assuming a monodisperse particle ensemble, the evaluation of the experimental curve in Figure 7 by only one exponential fit results in the mean monodisperse particle diameter $d_{p,mono} = 34$ nm. For this measurement, it differs only slightly from $d_{p,med}$, since the particle size distribution is rather narrow. Therefore, in this case $d_{p,mono}$ already represents a meaningful and reasonable result regarding primary particle sizing, even though the provision of information about higher moments of the particle size distribution is also valuable.

The choice of suitable evaluation intervals is determined by a compromise between signal strength and significance of the difference in signal decay times. The first interval should begin when heat conduction is the most important energy loss mechanism, i.e., after about 100 ns, and should last for 150–200 ns due to statistical reasons. For the choice of the second interval, the temporal development of the signal decay time must be considered, as it is shown in Figure 9 for the LII signal of two differently size distributed particle ensembles. It is observable that the significance of the differences in the signal decay times for different particle size distributions increases with the time after the laser pulse, so that a later second evaluation interval provides clearer results. However, the signal strength decreases with time, thus signal-to-noise ratio becomes worse. This leads to an increased uncertainty of the determined decay times. So, a suitable interval is to be specified in dependence on the experimental parameters. For high ambient temperatures and/or large particles, the evaluation can be accomplished 600–800 ns after the laser pulse. For other boundary conditions, a second

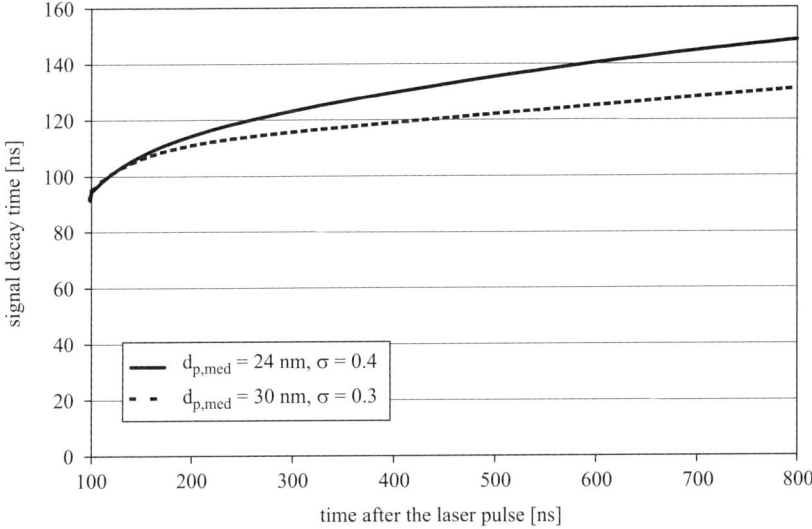

Figure 9 Calculated temporal evolution of the signal decay times for differently distributed particle ensembles (Dankers and Leipertz, 2004).

evaluation interval may cover 400–600 ns or even earlier times. The two evaluation intervals can in principle overlap.

An assessment of the sensitivity of the method is given in Table 1. The effect of a given deviation of the size distribution from starting values on signal decay times in different time intervals after the laser pulse is calculated. A change of $d_{p,med}$ mainly affects the signal decay time τ_1 and results only in a small change of the signal decay time τ_2 in later evaluation intervals. A change of the standard deviation of the size distribution σ, however, has greater influence on τ_2 than on τ_1.

A possible problem is the influence of the aggregate size on the signal decay, when primary particles are packed together to dense structures, which strongly affects the particle cooling (Liu et al., 2006a–c). As shown before (Figure 4), this is typically not the case for aggregates with a fractal dimension between 1.5 and 1.9. Therefore in the measurements presented in the following examples (Chapters III and IV), no shielding effects inside the aggregates are considered.

In principle, however, the effective surface area for heat transfer from the single monomer is decreased by the contact area between the particles. The fractal dimension of the aggregates varies in different systems and depends on the evolution process. For dense aggregates, heat transfer rates are decreased due to the reduced heat exchange area compared to primary particles only connected by point contact in a chain like structure (Liu et al., 2006a–c). This leads to an overestimation of

Table 1 Sensitivity of signal decay times on distribution parameters (Dankers and Leipertz, 2004)

Change of distribution parameter	Start values		Change of signal decay time (%)		
	$d_{p,med}$	σ	τ_1(100–300 ns)	τ_2(400–600 ns)	τ_2(600–800 ns)
$d_{p,med} \pm 5\%$	10	0.2	±5.2	±3.9	±3.6
	20	0.34	±4.1	±3.5	±3.0
	10	0.2	±3.0	±5.7	±6.5
$\sigma \pm 5\%$	20	0.34	±4.4	±6.0	±6.0

expected signal decay rates which results in an overestimation of the primary particle diameters. Basically, a dense aggregate structure can also lead to an alteration of optical properties which can be in principle considered by applying a Rayleigh–Debye–Gans (RDG) approach for the absorption and emission.

3. FLAME INVESTIGATIONS

Fundamental investigations of soot growth and oxidation were performed by many researchers in both diffusion and premixed flames with ethene and methane as fuels (see, e.g., Santoro and Shaddix, 2002; Schulz et al., 2006).

In exemplarily flame measurements conducted at the LTT-Erlangen (Will et al., 1996), flame temperatures were determined by emission spectroscopy or coherent anti-Stokes Raman scattering (CARS) thermometry depending on the maximum soot concentration. Typical temperatures are in the range of 1800 K in the middle of the flames and up to 2100 K in the outer regions where the reactions take place. A typical measurement setup for two-dimensional LII investigations is shown in Figure 10.

In the different flames, the measured concentrations cover a range from 1×10^{-7} to 6×10^{-6} and primary particle sizes range from 2 to 60 nm. Higher soot concentrations and larger particles were measured in the ethene diffusion flame. The resulting maximum number concentration of primary particles, however, is about one magnitude larger in the premixed methane flame. Distributions of the different soot quantities are shown exemplarily for the ethene flame in Figure 11. There is an annular structure in the soot volume concentration observable with maximum values in the outer regions. The concentration decreases in the upper part of the flame where mean primary particle sizes also decrease due to the increasing oxidation of the soot.

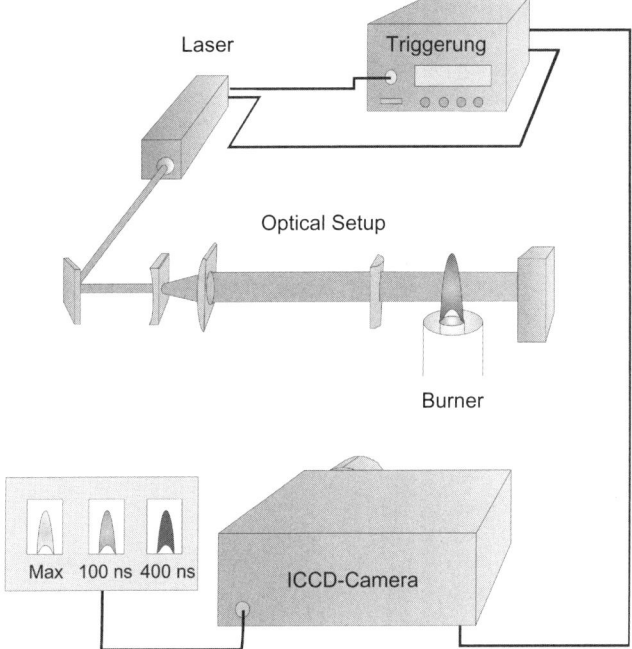

Figure 10 LII measurement setup for two dimensional flame investigations.

4. TECHNICAL APPLICATIONS

In this chapter technical applications are described which are of utmost interest for mechanical and chemical engineers working with particles in the nano size regime. These are the control of production processes of carbon blacks and of non-carbonous and metallic particles, the characterization of soot in automotive applications and the investigation of in liquids suspended particles.

4.1 Control of nanoparticle production processes

Characterization of carbon black primary particle size during the production process was realized at a furnace black production and a research plasma reactor (Dankers et al., 2003; Sommer et al., 2004, 2005).

In Figure 12 a typical LII measurement setup for such applications is shown, whereby here the detection of the enhanced radiation is in perpendicular direction relative to the incident laser beam.

4.1.1 Research plasma reactor

At first the spatially resolved investigation of carbon black primary particle sizes was carried out by realizing a LII setup in backscattering

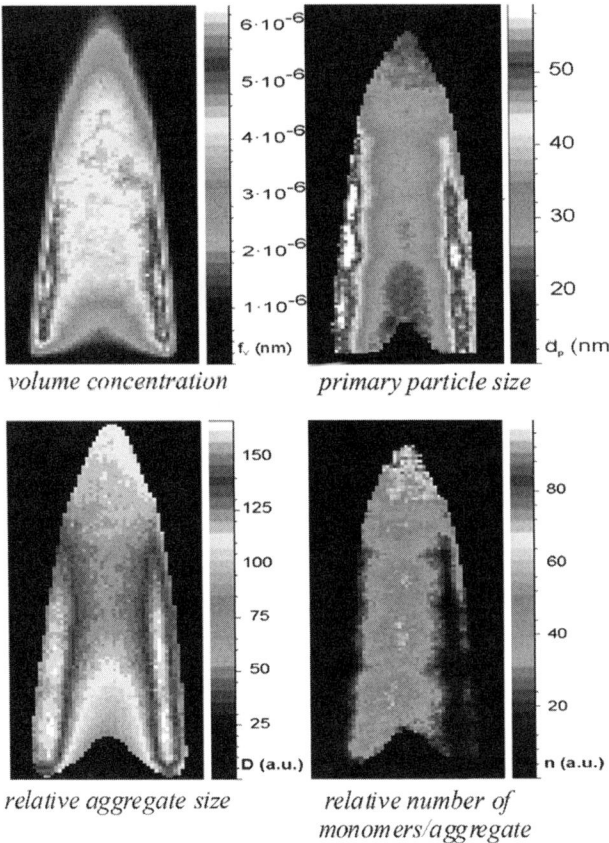

Figure 11 Characteristic properties of soot in a laminar diffusion flame (Will et al., 1996).

geometry within the research plasma reactor (Sommer et al., 2004, 2005). The basic principle of the plasma reactor is the cracking of precursor gas by means of electronically generated plasma and the formation of primary particles out of free molecules.

Unfortunately, at most of technical reactors, only one optical access is available, so LII setups in backscattering geometry have to be realized. Thereby, the carbon black particles are irradiated by a frequency-doubled Nd-YAG laser beam being led into the measurement volume by a mirror and a beam splitter, which are both highly reflective for 532 nm and transmittive for other wavelengths. The LII signal is detected within an appropriately selected spectral range by a fast PMT, fed to an oscilloscope, processed and evaluated by a computer. In the regarded plasma reactor optical access was possible at three different downstream positions (see Figure 13). To obtain an appropriate radial resolution, glass

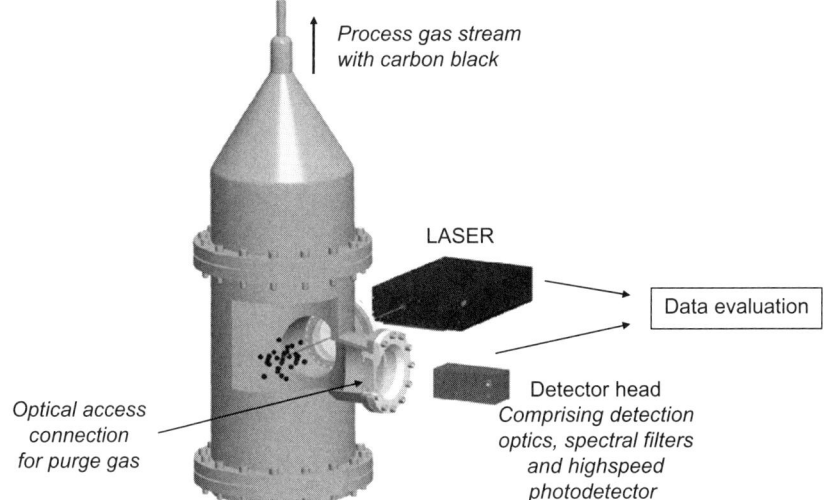

Figure 12 LII measurement setup at a carbon black reactor.

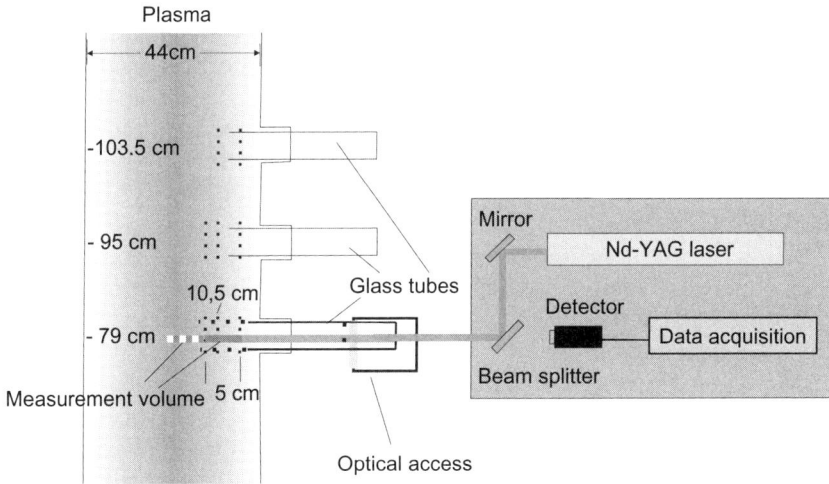

Figure 13 Backscattering configuration at a research plasma reactor (Sommer et al., 2004).

tubes were pushed into the reactor and kept free of carbon particles by means of purge gas. Absorption of the laser beam is so strong that the laser energy is sufficient to heat up the particles only at a short distance after entering the reactor, if the carbon black concentration is high enough. Thus, the outside edge of the LII measurement volume is determined by the end of the glass tubes, as the laser beam cannot irradiate particles until entering the reactor. However, in the course of the measurements it turned out that the carbon black concentrations were lower than expected, and absorption of the laser beam within a short distance (<1 cm) was not given. The laser irradiance was sufficient to induce incandescence over a long path within the reactor. Thus, the local allocation of the measuring volume became difficult as the signals are integrated over a certain radial distance. Furthermore, additional measurements with a 90°-setup configuration were carried out in order to compare these results with the results of the backscattering geometry. Thereby, the LII signal is detected through a second optical window perpendicular to the laser beam which could only be realized for the lowest local position inside the plasma reactor. The temperature distribution in the reactor was measured with thermocouples which is reasonable for the relatively low carbon black concentrations. The radial temperature gradient is steeper than the axial one which particularly holds in the upper part of the reactor.

For all examined reactor adjustments within the optically accessible range, temperatures do not exceed 1000°C, i.e., the particle formation is expected to be terminated. As the temperature is an important input parameter for evaluating the primary particle size, in case of the backscattering geometry the diameter can only be determined under consideration of the temperature range in the respective measurement downstream position. Nevertheless, these primary particle sizes reveal a good correlation with the results obtained in a perpendicular setup configuration and with calculated particle sizes based on the surface area from chemical analysis. Determinations in different downstream positions provided identical particle sizes for each reactor setting (Figure 14) which is in good agreement with the results of Brunauer-Emmett-Teller (BET) analysis (Table 2) (which is a standard evaluation technique for carbon black size determination) and with the expectation that primary particle formation has already been terminated.

As stated above (see Chapter II.B), LII signals also contain information about the size distribution. To compare the influence of different plasma powers on primary particle diameters, different ways of size evaluation have been accomplished. It could be shown by assuming a monodisperse distribution that the mean primary particle diameter is 31 nm for 30 kW and 33 nm for 70 kW. In contrast, under the assumption of a log-normal distribution and by applying the two-decay time evaluation, the determination yields a different result which can be seen in Figure 15. Size distributions with median sizes of 17 nm and 28 nm and standard deviations of 0.39 and 0.18 for 30 kW and 70 kW were observed, respectively. This indicates that in practical production systems, the evaluation of a mondisperse distribution is not sufficient. Unfortunately, the reconstruction of particle size distributions is relatively sensitive on

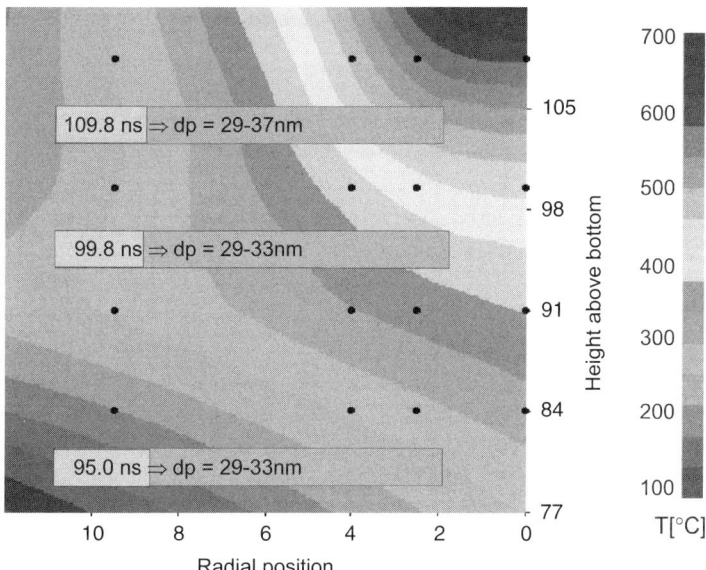

Figure 14 Temperature and calculated size distribution within the plasma reactor (Sommer et al., 2004).

Table 2 Primary particle size d_p derived from LII measurements and BET analysis (Sommer et al., 2004)

d_p (nm) 90°-Setup	d_p (nm) Backscattering geometry	BET (nm)
—	29–35	31
29	29–32	31
33	30–34	33

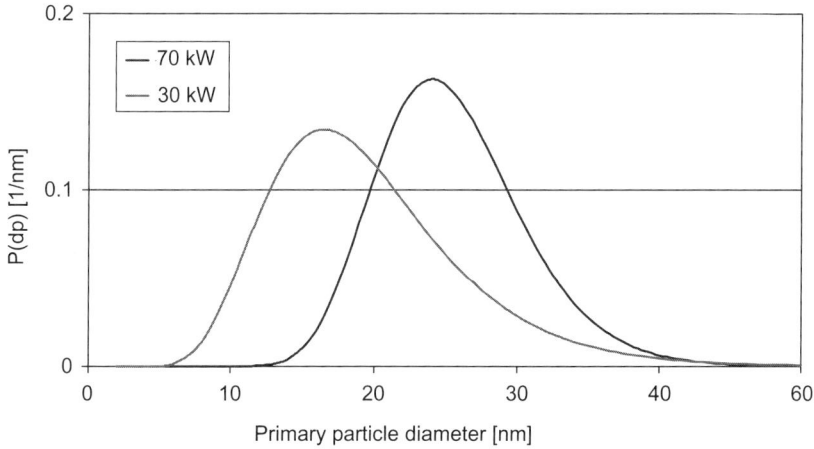

Figure 15 Primary particle size distributions for different plasma power (Sommer et al., 2004) (see Plate 18 in Color Plate Section at the end of this book).

the signal behavior, and therefore, a good signal-to-noise ratio is required.

4.1.2 Furnace black reactor

A similar experimental setup was used to perform online size characterization during the production process within a furnace black reactor (Dankers et al., 2003). During the first series of measurements at a production reactor, it was examined whether phased variations of the operating parameters of the reactor and the resulting changes of the particle size can be observed with the LII technique. It is shown (Figure 16) that the seven different reactor settings can be differentiated clearly by LII, and changes in the primary particle size are detected. The statistical fluctuations during one operation condition are comparatively small.

During the individual phases, carbon black was sampled in order to determine the iodine number. The iodine number results from an adsorption measurement and represents a measure for the specific surface which correlates with the primary particle diameter. Remarkably is that micro porosities of the surface of the particles influence the iodine number. For the comparison of the results from the laboratory analysis and from LII for each phase (duration approximately 20 min), an averaged diameter was determined from the LII signal decay times. Figure 17 shows the outstanding correlation of the results of the two methods for the reactor settings from Figure 16. The iodine number can give only an averaged information over one time period whereas LII

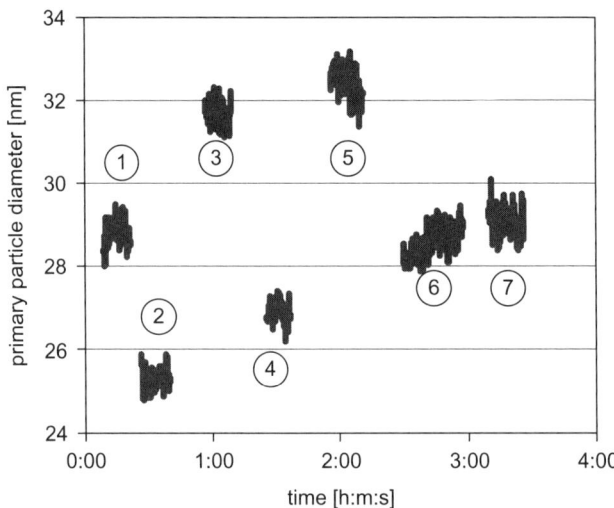

Figure 16 Primary particle size d_p from LII for seven different reactor settings (Dankers et al., 2003).

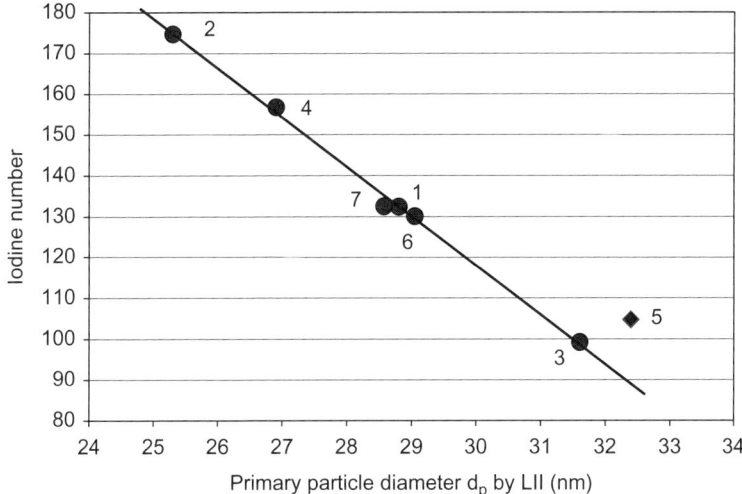

Figure 17 Comparison of the iodine numbers and the averaged d_p from LII with indication of the reactor phase (Dankers et al., 2003).

reacts practically immediately to changes of the measured variable. This highly time-resolved information, however, can be used only with simultaneous and similarly fast temperature measurement. In the measurements described here, the primary particle diameters were computed with the temperature averaged during one period.

Only the results during phase 5 do not fit into the correlation. This is probably caused by the fact that during this phase another position for the water quench to terminate the reaction in the production process had been selected, which affects also the surface porosity apart from the particle diameter. If one assumes that small holes in the surface do not affect the energy loss of the particles after the laser pulse crucially, because the thermal conduction is retarded there, LII determines an enveloping primary particle diameter independently of porosity. The iodine number, however, rises with porosity.

Only then a correlation of the LII results with the iodine number is to be expected if different specific surfaces are caused by variations in the primary particle size and not in porosity. The assumption that LII determines a measure for the enveloping surface without consideration of small porosities has been examined in further measurements at a test reactor of the Degussa AG with finer parameter variations. The aim was to manufacture carbon blacks of same primary particle size, but different surface structure.

Hereby, it was important that the reactor ran stationary. This is contrary to the measurements described before, where very strong parameter variations had been made and therefore small fluctuations of reactor stability had no significant influence. Now a reactor setting was maintained over several hours, in order to achieve stable operating conditions. Samples were taken for the determination of the iodine number (surface inclusive porosity), the Hexadecyltrimethylammoniumbromid (CTAB) value (represents particle size without consideration of small porosities) and the Dibutylphthalat (DBP) value (measure for the aggregate structure). An exact allocation of the LII results to the laboratory values determined from the samples makes an exact determination of the time interval necessary which the carbon black particles need to travel the way from the LII measurement location directly in the reactor to that place, where the sample is taken, i.e., at the end of the reactor. Since sampling of one probe takes several minutes, the LII signal decay time must be averaged during an appropriate time period. An average period of approximately 20 min that particles spend in the line had been determined already in former times empirically. Therefore, for the comparison of LII with the laboratory analyses, the average value of the LII results 20–15 min before sampling was calculated. The results of the variation of porosity are depicted in Table 3.

The first change leads to a reduction of the iodine number of about 10%, whereby the reduction of the CTAB number is smaller, which points out an increase of the primary particle size with change of porosity. In the next phase obviously larger primary particles (CTAB number decreases) with a more porous surface (iodine number remains constant) are produced. The last variation results in an increase of porosity with nearly constant CTAB number. The LII results correlate clearly with the CTAB number.

In a further series of measurements, the independence of the LII results from the aggregate structure, found in preliminary investigations in a measuring chamber (Chapter II.B), was checked. Differently aggregated particles of similar primary particle size were examined.

The Figures 18 and 19 contain the results of all examined operating points of both series of measurements. The primary particle size from LII does not show systematic dependence on the DBP value which again is a standard evaluation technique in carbon black production (Figure 18).

Table 3 Results of the measurements regarding the variation of porosity (Dankers et al., 2003)

Phase	Iodine number	CTAB value	Primary particle size by LII [nm]
1	206	157	27.0
2	179	155	27.9
3	179	148	29.9
4	190	149	29.5

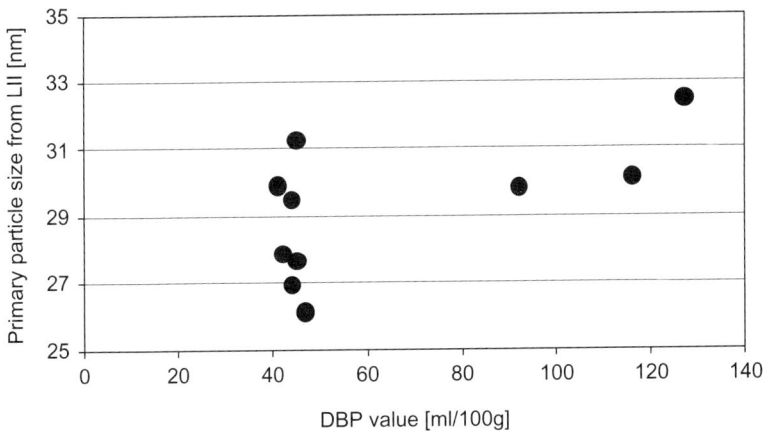

Figure 18 Primary particle size from LII as a function of the DBP value (Dankers et al., 2003).

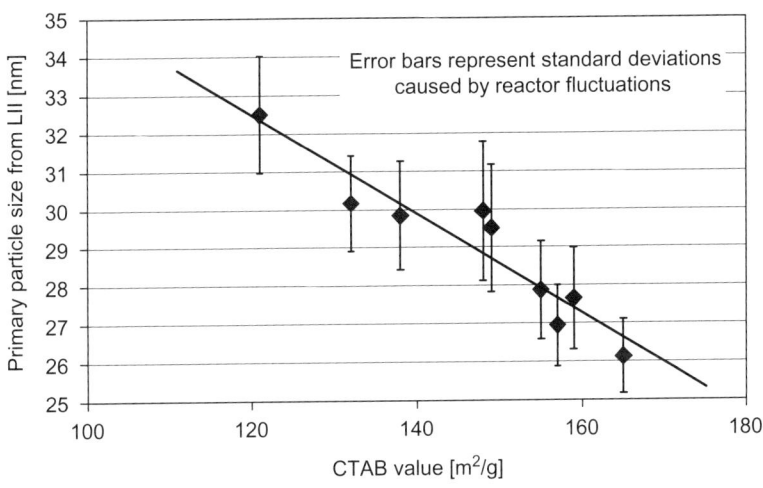

Figure 19 Correlation of primary particle size determined by LII and CTAB value (Dankers et al., 2003).

In Figure 19, however, a clear correlation of the primary particle diameter determined with LII and the CTAB number — the specific surface without consideration of micropores — is observable. The error bars result from the standard deviations of the average values.

It is to be noted that the size of the standard deviation is determined by fluctuations of the particle size, i.e., among other things by reactor fluctuations, and is not a characteristic of the measuring technique as it was already shown in laboratory tests at stationary objects free of doubts.

4.1.3 Metal and metal oxides production reactors

First investigations of metals and metal oxides indicated the principle applicability of LII also for these particles (Filippov et al., 1999; Vander Wal et al., 1999; Kock et al., 2005). Here problems may arise from the particle properties entering the energy balance, e.g., the complex refraction index, which are often unknown. In order to overcome these dependences, a new approach has been introduced by Snelling et al. (2002) and Lehre et al. (2005) using two different wavelengths (two-color-LII) which has already been described in Chapter II.A. Thereby, the temperature signal can be directly obtained from the fraction of both wavelength signals. For particle systems with unknown optical properties, this is an advantageous method for LII particle characterization.

4.1.3.1 Laser vaporization reactor.

At LTT-Erlangen, first investigations with other nanomaterials have been carried out in cooperation with the group of Staupendahl at the University of Jena in a laser vaporization reactor (LVR) (Staupendahl, 2003). The basic principle of this reactor is dispersing raw material by a fountain into the CO_2 laser beams, in which particles are vaporized and nanoparticles are formed by the subsequent condensation. As the LII measurement volume was located slightly above the vaporization zone, it was not possible to prevent coarse structures of material to occur inside the measurement volume (Figure 20).

Thereby, LII signals could be detected for ZrO_2 and TiO_2 by varying laser excitation wavelengths from 532 nm to 355 nm. It turned out that UV excitation yields a clearly improved LII signal for TiO_2 (Figure 21).

Thus, a quantitative evaluation of the particle size was impossible at that time, as coarse pieces had been constantly contributing to the detected LII signal. This is confirmed by the fact that the detected signal decay can be explained very well under the assumption of the superposition of two signal contributions, i.e., a fast dropping signal from nanoparticles and a long-lasting signal of coarse structures of the raw material (Figure 22).

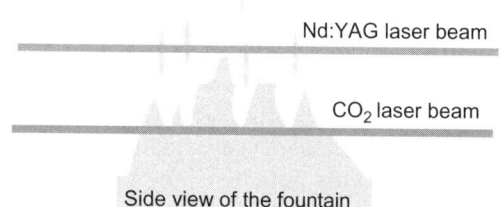

Figure 20 Side view of the dispersing fountain in the LVR (Sommer et al., 2004).

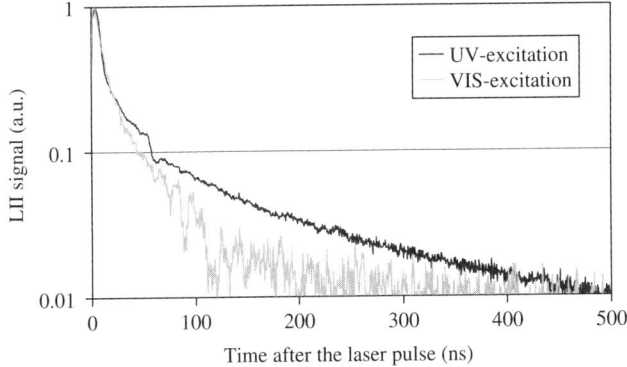

Figure 21 Typical LII courses of TiO_2 with different laser excitation wavelengths (Leipertz and Dankers, 2003).

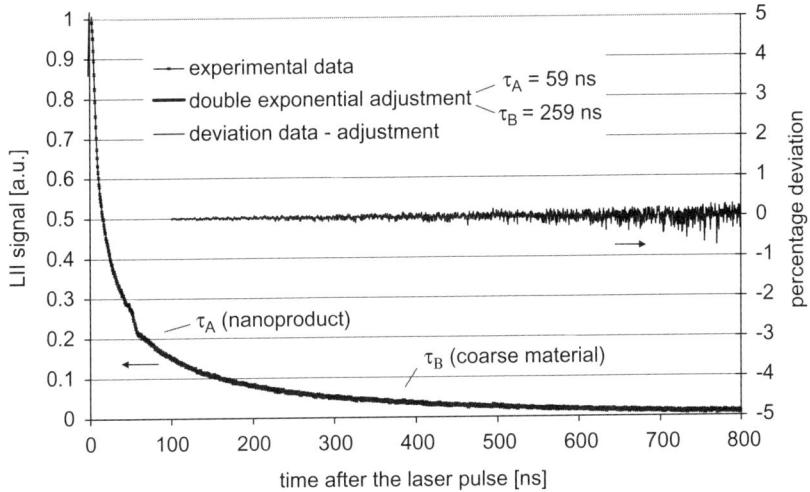

Figure 22 LII signal with double exponential adjustment (Sommer et al., 2005).

The changing mass ratios, however, do not allow a separation of the signals. To prevent these disturbances further investigations were carried out in backscattering geometry at a different LVR without a fountain (Figure 23). LII signals could be detected besides the already mentioned materials also for Fe_2O_3 and MnO_2. For these the influence of CO_2 laser power and pulse length could be examined. The influence of CO_2 laser power on LII signals of Fe_2O_3 is exemplarily shown for various laser powers in Figure 24. It could be shown that the signal maximum which is proportional to the mass concentration increases with rising CO_2 laser power, whereas the signal decay time decreases. Thus, higher concentrations of smaller primary particles were induced by high laser power.

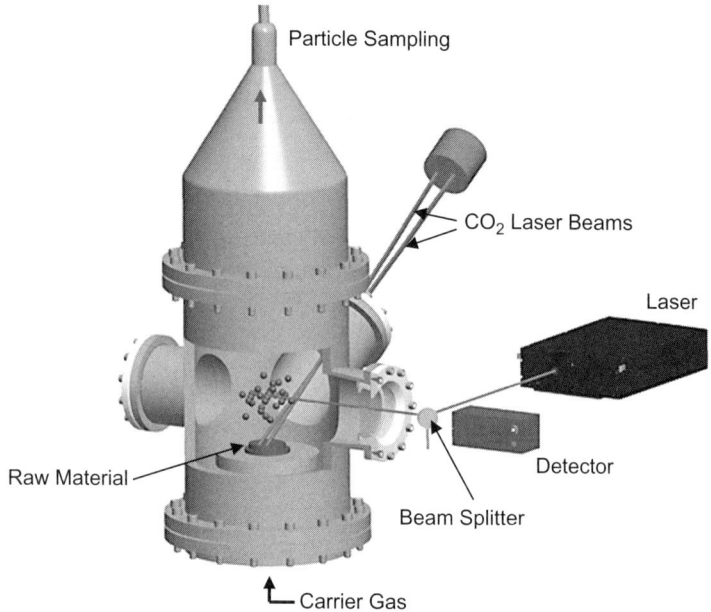

Figure 23 LII setup at the LVR.

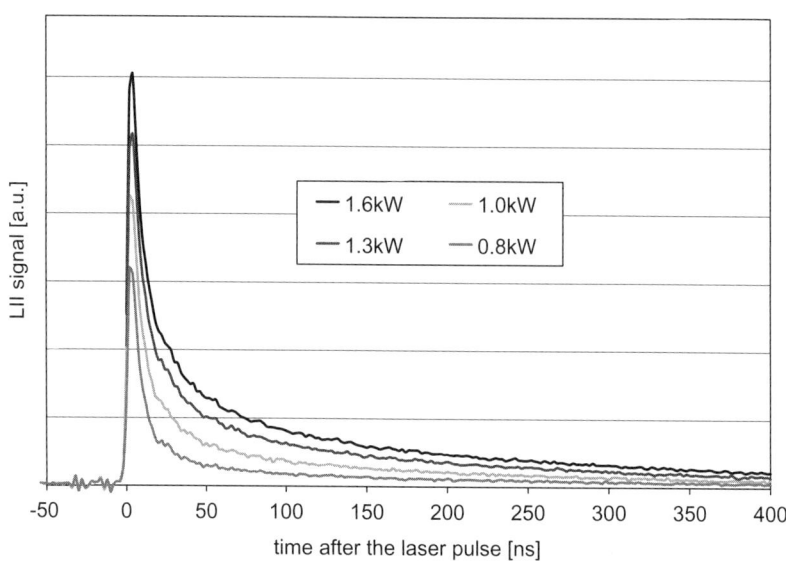

Figure 24 Influence of various CO_2 laser power on LII signals of Fe_2O_3 (Sommer et al., 2005).

Contrary to this only an increase of mass concentration but no alteration of primary particle size with longer pulse length could be observed. A similar behavior has been found for MnO_2. Although not real quantitative, the obtained information provides clear directions for further improvement of the production process.

4.1.3.2 Hot wall reactor. Another interesting and trendsetting investigation has currently been carried out in a hot wall reactor. Here, catalytic active iron particles are produced by thermal-induced metal-organic chemical vapor deposition (MOCVD) acting as formation basis for the growth of carbon nanotubes. In this case, a substrate (Al_2O_3 powder) was calcined over 2 hours at 400°C in a hot wall reactor by means of a fluidized bed under helium atmosphere (flow velocity: 265 ml/min). The actual production of the iron nanoparticles takes place by the decomposition of the deployed precursor from type [(arene)(diene)Fe(0)] at temperature of 150°C on the substrate surface (Michkova et al., 2006). Thereby, the precursor was vaporized at 80°C in advance. To monitor the growth of the iron particles, the LII measurement technique was applied at this reactor. A typical Fe particle deposited on the Al_2O_3 surface is shown in Figure 25.

Investigations were carried out by focusing the laser beam of a frequency doubled Nd:YAG laser in the upper part of the fluidized bed by means of a wavelength selective mirror at 532 nm which is described in more detail elsewhere (Leipertz et al., 2008).

In order to prevent vaporization effects, a low fluence approach was chosen. To consider the growth of the particles, the exponential signal decay time was determined. For this, the measurements were carried out

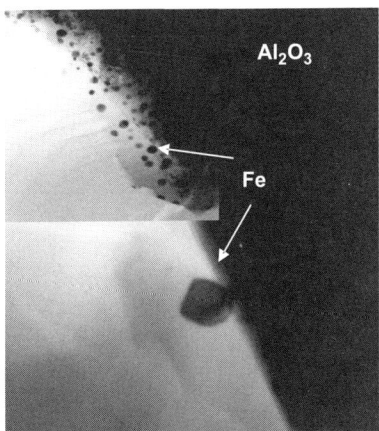

Figure 25 TEM: Fe on substrate (Leipertz et al., 2008).

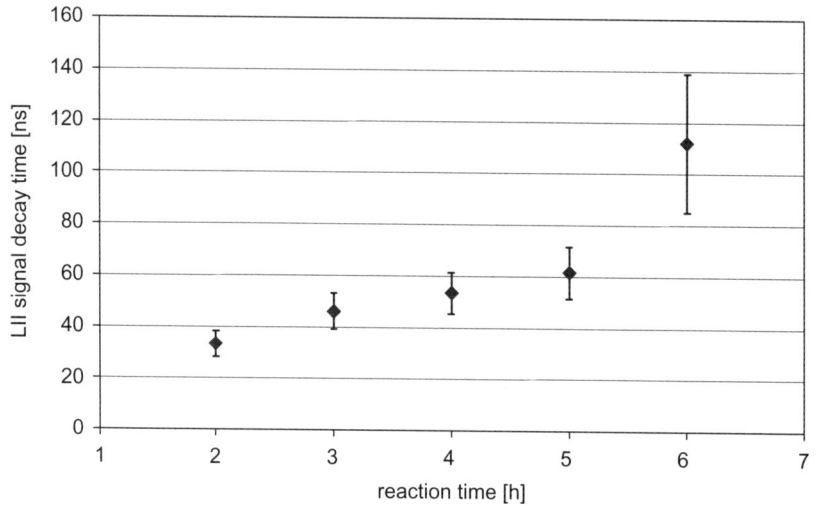

Figure 26 Experimental signal decay time in dependence of the reaction time (Leipertz et al., 2008).

every hour for 2 minutes by opening the heating coil which is coating the glass reactor.

An increase of the signal decay was found in the course of time, which is illustrated in Figure 26, indicating the particle surface growth. It was observable that up to 5 hours a permanent increase of the particles was given. After 6 hours a large rise of the signal decay time was found and a broadening of the experimental standard deviation was measured. This behavior can be explained by slightly removing the iron particles, thus altering the heat conduction, which can be directly seen in the LII signal decay time and was also found by TEM.

For size quantification of these particle systems, the underlying LII model has to be extended taking into consideration the optical metal particle properties (Vander Wal et al., 1999; Kreibig and Vollmer, 1995) on the one hand side and the contact surface area, on the other hand. The optical metal properties, which are in particular determined by the high imaginary part of the complex refraction index, show low absorption coefficients.

One important factor that influences the heat loss the most is the heat exchange area between the iron and substrate particle. For a spherical iron particle with a diameter of 120 nm and a gas temperature of about 423 K the temperature decay was calculated in dependence of the contact area which is denoted as the percentile value of the total iron particle surface, which is illustrated in Figure 27.

Smaller contact exchange surface areas cause a slow down of the temperature release due to the dominating gas phase conditions. Comparing the theoretical calculated signal decays in the time frame

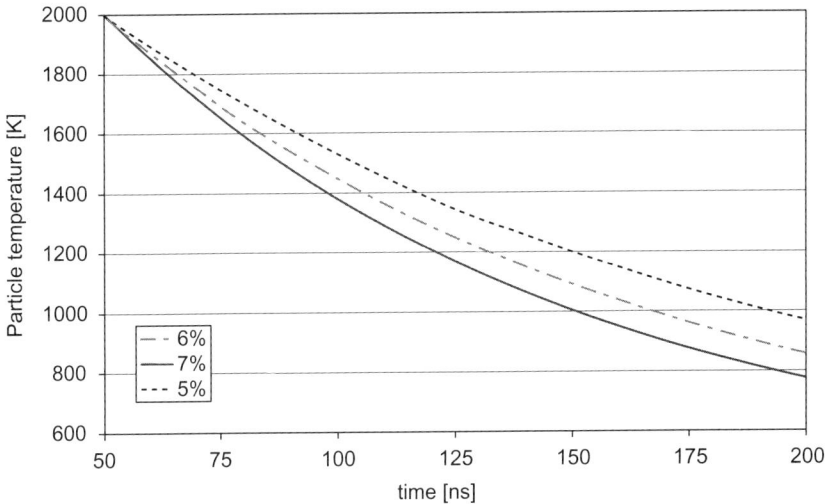

Figure 27 Calculated temperature decay for different percentile contact areas.

between 50 and 150 ns to the experimental ones for a mean primary particle size of 120 nm, which is measured by TEM, is in good agreement. Although the deviation of the LII experimentally determined signal decay times is broad after 6 hours, it is possible to describe these results by the LII model varying the involved parameters. Assuming a mean percentile contact area of 6% also over the whole formation process, the primary particle sizes can be directly determined from Figure 28.

4.2 Automotive soot investigations

Nanoparticles in form of soot play a prominent role in the automotive industry. Soot production, oxidation and emission is an important issue in diesel and gasoline direct-injection engine optimization as this is one of the pollutant components covered in legislation rules to be fulfilled by engine manufactures. TIRE-LII is a very suitable tool to control the soot distribution inside the engine combustion chamber and in the raw exhaust gas (Schraml et al., 1999). In combination with standard NOx measurement in the engine exhaust, TIRE-LII can be used for engine optimization (Schmid and Leipertz, 2006). In engine raw exhaust investigation, LII is only sensitive to elemental carbon (EC). Possible particle components such as ashes, soluble and volatile fractions are not determined, as they are vaporized by the high intense laser pulse.

4.2.1 LII exhaust gas sensor

For raw exhaust measurements in the automobile industry and fine dust investigations in the surrounding atmosphere, the TIRE-LII technique

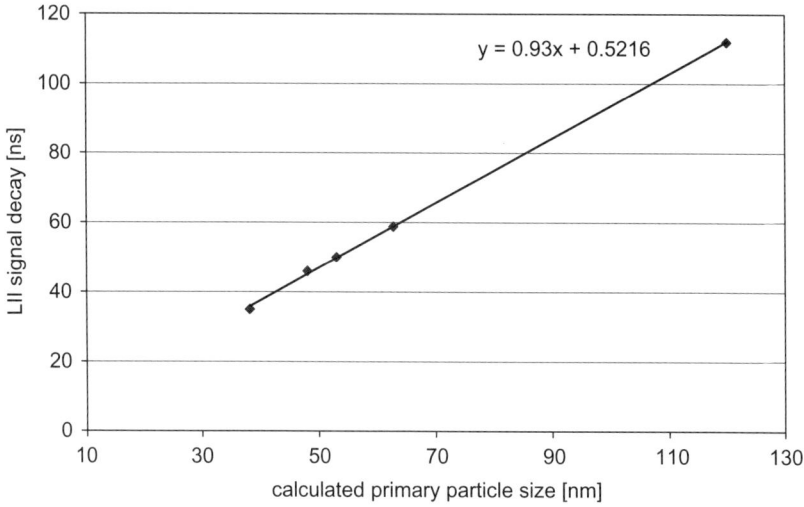

Figure 28 Correlation of experimental signal decay time and calculated primary particle size assuming a percentile contact area of 6%.

has been integrated in a compact sensor (Schraml et al., 2004; Sommer et al., 2005). Selective particle characterization is essential to appraise the performance characteristics of diesel exhaust after treatment systems and of specific engine optimization. For this, real-time investigations are important in order to resolve fast particle property changes.

The system consists of a water cooled ring adapter with purge air supply (Figure 29) and of a control and evaluation unit. The laser beam with an excitation wavelength of 1064 nm is focused in the measurement volume by a light fiber and the enhanced thermal radiation is captured by an appropriate detector head perpendicularly. It can be applied directly in the raw exhaust without dilution up to exhaust gas peak temperatures of 700°C. Its sensitivity ($3\,\mu g/m^3$) and variability for real-time (20 Hz) soot characterizing in different applications is shown.

As already published (Schraml et al, 2004), all the other components of the system are integrated into a compact 19″ rack cabinet. The embedded processing computer is responsible for control of the system parameters including laser and detector unit and allows fully automated operation of the system including data acquisition, online evaluation, visualization and storage. The detector gain is adjusted by a control voltage which is automatically controlled by the system processing unit to enable a broad dynamic range which is particularly important for highly transient tests. The signal is recorded by a fast analog-to-digital converter (PCI interface card) which additionally offers the possibility to enlarge the dynamic range. This can be done by adjusting the voltage resolution of the input signal. As the system software dynamically

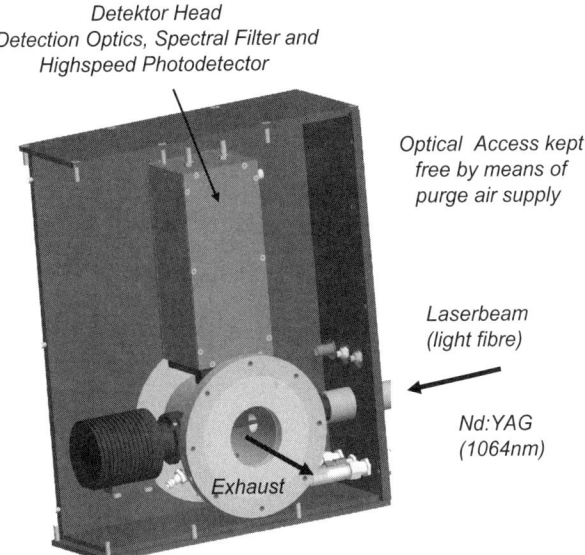

Figure 29 LII soot sensor (raw exhaust ring adapter).

adjusts detector gain and AD converter resolution, the dynamic range of a single setup is currently as high as nearly 10^5. This system is commercially available as Laser-Induced Soot Analyzer (LI²SA) being produced by the Spin-Off-Company of the Department of Engineering Thermodynamics of the University of Erlangen-Nuremberg. Similar systems are available on the market also from other companies, e.g., Atrium in the USA.

As mentioned before, in engine raw exhaust measurements, LII is only sensitive to EC. This was shown in many previous studies in comparison with conventional standard reference methods such as coulometry, gravimetry or opazimetry (Schraml et al., 2004, and references therein). The only comparable technique that determines only the EC content is coulomteric filter analysis. Therefore, the mass concentration calibration for the sensor is done in comparison with coulometry at a combustion aerosol standard (CAST), which is commercially available from Mattern Engineering in Switzerland and is basically a laminar diffusion flame quenched by a nitrogen flow and equipped with a dilution unit. The soot particles of the CAST system are considered to be very similar to diesel soot in both size and composition if appropriate CAST settings are chosen. By a variation of the dilution air flow different EC mass concentrations can be adjusted in a broad concentration range. In Figure 30 the correlation of the maximal peak LII signal (uncalibrated voltage signal) and the CAST EC concentration setting is depicted. The

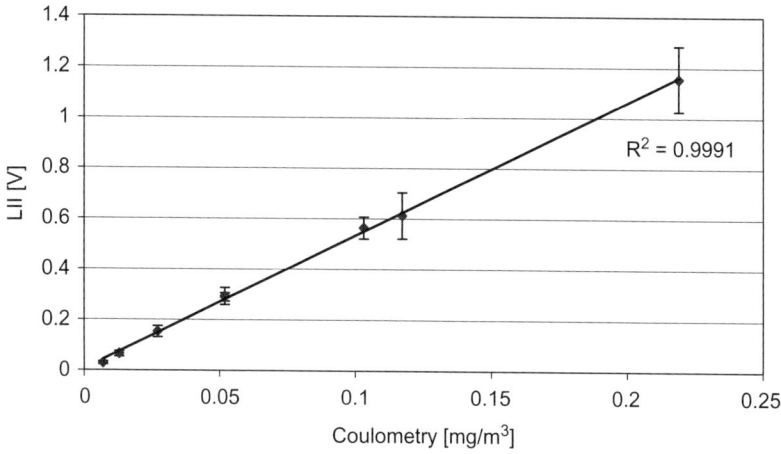

Figure 30 Correlation of the maximal peak LII signal to coulometric filter analysis obtained at a CAST (Schraml et al., 2004).

latter one has been determined by a coulometric analysis of the aerosol flow further taking into account the different dilution ratios.

The error bars of the LII signal are indicating the signal variation (mainly due to fluctuations of the aerosol generator) during the time interval of signal averaging (typically 5 min corresponding to 6000 single measurements). The uncertainty in the coulometric analysis can roughly be estimated with about $\pm 10\%$. Additionally, the figure also gives a first impression about the high sensitivity of the system. From the slope of the graph, a system-specific calibration constant can be deduced.

To verify the results of the determination of the mean primary particle size, the TEM has been applied. To verify the engine exhaust gas measurements in a more adjusted application here additional measurements have been performed again using the CAST aerosol generator operated in different modes. Although just 100–200 single primary particles have been evaluated per test with the TEM method the results are in reasonable agreement with the LII values. It should be pointed out in this context that the statistical error of these measurements is generally much lower than that of TEM as typically up to 10^8 particles (depending on the mass concentration) contribute to the size evaluation. Besides this statistical error, the uncertainties in Table 4 also include temporal fluctuations of the aerosol generator.

4.2.2 Raw exhaust gas measurements

The system has been applied in several configurations on various test benches within different projects, which are here just briefly reported to

Table 4 Comparison of primary particle diameters of TEM and LII obtained with the CAST aerosol generator (Schraml et al., 2004)

Test (diffusion diameter)	d_p TEM (nm)	d_p LI^{2}SA (nm)
CAST 1 (55 nm)	16 ± 2	14 ± 4
CAST 2 (80 nm)	18 ± 3	17 ± 1
CAST 3 (110 nm)	19 ± 3	17 ± 1
CAST 4 (220 nm)	38 ± 2	44 ± 1
CAST 6 (110 nm)	17 ± 3	16 ± 1
CAST 8 (55 nm)	16 ± 3	13 ± 3

illustrate the broad applicability and performance of the TIRE-LII technique in form of the LI^{2}SA sensor.

One possible application of the LI^{2}SA system is the investigation of strategies for internal pollutant reduction. Exemplarily, this can be performed by a variation of the injection timing of heavy duty engine (MAN D08). The used common-rail-system enables an injection which is independent of the engine speed and the free choice of single injection times.

As shown in Figure 31, by adjusting the main injection positioning and the injection pressure, the soot concentrations could be reduced from about $40\,\text{mg}/\text{m}^3$ at the base point to $2.5\,\text{mg}/\text{m}^3$ and a simultaneous reduction of NO$_x$ emissions. Moreover, the pre-injection at first showed increased particle emissions.

For the engine optimization, some injection parameters have been changed, which is indicated as part of Figure 31. Thereby, the advantage of this sensor technology for research and development applications could be shown. Because of the high sensitivity of the LII sensor also very low concentrations of the emitted soot mass concentration can be accurately measured. The NO$_x$ concentration was measured by means of conventional exhaust measurement devices. A typical method for NO$_x$ reduction is the variation of the injection start, by shifting it to later injection times. It could be shown that the known correlation between NO$_x$ formation and particle emission is valid also for modern engine combustion processes. The retarding of the injection start leads to a significant increase, whereas the NO$_x$ concentration is decreasing (Figure 32).

Another important research field is the investigation of exhaust gas after treatment systems. Besides its high sensitivity, which is important for the characterization of diesel particle filters and fine dust monitoring, also the fast in situ measurement up to 20 Hz makes it possible to optimize filter regeneration strategies or engine combustion even under highly transient conditions. Especially, the investigation of exhaust after treatment system behavior during different test cycles is a very important

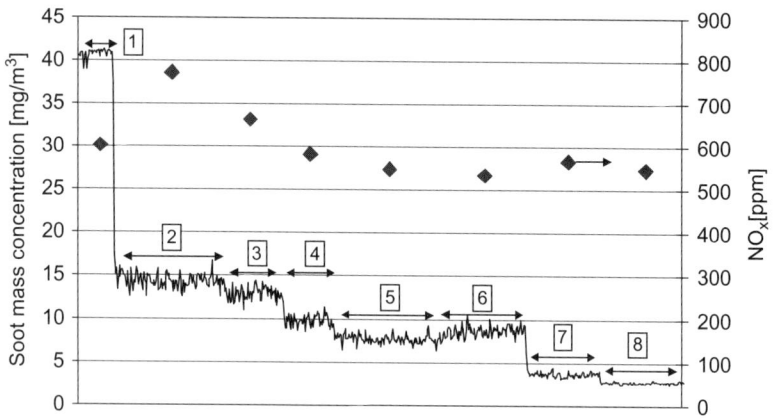

1: Base p_R = 860 bar, SB 3.5° before TDC
2→ 3: increase of injection pressure 860 bar → 1150 bar
3→ 4: positioning of injection 3.5° before TDC → 1.5° after TDC
4→ 5: 1.5° before TDC → 0.5° after TDC
5→ 6: decrease of injection pressure 1150 bar → 1100 bar
6→ 7: without pre-injection
7→ 8: 1.5° after TDC 2.5° after OT

Figure 31 Optimization of an engine operation point (1800 rpm, 605 Nm).

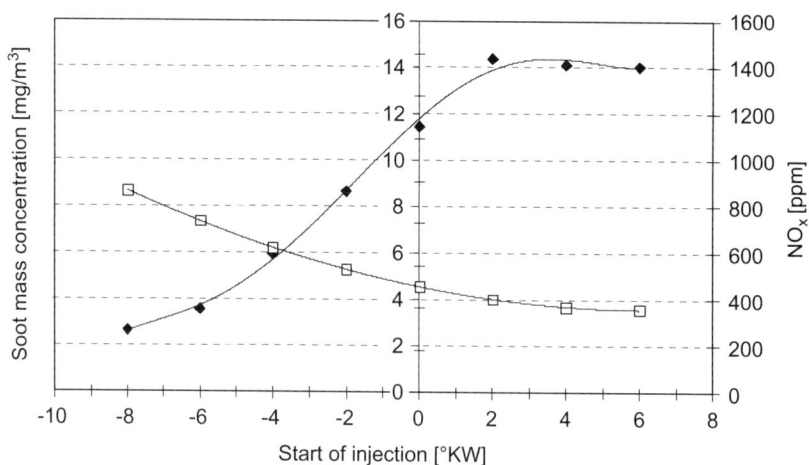

Figure 32 Variation of the injection start (1450 1/min, 313 Nm).

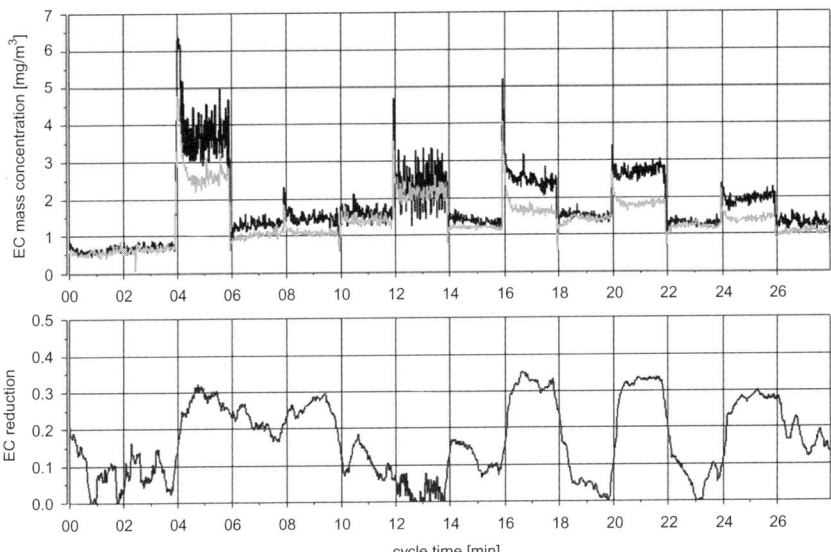

Figure 33 EC mass concentration during ESC test cycle (top) in front of (dark curve) and after (grey curve) a SINOx catalyst and the resulting EC mass reduction (bottom).

task in research and development. In Figure 33 the temporal course of the EC mass concentration is depicted for two different measurement locations during an European Stationary Cycle (ESC) test cycle of a medium duty truck engine equipped with an SINOx© SCR catalyst (SCR stands for selective catalytic reduction) together with the deduced EC mass reduction. The first measurement location was directly in front of the catalyst, whereas the second was immediately afterwards, each utilizing a full-flow sensor head. It could be observed that the EC mass reduction is significantly different for the individual operation points during this 13-stage stationary test cycle. Although the SINOx system has been designed to reduce the nitrogen oxide emission by a selective catalytic reaction after urea injection, an EC mass reduction of about 10–30% was also found. This reduction is strongly dependent on the individual operating conditions, e.g., a significant higher mass reduction occurs under high load conditions. Amazingly, no simultaneous change was observed in specific particle surface or primary particle size (not depicted), respectively. Thus, the dominating reduction process cannot be caused by a particle size reduction in this case, but in particle number. It seems to be most likely that some of the particles passing the SCR catalyst are completely oxidized, whereas other ones pass it entirely unaffected. A possible mechanism is impaction and subsequent oxidization of these particles on the catalyst surface. This example shows that by selective particle characterization of the mass concentration as well as the primary particle size conclusions can be drawn for the underlying reduction processes in such systems.

Especially the application of particulate filter systems is broadly considered to be the most efficient way for a significant reduction of the soot emission for diesel engines. Consequently, these systems also make high demands on the measurement systems, especially in sensitivity.

In order to simulate decreasing emission levels a truck engine of Euro III emission level (6 cylinder Volvo, displacement 7 liters) has been equipped with a continuously regeneration trap (CRT) filter system and an additional bypass to this filter by the Swiss EMPA institute. By a variation of the flow ratio of filtered/unfiltered exhaust gas, an emission level of about 60% of Euro IV has been adjusted (Mohr and Lehmann, 2003). The temporal courses of EC mass concentration and EC-specific surface area are depicted in Figure 34. In this graph, the mass and surface emissions of the 13 operation points within this test yield a different behavior. Most pronounced is the small specific surface area for idle operation which is caused by relatively large particles and in agreement to the results of previous studies.

Moreover, a CRT-equipped heavy-duty diesel engine was investigated utilizing the partial flow sensor head applied to the secondary dilution tunnel of a full flow CVS system. (Anderson, 2003). In Figure 35 the EC mass concentration during a ETC test cycle is depicted for four consecutive tests. From these measurements, it could be seen that LII is feasible to detect EC mass concentrations with high signal dynamics and

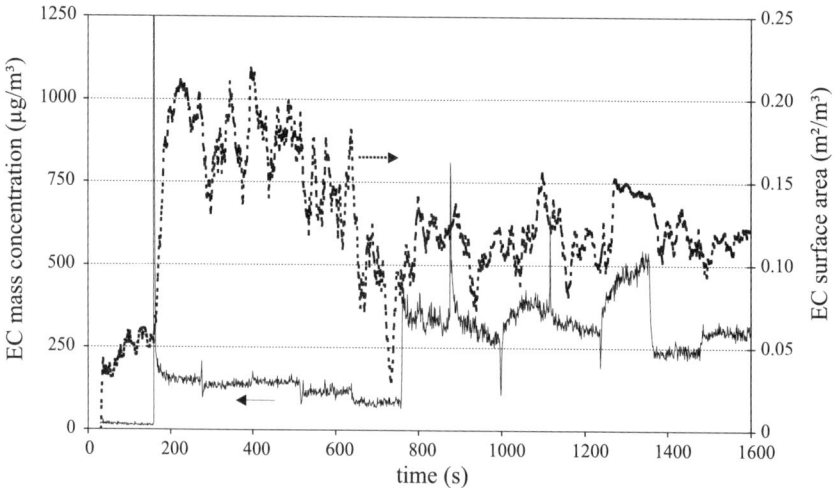

Figure 34 EC mass concentration (dark curve) and EC specific surface area (light curve) during ESC test cycle of a heavy-duty truck engine of Euro IV emission level.

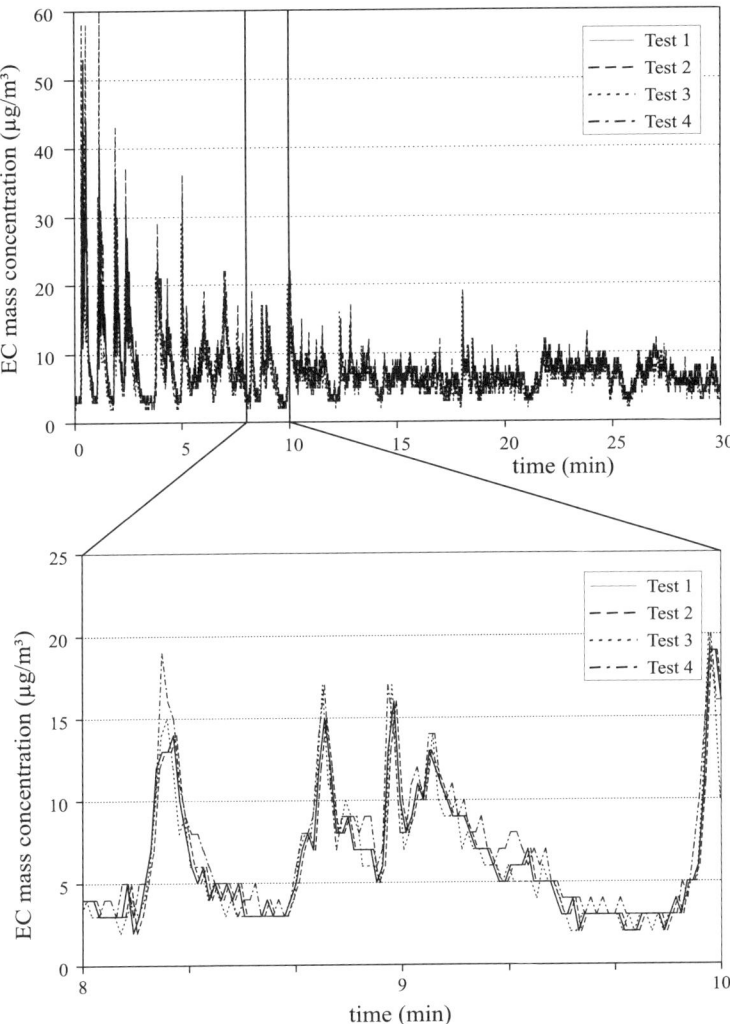

Figure 35 EC mass concentration during four consecutive runs of a ETC (European test cycle) of a CRT equipped heavy-duty truck engine (top: full cycle, bottom: cutout).

high temporal resolutions even at very low, near ambient emission levels. In this application, the very good reproducibility of engine operation conditions for the first time allowed a detailed investigation of the system repeatability under ultra low emission vehicle (ULEV) conditions which has been proven to be excellent (figure, bottom). The discrete steps that are obvious in Figure 34 are just caused by the limitation of the current setting of the measurement resolution ($1\,\mu g/m^3$, given by number of digits stored in the recorded data file) rather than by the bit resolution and are therefore no limitation of the dynamic of the AD converter which was automatically adjusted, as discussed above.

Besides measurements on engine test beds, tests on chassis dynamometers are also feasible either with the partial flow or even with a full flow version of the sensor head. Additional to the investigation of diesel vehicles, tests with gasoline-driven vehicles have also been performed (Feest et al., 2003). Here, particular attention has to be paid to modern gasoline direct injection concepts (GDI engines). As a specific example the EC mass concentration of an Audi A4 2.0 liter FSI during a hot federal test procedure (FTP) test cycle is depicted in Figure 36, underlying the applicability of LII also for these concepts.

From Figure 36 it can be observed that the EC emission level from this GDI engine is, speaking in terms of concentration, approximately of the same order than the emission level of a Euro-IV heavy duty truck engine (see e.g., Figure 34) and even one order of magnitude higher if the latter one is equipped with a trap . This is particularly interesting as particulate emission from GDI engines is currently not regulated at all.

4.2.3 Ambient Air Studies

With respect to emission legislation, many European cities transgressed the threshold value of particulate matter exposure. Unfortunately, only the total mass of particles smaller than 10 μm is restricted without distinguishing between different components. Epidemiological studies

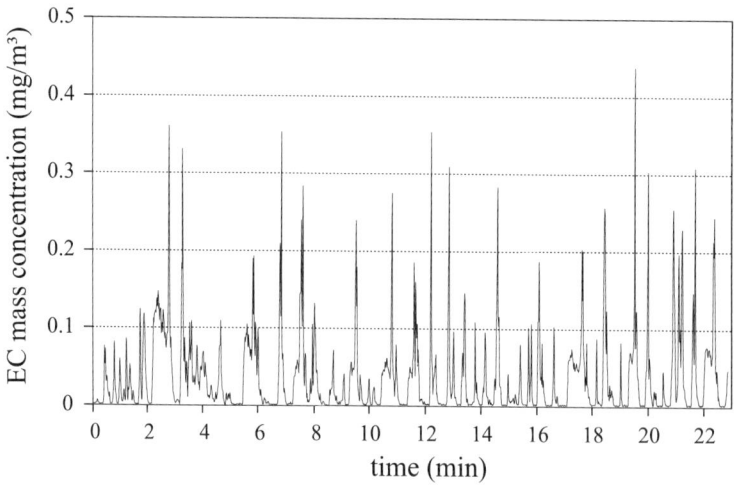

Figure 36 EC mass concentration during a hot FTP test cycle of a Audi A4 FSI equipped with a direct injection gasoline engine (CVS sampling with partial flow sensor head).

indicate that ultra fine particles of definite nature are significantly more dangerous than larger ones. Thus, simultaneous selective determination of mass and specific surface area is necessary to better appraise the exposure composition.

First investigations were carried out next to two different measurement stations, one located at a busy urban road, the other one based at a residential area next to a forest. In comparison with measurements using a tapered element oscillating microbalance (TEOM), which determine the total mass of PM10 per hour, LI^{2}SA characterized the EC fraction averaged in the same time scale. Thereby, at the busy street the mean concentration of EC was always between 7 and $12\,\mu g/m^3$. Beyond this, due to the deviation of the non-traffic-related particles, the EC fraction lies always between 20 and 40% of the total mass. By means of time-resolved measurements (5 Hz), short but high peak concentrations were detectable and unambiguous correlation to single vehicles were found when comparing to simultaneously detected video tapes (Figure 37).

Other investigations were conducted at school bus stops to appraise the exposure on pupils. It was found that short (2–6 s) but very high EC mass concentrations of about several hundred $\mu g/m^3$ occurred (Figure 38). This underlines the need for comprehensive equipping of buses with particulate traps.

Mobile studies on the road showed the exposure on drivers inside and outside the car (Table 5), whereas the mean EC mass concentration inside the vehicle was approximately doubled. Thereby, specific surface area and primary particle sizes are in the same range as expected. Furthermore, high concentrations at traffic congestions or start ups at

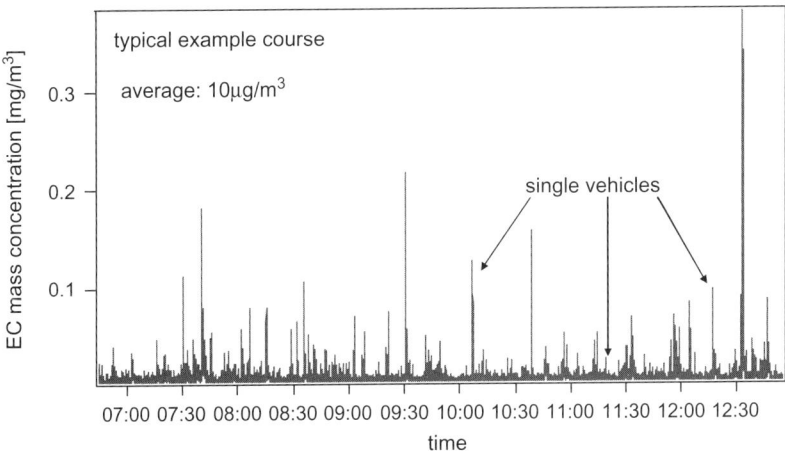

Figure 37 EC concentration at roadside.

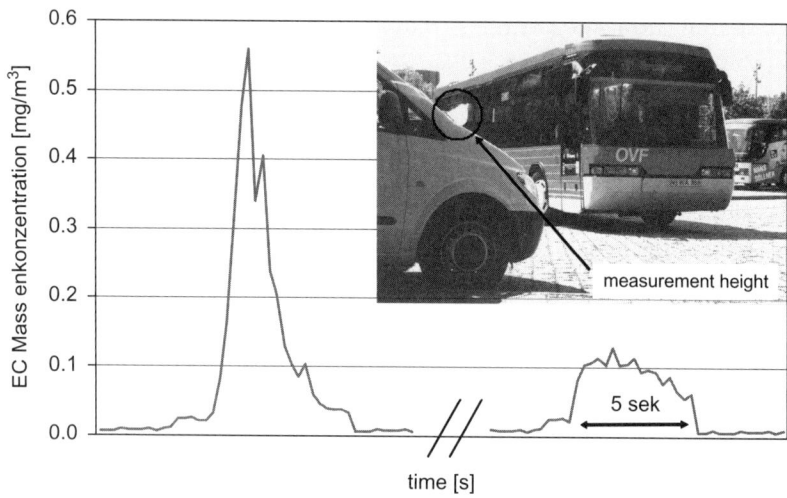

Figure 38 EC concentration at a bus stop.

Table 5 Soot measurements inside and outside the car

	Outside the car	Inside the car
Mean EC mass concentration ($\mu g/m^3$)	8–10	19–28
Maximum EC mass concentration ($\mu g/m^3$)	up to 820	max. 120
Specific surface area (m^2/mg)	0.361–0.067	0.279–0.071
Primary particle size (nm)	11–50	13–47

traffic lights and street crossings can be seen, as well as typical concentration courses inside and after a tunnel.

4.3 Particle suspensions

So far LII has only been applied for aerosol processes without the consideration of particles dispersed in liquids. First, investigations were carried out with re-dispersed carbon blacks. Besides furnace blacks (Printex A, G, 25, 35, and 55), various gas carbon black particles (FW 18, Colour Black S160 and S170, Printex U and U140) were also considered. The particles were suspended in different liquids and dispersed by ultrasonic excitation. The stability of the suspension was recorded by measuring the aggregate size distribution (diffusion diameter) with DLS. Moreover, this was done before and after the LII measurements in order to control the stability of the particle suspension. To achieve LII excitation, the second harmonic of a Nd:YAG laser at 532 nm having a pulse repetition rate of 20 Hz and a pulse length of 6 ns was used. In order to get a nearly homogenous beam profile, the edges of the laser beam were cut off by passing a 2 mm pinhole. A part of the incident laser beam was reflected by a beam splitter onto the sensor head of a power meter to monitor the initial laser pulse energy. Subsequently, the transmitted laser beam was used for particle heating inside a quartz cell. The excitation energy in the measurement volume was calculated by considering the absorptivities of the quartz cell and of the solvent, the values of which had been predetermined in a test series. The LII signal was collected and imaged on the surface of a fast PMT with a time response of 0.65 ns. In this step elastically scattered laser light was suppressed by using a holographic notch filter. Furthermore, a shortpass filter with a cut-off wavelength of 450 nm and absorptive neutral density filters were inserted for spectrally filtering and signal attenuation, respectively. The PMT signal was digitized by a fast 8-bit analog to digital converter with 250 MHz bandwidth and 1 GS/s sampling rate. Since the temperature of the solvent influences the signal decay, it was determined before and after the LII measurements.

In the course of our investigations, it was possible to detect LII signals for all particle suspensions studied. First, the dependence of the maximal incandescence signal on the laser fluence is documented. The initial increase in laser fluence leads to a rapid rise in the peak LII signal until the sublimation-limited region is reached from which no significant signal change is observable. These results are comparable to those from aerosols and flames as mentioned earlier. Fluence dependency as described here was representative for all carbon blacks tested in this study (Sommer and Leipertz, 2007).

To evaluate the relation of the exponential signal decay time, which is determined 100–250 ns after the laser pulse, and the size of the primary particles, which has been measured by TEM, the laser fluence was set at $0.17 J/cm^2$. This corresponds to the beginning of the plateau region discussed in more detail in Chapter II.A. A linear correlation between both quantities was found, which is illustrated in Figure 39. No influence of the aggregate size could be found in these investigations. A comparison of Printex U and Printex 55 in particular, which have the same TEM primary particle diameter but different aggregate sizes, confirms this observation, which also has been found before (e.g., Figure 4).

The LII signal decay time was almost the same for both, showing that for the considered carbon black suspensions, the exponential decay time is only a measure of the primary particle size.

In this context, the influence of different solvents on the signal decay behavior was also considered. Despite different thermal conductivities, the signal decay time was not influenced by the solvent in question. This

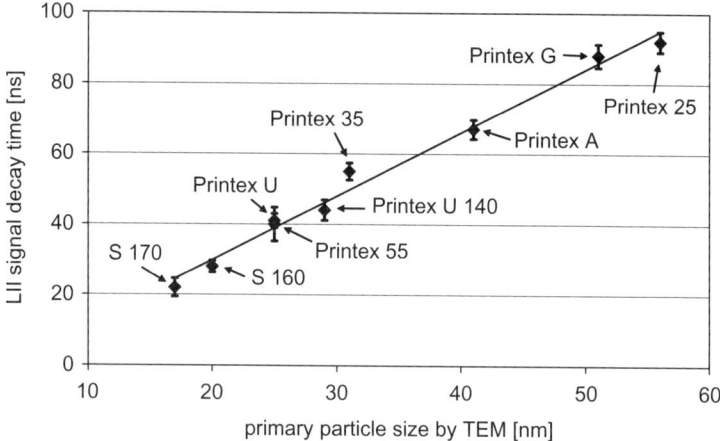

Figure 39 Correlation of LII signal decay time and primary particles size measured by TEM for different carbon blacks dispersed in ethanol (Sommer and Leipertz, 2007).

factor and the signal relation to the laser fluence can possibly be described by vapor layers enclosing the particles, which is described in more detail elsewhere (Sommer and Leipertz, 2007).

Regarding the dependence of the signal decay time on laser energy, a typical course is shown in Figure 40 for Printex G dispersed in ethanol. At first, the signal decay time rises with increasing laser fluence, achieves constant values of about 89 ns in the plateau regime, and subsequently increases. This behavior is characteristic for all carbon blacks and is in contrast to results obtained in a gaseous environment. A possible explanation might be found in solvent vaporization and the formation of a gaseous layer around the particles which directly affects particle cooling. Therefore, the initial increase in the decay time with higher laser fluence could be attributed to a thickening of the vapor layer which leads to an altered heat loss from the particles to the bulk fluid. At higher laser excitations, a gas phase might consist of vaporized solvent and sublimated carbon black. Interestingly, in all studied cases, there is a slight slow down in the signal despite decreasing concentration. This latter behavior is caused in a gaseous environment mainly by particle surface sublimation as mentioned above. However, due to smaller particles, this normally leads to decreasing signal decay times. Increasing decay times, on the contrary, might be referred to particle merging or to a continuous thickening of the vapor layer.

As mentioned before the influence of the liquid on the signal decay behavior was also considered. For this purpose, we worked not only with ethanol but also with isopropanol and methoxy-nonafluorobutane (HFE-7100) because of their different thermal conductivities. Solvent

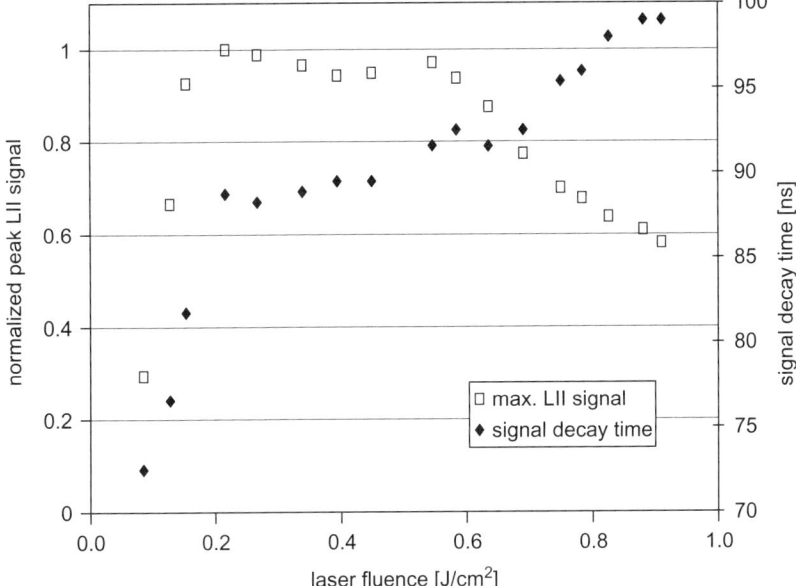

Figure 40 Dependences of the signal decay time and the peak LII signal on laser fluence for Printex G/ethanol suspension.

Table 6 Comparison of LII signal decay times of Printex G for different laser fluences and solvents (Sommer and Leipertz, 2007)

	Decay time (ns)	Ethanol	Isopropanol	HFE-7100
mJ/cm^2	200	89	91	92
	250	90	91	92
	300	90	89	90
	350	92	92	91
	400	91	90	92

temperature was the same before and after the LII measurements. Surprisingly, signal decay time is not affected by the solvent in the plateau region, which can be seen in Table 6. Comparable results are shown for different laser fluences confirming the theory of gaseous layers being formed around the particles.

5. CONCLUSIONS

In this paper the theoretical background and the potentials of TIRE-LII for the determination of the size and mass concentration of nanoscaled particles has been shown. TIRE-LII has successfully been used for the investigation of soot formation and oxidation in basic combustion research and in two different important technical application fields. The technique has been applied both at different nanoparticle production reactors and directly in engine raw gas exhaust. An online determination of the mean primary particle size of carbon blacks during the particle formation has been carried out and compared to standard chemical analysis indicating very good agreement. Furthermore, primary particle size distributions for different reactor settings were reconstructed by the determination of two signal decay times from the experimental LII signals in comparison with numerically calculated ones. Moreover, the applicability for metal and metal oxide process control was shown in a laser vaporization and a hot wall reactor.

For automotive investigations one of the main benefits of LII, in comparison with conventional particle measurement systems, lies in the exceptional high sensitivity and the possibility of a simultaneous evaluation of EC mass concentration and primary particle size (respectively size of particle surface area). This feature enables a convenient investigation of several interesting effects, which up to now, have not been observed in detail — either because of the lack of possibility or because of the complexity of alternative solutions. Furthermore, the high dynamic range and temporal resolution qualifies LII as an ideal tool for research and engine development as it enables the user to optimize the engine with respect to soot emission even for transient operation conditions. This is particularly important as time-consuming data evaluation is not necessary with this system and so extensive parameter studies can be performed very fast. Therefore, LII is able to comply with all forthcoming developments of new combustion concepts, traps or other exhaust gas after treatment systems. Furthermore, with this measurement technique, it is possible to determine very low soot concentration also in ambient air.

For the first time, TIRE-LII has been successfully applied to the characterization of in liquids suspended nanoparticles. Re-dispersed carbon blacks were investigated in different solvents, whereby a linear correlation between the exponential LII signal decay time and the primary particle size determined by transmission-electron microscopy was found.

TIRE-LII has been found to be a powerful tool for the characterization of nanoscaled particles for basic studies on particle formation and for the investigation of technical nanoparticle applications in mechanical and chemical engineering.

NOMENCLATURE

NOTATION

C_{abs}	absorption efficiency
c_P	specific heat
d_p	particle diameter
E_i	laser energy
ΔH_V	sublimation enthalpy
M	molar mass
M_λ^b	spectral energy density
T	particle temperature
T_0	temperature of the surrounding

GREEK LETTERS

E	emission coefficient
Λ	heat transfer coefficient
λ	Wavelength
P	mass density
σ	size distribution width
σ_g	geometric standard deviation
τ	signal decay

SUBSCRIPTS AND SUPERSCRIPTS

Med	Median
Mono	Monodisperse

ACKNOWLEDGEMENTS

The authors gratefully acknowledge the financial support for parts of the work by the German National Science Foundation (DFG) and the German Federal Ministry for Education, Science, Research und Technology (BMBF). They additionally thank Degussa AG for support in providing and running the reactors investigated, and furthermore the RWTÜV Institute in Essen and the companies Robert-Bosch GmbH, Argillon GmbH, MAN Nutzfahrzeuge GmbH in Germany within the frame of different projects for enabling the test bench measurements. The LI^{2}SA system has been provided for the measurements by the company ESYTEC Energie und Systemtechnik GmbH in Erlangen, Germany. The Erlangen Graduate School in Advanced Optical Technologies is funded by the DFG in the framework of the Excellence Initiative of the German Federal and State Governments to Promote Science and Research at German Universities.

REFERENCES

Anderson, J. D., "UK particle measurement programme – heavy duty methology development, Final Report Phase 2, Ricardo Consulting Engineers". Shoreham-by-Sea, Great Britain (2003).
Bladh, H., and Bengtsson, P.-E. *Appl. Phys. B* **78**, 241–248 (2004).
Bladh, H., Johnsson, J., and Bengtsson, P.-E. *Appl. Phys. B* **90**, 109–125 (2008).
Bockhorn, H., Geitlinger, H., Jungfleisch, B., Lehre, T., Schon, A., Streibel, T., and Suntz, R. *Phys. Chem.* **4**, 3780–3793 (2002).
Dankers, S., and Leipertz, A. *Appl. Opt.* **43**, 3726–3731 (2004).
Dankers, S., Schraml, S., Will, S., and Leipertz, A. *Chem. Eng. Technol.* **25**, 1160–1164 (2002).
Dankers, S., Leipertz, A., Will, S., Arndt, J., Vogel, K., Schraml, S., and Hemm, A. *Chem. Eng. Technol.* **26**, 966–969 (2003).
Dasch, C. J. *Proc. Comb. Inst.* 231–237 (1984).
Daun, K. J., Stagg, B. J., Liu, F., Smallwood, G. J., and Snelling, D. R. *Appl. Phys. B* **87**, 363–372 (2007).
Feest, E. A., Marshall, I. A., Norris, J. O. W., Reading, A. H., Sandbach, E. L., AEA Technology, Didcot, UK Particle Measurement Programme – Light Duty Methology Development, Final Report Phase 2, Great Britain (2003).
Filippov, A. V., Markus, M. W., and Roth, P. *J. Aerosol. Sci.* **30**, 71–87 (1999).
Hiers, R. *Appl. Phys. B* **92**, 635–641 (2008).
Kock, B. F., Kayan, C., Knipping, J., Orthner, H. R., and Roth, P. *Proc. Comb. Inst.* **30**, 1689–1697 (2005).
Kreibig, U., and Vollmer, M., "Optical Properties of Metal Clusters". Springer Verlag, Heidelberg (1995).
Lehre, T., Jungfleisch, B., Suntz, R., and Bockhorn, H. *Appl. Opt.* **42**, 2021–2030 (2003).
Lehre, T., Suntz, R., and Bockhorn, H. *Proc. Comb. Inst.* **30**, 2585–2593 (2005).
Leipertz, A., and Dankers, S. *Part. Part. Syst. Charact.* **20**, 81–93 (2003).
Leipertz, A., Ossler, F., and Alden, M., *in* "Appl. Comb. Diagns." (K. Kohse-Höinghaus, and J. Jeffries Eds.), pp. 359–382. Taylor & Francis, New York (2002).
Leipertz, A., Sommer, R., Danova, K., Popovska, N., *Proc. CARBON 2008* (2008).
Liu, F., Daun, K. J., Snelling, D. R., and Smallwood, G. J. *Appl. Phys. B* **83**, 355–382 (2006a).
Liu, F., Stagg, B. J., Snelling, D. R., and Smallwood, G. J. *Int. J. Heat Mass Transfer* **49**, 777–788 (2006b).
Liu, F., Yang, M., Hill, F. A., and Snelling, D. R. *Appl. Phys. B* **83**, 383–395 (2006c).
Melton, L. A. *Appl. Opt.* **48**, 4473–4479 (1984).
Mewes, B., and Seitzman, J. M. *Appl. Opt.* **36**, 709–717 (1997).
Michelsen, H. A. *J. Chem. Phys.* **118**, 7012–7045 (2003).
Michelsen, H. A., Liu, F., Kock, B. F., Bladh, H., Boiarciuc, A., Charwath, M., Dreier, T., Hadef, R., Hofmann, M., Reimann, J., Will, S., Bengtsson, P. E., Bockhorn, H., Foucher, F., Geigle, K. P., Mounaim-Rousselle, C., Schulz, C., Stirn, R., Tribalet, B., and Suntz, R. *Appl. Phys. B* **87**, 503–521 (2007).
Michelsen, H. A., Linne, M. A., Kock, B. F., Hofmann, M., Tribalet, B., and Schulz, C. *Appl. Phys. B* **93**, 654–656 (2008).
Michkova, K., Schneider, A., Gerhard, H., Popovska, N., Jipa, I., Hofmann, M., and Zeneck, U. *Apll. Cataly.* **315**, 83–90 (2006).
Mohr, M., Lehmann, U., Particle Measurement Programme (GRPE-PMP) – Comparison Study of Particle Measurement Instruments for Future Type Approval Applications, EMPA Report, Dübendorf, Switzerland, (2003).
Reimann, J., Kuhlmann, S.-A., and Will, S. *Comb. Flame* **153**, 650–654 (2008).
Roth, P., and Filippov, A. V. *J. Aerosol. Sci.* **27**, 95–104 (1996).

Santoro, R. J., and Shaddix, C. R., *in* "Appl. Comb. Diagns." (K. Kohse-Höinghaus, and J. Jeffries Eds.), pp. 252–286. Taylor & Francis, New York (2002).

Schmid, M., and Leipertz, A. *Int. J. Engine Design* **41**, 188–205 (2006).

Schraml, S., Will, S., Leipertz, A., *SAE Technical Paper Series*, 1999-01-0146 (1999).

Schraml, S., Dankers, S., Bader, K., Will, S., and Leipertz, A. *Comb. Flame* **120**, 439–450 (2000).

Schraml, S., Kremer, H., Sommer, R., and Leipertz, A., Proceedings of the 8th International Symposium on Diagnostics and Modeling of Combustion, *in* "Int. Comb. Eng. (COMODIA)" Yokohama, Japan (2004).

Schulz, C., Kock, B. F., Hofmann, M., Michelsen, H., Will, S., Bougie, B., Suntz, R., and Smallwood, G. J. *Appl. Phys. B* **83**, 333–354 (2006).

Shaddix, C. R., and Smyth, K. C. *Comb. Flame* **107**, 418–452 (1996).

Snelling, D. R., Liu, F., Smallwood, G. J., and Gülder, Ö. L., "Proceedings of the 34th National Heat Transfer Conference". Pittsburgh, PA, USA, NHTC2000-12132 (2000).

Snelling, D. R., Smallwood, G. J., Gülder, O. L., Liu, F., and Bachalo, W. D., "Second Joint Meeting U.S. Sections Combustion Institute". Oakland, CA, USA (2001).

Snelling, D. R., Thomson, K. A., Smallwood, G. J., Gülder, O. L., Weckman, E. J., and Fraser, R. A. *AIAA Journal* **40**, 1789–1795 (2002).

Sommer, R., and Leipertz, A. *Opt. Lett.* **32**, 1947–1949 (2007).

Sommer, R., Dankers, S., Will, S., and Leipertz, A., *in* "Handling of Highly Dispersed Powders" (E. Müller, and C. Oestreich Eds.), pp. 32–39. Shaker, Aachen (2004).

Sommer, R., Dankers, S., and Leipertz, A. *Chemie Ingenieur Technik* **77**, 214–223 (2005).

Sommer, R., Kaste, A., and Leipertz, A., "Proceedings of the 9th International ETH Conference on Combustion Generated Nanoparticles". Zürich, Switzerland (2005).

Staupendahl, G., Kurland, H.-D., and Garbow, J., IEEE LEOS Newsletter 17 (2003).

Tait, N. P., and Greenhalgh, D. A. *Phys. Chem.* **97**, 1619–1625 (1993).

Vander Wal, R. L., and Choi, M. Y. *Comb. Flame* **102**, 200–204 (1995).

Vander Wal, R. L., and Choi, M. Y. *Carbon* **37**, 231–238 (1999).

Vander Wal, R. L., and Weiland, K. J. *Appl. Phys. B* **59**, 445–452 (1994).

Vander Wal, R. L., Ticich, T. M., and Stephenson, A. B. *Appl. Phys. B* **67**, 115 (1998).

Vander Wal, R. L., Ticich, T. M., and West, J. R. *Appl. Opt.* **38**, 5867–5879 (1999).

Will, S., Schraml, S., and Leipertz, A. *Opt. Lett.* **20**, 2342–2344 (1995).

Will, S., Schraml, S., and Leipertz, A. *Proc. of the Comb. Inst.* **26**, 2277–2284 (1996).

Will, S., Schraml, S., Bader, K., and Leipertz, A. *Appl. Opt.* **37**, 5647–5657 (1998).

SUBJECT INDEX

ABACUS, 59
ADAC Forte, 154
Adhesion strength, measurement of
  of bacteria, 72–73
  food fouling, 75
Advantose™ 100, 63, 64, 65
AFM. *See* Atomic force microscopy (AFM)
Algebraic reconstruction techniques (ART), 182
Alginate, 58
Alginate–chitosan microcapsules, 58–59
Ambient air studies, in automotive soot investigations, 260–262
Animal cells, in suspension culture, 51–53
Ar laser, 94, 138
Aspergillus nidulans, 56
Atomic force microscopy (AFM), 33–35, 54, 56, 70–71, 73, 74
Automotive soot investigations
  ambient air studies, 260–262
  inside and outside the car, 261–262
  LII applications, 251–262
  LII exhaust gas sensor, 251–254
  raw exhaust gas measurements, 254–260
A_{wet}. *See* Liquid phase (A_{wet})
Axial dispersion coefficient, D, 165

Bacillus subtilis, 37, 55
"Back-to-back" γ-rays, 151
Bacterial cells, 54–55
  adhesion strength of, measurement of, 72–73
Barcroft™ CS90 calcium carbonate, 63
Biocompatible particles, 58–59. *See also* Micromanipulation, in mechanical characterisation of single particles

Biological particles. *See also* Micromanipulation, in mechanical characterisation of single particles
  animal cells, in suspension culture, 51–53
  chondrocytes, 53–54
  filamentous microorganisms, 56
  plant cells, 56–57
  yeast and bacterial cells, 54–55
Biomass and biofilm formation, fouling of surfaces by, 72–74
Birmingham positron camera, 154, 171
Boltzmann integral expression, 43
BP's Hull Research and Technology Centre, 173
Brunauer-Emmett-Teller (BET) analysis, 240–241
Bubbly flow, in gas-liquid two-fluid flows measurement, 121, 125–136

Calcium alginate microspheres, 58, 59
Calcium–shellac microsphere, 59
CARPT. *See* Computer-automated radioactive particle tracking technique (CARPT)
Cartesian coordinates, 18
Cell poking technique, 33–35
CFB. *See* Circulating fluidised beds (CFB)
Chitosan, 58, 59
Chondrocytes, 53–54
Chondrons, defined, 53
CIP. *See* Cleaning-in-place (CIP)
Circulating fluidised beds (CFB), 156, 159–160
  ECT applications in, 186–190
Cleaning-in-place (CIP), 74
Coherent anti-stokes Raman scattering (CARS) thermometry, 236

271

Compression testing, by micromanipulation. See Diametrical compression
Computed tomography (CT), 182
Computer-automated radioactive particle tracking technique (CARPT), 150–151
Confocal μ-PIV technique, 104–105, 120
Continuously regeneration trap (CRT) filter system, 258
Continuous wave (CW) lasers, 94–95
Core/shell structure of microparticles, models for, 44–51
Coriolis flow meter, 11–12
Correlation analysis for gas–liquid interface heights, 15–16
Correlation coefficient, defined, 15
Cross-correlation analysis, for PIV technique, 97–98
CTI ECAT931/08, 171–172
^{61}Cu (half-life 3.4 h), 154

DCPD. See Dicyclopentadiene (DCPD)
Defocusing concepts, 111
Defocusing particle image velocimetry (DPIV) technique, 111–112
DEM. See Discrete element modelling (DEM)
Diametrical compression, 37–51
  experimental setup, 37–40
  microcapsules and, 65–67
  of single particles, mathematical modelling of
    Hertz model, 40–41
    microparticles with core/shell structure, models of, 44–51
    Tatara analysis, 41–42
    viscoelastic model, 42–44
  of two-week-old suspension-cultured tomato cells, 57
Dicyclopentadiene (DCPD), 67
Dielectric particle, optical trapping of, 35–36
Dipalmitoyl phosphatidyl choline (DPPC), 70
Discrete element modelling (DEM), 164
Disc-shearing device, 72–73
Dispersion of particle pulse in gas-solid fluidized beds, PET applications in, 212–213
Dispersion time, defined, 167
DLVO theory, 68

Doppler velocimetry, 2–3
  laser, 2
  ultrasonic, 3
DPPC. See Dipalmitoyl phosphatidyl choline (DPPC)
Dual-plane PIV, 117–118
Dynamic light scattering (DLS), 224, 262
Dynamic PIV, 115

Elastic membrane model, 44–47, 57
Electrical capacitance tomography (ECT), 180–196
  applications, 186–196
    circulating fluidized beds (CFB), 186–190
    hydrodynamic behaviors in bubble and slurry bubble columns, 191–196
    pneumatic solid conveying, 190–191
  principle of, 183–186
Electrical capacitance volume tomography (ECVT), 192–194, 196–197, 215
Electrical impedance tomography (EIT), 198, 200–202
  strategies employed in, 200–202
Electrical magnetic tomography (EMT), 181
Electrical resistance tomography (ERT), 196–209
  applications, 204–209
    high-speed flow imaging in slurry conveying, 205–207
    hydrocyclone flow visualization and comparison with computational fluid dynamics, 204–205
    visualization of dispersions in an oscillatory baffled reactor, 207–209
  principle of, 198–203
Electromagnetic flow meters, 11–12
Electron microscopy, 76
Energy balance, determination and LII signal, 225–228
Environmental scanning electron microscope (ESEM), 76–77
EPS. See Exopolysaccharides (EPS)
Escherichia coli, 55
ESEM. See Environmental scanning electron microscope (ESEM)

Eudragit® L100-55, 62–65
Eudragit microparticles, 39
Eudragit® S100, 63
3D Euler–Lagrangian hardsphere discrete particle model (DPM), 160
Excipients, pharmaceutical, 61–64
Exopolysaccharides (EPS), 72

^{18}F (half-life 110 min), 153–154
Filamentous microorganisms, 56
Flame investigations, by LII, 236–237
Fluidised beds, PEPT used in
  circulating fluidised beds (CFBs), 159–160
Fluidized beds, PEPT used in
  computational models, validation of, 160–162
  motion close to surfaces, 158–159
  solids motion studies, 156–158
Food fouling deposits, 74–75
Force spectroscopy, 34
Fouling, of surfaces
  biomass and biofilms, 72–74
  food fouling deposits, 74–75
Four-pulse ultrasound wave, at gas–liquid interface, 7
Free-surface flow, in gas-liquid two-fluid flows measurement, 121–125
Furnace black reactor, 242–245
Fusarium graminearum, 56

^{66}Ga (9.3 h), 154
Gas–liquid interface
  four-pulse ultrasound wave at, reflected, 7
  liquid velocity and
    experimental approach, 6–10
    experimental setup, 4–5
    proposed experimental method, validation of, 10–11
    peak ultrasound echo intensity and, 3, 11
    experimental approach, validation of, 18–24
    experimental setup, 12–13
    method, experimental, 13–18
Gas-liquid two-fluid flows measurement using PIV technique, 121–137
  bubbly flow, 125–136
  free-surface flow, 121–125
  gas–liquid two-phase flows in microchannels, 136–137

Gas–liquid two-phase flows in microchannels, in gas-liquid two-fluid flows measurement, 136–137
γ-ray tomography (GRT), 181
Green strain, defined, 49

Hencky/true strain, defined, 49–50
He–Ne laser, 94
Hertz equation, 41
Hertz model, 40–41
High-speed flow imaging in slurry conveying, ERT applications in, 205–207
Holographic particle image velocimetry (HPIV) technique, 109–111
Hooke's law strain energy function, 45, 49
Hot wall reactor, 249–251
Hot-wire anemometry, 2
Hydrocyclone flow visualization and comparison with computational fluid dynamics, ERT applications in, 204–205
Hydrodynamic behaviors in bubble and slurry bubble columns, ECT applications in, 191–196

Imaging tomography, 2
Instantaneous shear modulus (G_0), 43
Ion-exchange resin particle, 38–39
Iterative linear back projection (ILBP), 185–187

JKR theory, 69
Jurket T lymphomas cells, 35

Kelvin–Voigt element, 42

Lactose, particle–surface interactions of AFM for, 70
Laplacian filter, 7–8
Lardner and Pujara's model, 67
Laser Doppler velocimetry, 2
Laser-induced fluorescence (LIF) technique, 92, 119, 121, 127–137, 139–140

Laser-induced incandescence (LII), 224
  applications, 237–265
    automotive soot investigations, 251–262
    nanoparticle production processes control, 237–251
    particle suspensions, 262–265
  flame investigations, 236–237
  signal and determination of energy balance, 225–228
  primary particle size and its distribution, 228–236
Laser-Induced Soot Analyzer (LI^{2}SA), 253, 255
Laser trapping, 68–70
Laser tweezers. See Optical trapping method
Laser vaporization reactor (LVR), 246–249
LII exhaust gas sensor, 251–254
Linear back projection (LBP) technique, 185–187, 203
Linear forward projection (LFP) technique, 184
"Liquid-drop" model, 44
Liquid–liquid two-fluid flows measurement, using PIV technique, 119–121
Liquid phase (A_{wet}), calculation of portion of pipe occupied by, 16–18
Liquid velocity
  gas-liquid interface inferred from experimental approach, 6–10
  experimental setup, 4–5
  proposed experimental method, validation of, 10–11
Loading and unloading curves, 39–40
Local-field correlation particle image velocimetry (LFCPIV) method, 99
Long-term shear modulus (G_∞), 43
Lycopersicon esculentum, 56

Markov chain Monte Carlo (MCMC) method, 202
MATLAB ode45 solver, 50
Maxwell model, 42
ME. See Mixer effectiveness (ME)
Mean von Mises stress, 54
Mechanical characterisation of single particles, micromanipulation in. See Micromanipulation, in mechanical characterisation of single particles

Mechanotransduction, 53
Melamine formaldehyde (MF)
  microcapsules, 67
  microparticles, 71
Metal and metal oxides production reactors, 246–251
  hot wall reactor, 249–251
  laser vaporization reactor, 246–249
MF. See Melamine formaldehyde (MF)
Microcapsules, 65–67
Micromanipulation, in mechanical characterisation of single particles, 29–85
  status and applications
    biocompatible particles, 58–59
    biological particles, 51–57
    fouling deposits on surfaces, 72–75
    non-biological particles, 59–67
    particle adhesion to surface, 70–72
    particle–particle adhesion, 68–70
    sub-micron/nanoparticles, nanomanipulation of, 75–77
  techniques for
    cell poking and atomic force microscopy (AFM), 33–35
    diametrical compression. See Diametrical compression
    micropipette aspiration, 32–33
    optical trapping, 35–37
    pressure probe, 31–32
Micropipette aspiration, 32–33, 53, 68
Microspheres, 59–65
  chromatographic resins, 60–61
  pharmaceutical excipients, 61–65
Mixer effectiveness (ME), 167
Mooney–Rivlin model, 45

Nanomanipulation, of sub-micron/nanoparticles, 75–77
Nanoparticle production processes control, 237–251
  furnace black reactor, 242–245
  LII applications to, 237–251
  metal and metal oxides production reactors, 246–251
    hot wall reactor, 249–251
    laser vaporization reactor, 246–249
  research plasma reactor, 237–242
Nd:YAG lasers, 94, 103–104, 115, 117, 136, 238–239, 249, 253, 263
Neo-Hookean equations, 67

Neural network multi-criterion image reconstruction technique (NN-MOIRT), 185–188
Newton-Raphson method (NRM), 203
N-methylmorpholine-N-oxide (NMMO), 71
NMMO. See N-methylmorpholine-Noxide (NMMO)
Non-biological particles. See also Micromanipulation, in mechanical characterisation of single particles
  microcapsules, 65–67
  microspheres, 60–61
    chromatographic resins, 59–61
    pharmaceutical excipients, 61–64

Optical trapping method, 35–37
Optimization reconstruction techniques (ORT), 182
Orthogonal-plane PIV technique, 118

"Packet model", 158
Particle adhesion, to surface, 70–72
Particle collision dynamics, 160
Particle image accelerometry technique (PIA), 139
Particle image velocimetry (PIV) technique, 2
  fundamentals, 90–102, 113
    analysis, 95–101
    cross-correlation analysis, 97–98
    error elimination and accuracy improvement, 100–101
    illumination and image recording, 92, 94–95
    post-processing of velocity vectors, 101–102
    resolution improvement, 98–100
    seeding flow, 91–93
  multiphase flow measurement using, 118–140
    gas–liquid two-fluid flows, 90, 92, 119–137
    liquid–liquid two-fluid flows, 119–121
    particle-laden multiphase flows, 137–140
  seeding particles used for, 93
  types, 102–118
    2D-2C PIV techniques, 103–105
    2D-3C PIV techniques, 105–108
    3D-3C PIV techniques, 109–115

  dual-plane PIV, 117–118
  dynamic PIV, 94–95, 103, 109, 115
  orthogonal-plane PIV technique, 118
  scanning PIV, 115–116
Particle-laden multiphase flows measurement, using PIV technique, 137–139
Particle–particle adhesion, 68–70
Particle suspensions, LII applications to, 262–265
Particle tracking velocimetry (PTV) algorithms, 98, 107, 112, 133–135, 140
PCM. See Pericellular matrix (PCM)
Peak-locking error, 100–101
Peak ultrasound echo intensity
  gas–liquid interface inferred from, 11
  experimental approach, validation of, 18–24
  experimental setup, 12–13
  method, experimental, 13–18
PEPT. See Positron emission particle tracking (PEPT)
Pericellular matrix (PCM), 53
PET. See Positron emission tomography (PET)
Piezoelectric scanner, 34
Piezoelectric stack, for compression method, 38
PIV. See Particle image velocimetry (PIV)
2D-2C PIV techniques, 94, 103–105
  macro scale, 103
  micro scale, 103–105
2D-3C PIV techniques, 105–108
  macro scale, 105–108
  micro scale, 108
3D-3C PIV techniques, 109–114
  DPIV technique, 111–112
  HPIV technique, 109–111, 114
  macro scale, 109–114
  micro scale, 114
  TPIV technique, 112–114
Plane stress, defined, 45
Plant cells, 56–57
Plasmodium falciparum, 52
Pluronic F68, 52
PMMA nanoparticles. See Polymethylmethacrylate (PMMA) nanoparticles
Pneumatic solid conveying, ECT applications in, 190–191

Poisson ratio, 33
Polymethylmethacrylate (PMMA) nanoparticles, 76
Polyurethane microcapsules, 67
Portable PEPT, 168–174. *See also* Positron emission particle tracking (PEPT)
Positron emission particle tracking (PEPT), 151–153
  applications of, 169
    fluidised beds, 156–160
    rotating drums and kilns, 162–163
    solids mixing, 163–168
  detectors, 154–155
  and PET, difference between, 151
  portable, 168, 171–173
  positron-emitting tracers, 153–154
  principles of, 151–152
  technique development, 155
Positron emission tomography (PET), 180–182, 209–215
  applications, 211–215
    dispersion of particle pulse in gas-solid fluidized beds, 212–214
    slurry mixtures in stirred tanks, 211–212
    visualization of multi-phase fluids through sudden expansions, 213–214
  principle, 209–211
Positron emission tomography (PET) and PEPT, difference between, 151
Positron-emitting tracers, 153–154
Positron Imaging Centre at Birmingham, 154
Pressure probe, 31–32, 57
Primary particle size and its distribution, LII signal and determination of, 228–236
Pseudomonas fluorescens, 73
Pulsed lasers, 90, 91, 94, 115

Radioactive particle tracking (RPT), 181
Raman spectroscopy, 36
Ramp correction factor (RCF*i*), 43
Raw exhaust gas measurements, in automotive soot investigations, 254–260
Rayleigh-Debye-Gans (RDG) approach, 236
RBCs. *See* Red blood cells (RBCs)
RCF*i*. *See* Ramp correction factor (RCF*i*)

Red blood cells (RBCs), 32
  stretching of, using optical trapping method, 36
Research plasma reactor, 237–242
Residence time, defined, 158–159
Resins, chromatographic, 60–61
Rotating drums and kilns, 162–163
Runge–Kutta method, 50

Saccharomyces cerevisiae, 35, 54–55
Saccharopolyspora erythraea, 56
Scanners, for medical PET, 155
Scanning electron microscopes (SEMs), 76
Scanning mobility particle sizers (SMPS), 224
Scanning PIV, 115–116
SEMs. *See* Scanning electron microscopes (SEMs)
Sensitivity conjugate gradients (SCG) method, 203
Shadow image technique (SIT), 127–136, 140
SIRT, 186–187
Skalak–Tozeren–Zarda–Chien (STZC) material relationship, 45
Slurry mixtures in stirred tanks, PET applications in, 211–212
Sobel filter, 8, 9
Solids mixing, PEPT and PET used in, 163–168
Sound, use of, in science and technology, 2
Speckle correlation velocimetry, 3
Standard stirred tank reactors (STR), 207, 211–212, 216
Starlac™, 63
Stereoscopic μ-PIV technique, 108
Stereoscopic PIV system, 106–107
Strain energy, defined, 45
Stretch ratios, defined, 45
STZC. *See* Skalak-Tozeren-Zarda-Chien (STZC)
Sub-micron/nanoparticles, nanomanipulation of, 75–77

Tapered element oscillating microbalance (TEOM), 261
Tatara analysis, 41–42, 61
Time-resolved laser-induced incandescence (TIRE-LII), 223–266
Tomato. *See Lycopersicon esculentum*

Tomographic particle image velocimetry (TPIV) technique, 112–115
4V transistor–transistor logic signal, 13
Transmission electron microscopy (TEM), 224
Turgor (hydrostatic) pressures, within cells, 32

Ultrasonic Doppler velocimetry, 3
Ultrasonic velocity profiler (UVP) measurements, 1, 2, 4. *See also* Gas–liquid interface
Ultrasound echo intensity, 10–11, 13–15, 20
Ultrasound tomography (UST), 205
Ultrasound transducer arrangement, 14
Urea–formaldehyde microcapsules, 67
UVP–DUO systems, 2, 7, 12–13
UVP measurements. *See* Ultrasonic velocity profiler (UVP) measurements

Video microscopy, 32
Viscoelastic model, 42–44, 53, 59

Visualization of dispersions in an oscillatory baffled reactor, ERT applications in, 207–209
Visualization of multi-phase fluids through sudden expansions, PET applications in, 214
V-mixer, 163
Volumetric elastic modulus, 32

We. *See* Weber number (We)
Weber number (We), 6
Window displacement iterative multigrid (WIDIM) interrogation method, 99

XPS. *See* X-ray photoelectron spectroscopy (XPS)
X-ray photoelectron spectroscopy (XPS), 72

Yeast. *See Saccharomyces cerevisiae*
Young's modulus, 33, 35, 53–54

Contents of Volumes in This Serial

Volume 1 (1956)

J. W. Westwater, *Boiling of Liquids*
A. B. Metzner, *Non-Newtonian Technology: Fluid Mechanics, Mixing, and Heat Transfer*
R. Byron Bird, *Theory of Diffusion*
J. B. Opfell and B. H. Sage, *Turbulence in Thermal and Material Transport*
Robert E. Treybal, *Mechanically Aided Liquid Extraction*
Robert W. Schrage, *The Automatic Computer in the Control and Planning of Manufacturing Operations*
Ernest J. Henley and Nathaniel F. Barr, *Ionizing Radiation Applied to Chemical Processes and to Food and Drug Processing*

Volume 2 (1958)

J. W. Westwater, *Boiling of Liquids*
Ernest F. Johnson, *Automatic Process Control*
Bernard Manowitz, *Treatment and Disposal of Wastes in Nuclear Chemical Technology*
George A. Sofer and Harold C. Weingartner, *High Vacuum Technology*
Theodore Vermeulen, *Separation by Adsorption Methods*
Sherman S. Weidenbaum, *Mixing of Solids*

Volume 3 (1962)

C. S. Grove, Jr., Robert V. Jelinek, and Herbert M. Schoen, *Crystallization from Solution*
F. Alan Ferguson and Russell C. Phillips, *High Temperature Technology*
Daniel Hyman, *Mixing and Agitation*
John Beck, *Design of Packed Catalytic Reactors*
Douglass J. Wilde, *Optimization Methods*

Volume 4 (1964)

J. T. Davies, *Mass-Transfer and Inierfacial Phenomena*
R. C. Kintner, *Drop Phenomena Affecting Liquid Extraction*
Octave Levenspiel and Kenneth B. Bischoff, *Patterns of Flow in Chemical Process Vessels*
Donald S. Scott, *Properties of Concurrent Gas–Liquid Flow*
D. N. Hanson and G. F. Somerville, *A General Program for Computing Multistage Vapor–Liquid Processes*

Volume 5 (1964)

J. F. Wehner, *Flame Processes—Theoretical and Experimental*
J. H. Sinfelt, *Bifunctional Catalysts*
S. G. Bankoff, *Heat Conduction or Diffusion with Change of Phase*
George D. Fulford, *The Flow of Lktuids in Thin Films*
K. Rietema, *Segregation in Liquid–Liquid Dispersions and its Effects on Chemical Reactions*

Volume 6 (1966)

S. G. Bankoff, *Diffusion-Controlled Bubble Growth*
John C. Berg, Andreas Acrivos, and Michel Boudart, *Evaporation Convection*
H. M. Tsuchiya, A. G. Fredrickson, and R. Aris, *Dynamics of Microbial Cell Populations*
Samuel Sideman, *Direct Contact Heat Transfer between Immiscible Liquids*
Howard Brenner, *Hydrodynamic Resistance of Particles at Small Reynolds Numbers*

Volume 7 (1968)

Robert S. Brown, Ralph Anderson, and Larry J. Shannon, *Ignition and Combustion of Solid Rocket Propellants*
Knud Østergaard, *Gas–Liquid–Particle Operations in Chemical Reaction Engineering*
J. M. Prausnilz, *Thermodynamics of Fluid–Phase Equilibria at High Pressures*
Robert V. Macbeth, *The Burn-Out Phenomenon in Forced-Convection Boiling*
William Resnick and Benjamin Gal-Or, *Gas–Liquid Dispersions*

Volume 8 (1970)

C. E. Lapple, *Electrostatic Phenomena with Particulates*
J. R. Kittrell, *Mathematical Modeling of Chemical Reactions*
W. P. Ledet and D. M. Himmelblau, *Decomposition Procedures foe the Solving of Large Scale Systems*
R. Kumar and N. R. Kuloor, *The Formation of Bubbles and Drops*

Volume 9 (1974)

Renato G. Bautista, *Hydrometallurgy*
Kishan B. Mathur and Norman Epstein, *Dynamics of Spouted Beds*
W. C. Reynolds, *Recent Advances in the Computation of Turbulent Flows*
R. E. Peck and D. T. Wasan, *Drying of Solid Particles and Sheets*

Volume 10 (1978)

G. E. O'Connor and T. W. F. Russell, *Heat Transfer in Tubular Fluid–Fluid Systems*
P. C. Kapur, *Balling and Granulation*
Richard S. H. Mah and Mordechai Shacham, *Pipeline Network Design and Synthesis*
J. Robert Selman and Charles W. Tobias, *Mass-Transfer Measurements by the Limiting-Current Technique*

Volume 11 (1981)

Jean-Claude Charpentier, *Mass-Transfer Rates in Gas–Liquid Absorbers and Reactors*
Dee H. Barker and C. R. Mitra, *The Indian Chemical Industry—Its Development and Needs*
Lawrence L. Tavlarides and Michael Stamatoudis, *The Analysis of Interphase Reactions and Mass Transfer in Liquid–Liquid Dispersions*
Terukatsu Miyauchi, Shintaro Furusaki, Shigeharu Morooka, and Yoneichi Ikeda, *Transport Phenomena and Reaction in Fluidized Catalyst Beds*

Volume 12 (1983)

C. D. Prater, J, Wei, V. W. Weekman, Jr., and B. Gross, *A Reaction Engineering Case History: Coke Burning in Thermofor Catalytic Cracking Regenerators*
Costel D. Denson, *Stripping Operations in Polymer Processing*
Robert C. Reid, *Rapid Phase Transitions from Liquid to Vapor*
John H. Seinfeld, *Atmospheric Diffusion Theory*

Volume 13 (1987)

Edward G. Jefferson, *Future Opportunities in Chemical Engineering*
Eli Ruckenstein, *Analysis of Transport Phenomena Using Scaling and Physical Models*
Rohit Khanna and John H. Seinfeld, *Mathematical Modeling of Packed Bed Reactors: Numerical Solutions and Control Model Development*
Michael P. Ramage, Kenneth R. Graziano, Paul H. Schipper, Frederick J. Krambeck, and Byung C. Choi, *KINPTR (Mobil's Kinetic Reforming Model): A Review of Mobil's Industrial Process Modeling Philosophy*

Volume 14 (1988)

Richard D. Colberg and Manfred Morari, *Analysis and Synthesis of Resilient Heat Exchange Networks*
Richard J. Quann, Robert A. Ware, Chi-Wen Hung, and James Wei, *Catalytic Hydrometallation of Petroleum*
Kent David, *The Safety Matrix: People Applying Technology to Yield Safe Chemical Plants and Products*

Volume 15 (1990)

Pierre M. Adler, Ali Nadim, and Howard Brenner, *Rheological Models of Suspenions*
Stanley M. Englund, *Opportunities in the Design of Inherently Safer Chemical Plants*
H. J. Ploehn and W. B. Russel, *Interations between Colloidal Particles and Soluble Polymers*

Volume 16 (1991)

Perspectives in Chemical Engineering: Research and Education
Clark K. Colton, *Editor*
Historical Perspective and Overview
L. E. Scriven, *On the Emergence and Evolution of Chemical Engineering*
Ralph Landau, *Academic—industrial Interaction in the Early Development of Chemical Engineering*

James Wei, *Future Directions of Chemical Engineering*
Fluid Mechanics and Transport
L. G. Leal, *Challenges and Opportunities in Fluid Mechanics and Transport Phenomena*
William B. Russel, *Fluid Mechanics and Transport Research in Chemical Engineering*
J. R. A. Pearson, *Fluid Mechanics and Transport Phenomena*
Thermodynamics
Keith E. Gubbins, *Thermodynamics*
J. M. Prausnitz, *Chemical Engineering Thermodynamics: Continuity and Expanding Frontiers*
H. Ted Davis, *Future Opportunities in Thermodynamics*
Kinetics, Catalysis, and Reactor Engineering
Alexis T. Bell, *Reflections on the Current Status and Future Directions of Chemical Reaction Engineering*
James R. Katzer and S. S. Wong, *Frontiers in Chemical Reaction Engineering*
L. Louis Hegedus, *Catalyst Design*
Environmental Protection and Energy
John H. Seinfeld, *Environmental Chemical Engineering*
T. W. F. Russell, *Energy and Environmental Concerns*
Janos M. Beer, Jack B. Howard, John P. Longwell, and Adel F. Sarofim, *The Role of Chemical Engineering in Fuel Manufacture and Use of Fuels*
Polymers
Matthew Tirrell, *Polymer Science in Chemical Engineering*
Richard A. Register and Stuart L. Cooper, *Chemical Engineers in Polymer Science: The Need for an Interdisciplinary Approach*
Microelectronic and Optical Material
Larry F. Thompson, *Chemical Engineering Research Opportunities in Electronic and Optical Materials Research*
Klavs F. Jensen, *Chemical Engineering in the Processing of Electronic and Optical Materials: A Discussion*
Bioengineering
James E. Bailey, *Bioprocess Engineering*
Arthur E. Humphrey, *Some Unsolved Problems of Biotechnology*
Channing Robertson, *Chemical Engineering: Its Role in the Medical and Health Sciences*
Process Engineering
Arthur W. Westerberg, *Process Engineering*
Manfred Morari, *Process Control Theory: Reflections on the Past Decade and Goals for the Next*
James M. Douglas, *The Paradigm After Next*
George Stephanopoulos, *Symbolic Computing and Artificial Intelligence in Chemical Engineering: A New Challenge*
The Identity of Our Profession
Morton M. Denn, *The Identity of Our Profession*

Volume 17 (1991)

Y. T. Shah, *Design Parameters for Mechanically Agitated Reactors*
Mooson Kwauk, *Particulate Fluidization: An Overview*

Volume 18 (1992)

E. James Davis, *Microchemical Engineering: The Physics and Chemistry of the Microparticle*
Selim M. Senkan, *Detailed Chemical Kinetic Modeling: Chemical Reaction Engineering of the Future*
Lorenz T. Biegler, *Optimization Strategies for Complex Process Models*

Volume 19 (1994)

Robert Langer, *Polymer Systems for Controlled Release of Macromolecules, Immobilized Enzyme Medical Bioreactors, and Tissue Engineering*
J. J. Linderman, P. A. Mahama, K. E. Forsten, and D. A. Lauffenburger, *Diffusion and Probability in Receptor Binding and Signaling*
Rakesh K. Jain, *Transport Phenomena in Tumors*
R. Krishna, *A Systems Approach to Multiphase Reactor Selection*
David T. Allen, *Pollution Prevention: Engineering Design at Macro-, Meso-, and Microscales*
John H. Seinfeld, Jean M. Andino, Frank M. Bowman, Hali J. L. Forstner, and Spyros Pandis, *Tropospheric Chemistry*

Volume 20 (1994)

Arthur M. Squires, *Origins of the Fast Fluid Bed*
Yu Zhiqing, *Application Collocation*
Youchu Li, *Hydrodynamics*
Li Jinghai, *Modeling*
Yu Zhiqing and Jin Yong, *Heat and Mass Transfer*
Mooson Kwauk, *Powder Assessment*
Li Hongzhong, *Hardware Development*
Youchu Li and Xuyi Zhang, *Circulating Fluidized Bed Combustion*
Chen Junwu, Cao Hanchang, and Liu Taiji, *Catalyst Regeneration in Fluid Catalytic Cracking*

Volume 21 (1995)

Christopher J. Nagel, Chonghum Han, and George Stephanopoulos, *Modeling Languages: Declarative and Imperative Descriptions of Chemical Reactions and Processing Systems*
Chonghun Han, George Stephanopoulos, and James M. Douglas, *Automation in Design: The Conceptual Synthesis of Chemical Processing Schemes*
Michael L. Mavrovouniotis, *Symbolic and Quantitative Reasoning: Design of Reaction Pathways through Recursive Satisfaction of Constraints*
Christopher Nagel and George Stephanopoulos, *Inductive and Deductive Reasoning: The Case of Identifying Potential Hazards in Chemical Processes*
Keven G. Joback and George Stephanopoulos, *Searching Spaces of Discrete Solutions: The Design of Molecules Processing Desired Physical Properties*

Volume 22 (1995)

Chonghun Han, Ramachandran Lakshmanan, Bhavik Bakshi, and George Stephanopoulos, *Nonmonotonic Reasoning: The Synthesis of Operating Procedures in Chemical Plants*
Pedro M. Saraiva, *Inductive and Analogical Learning: Data-Driven Improvement of Process Operations*
Alexandros Koulouris, Bhavik R. Bakshi and George Stephanopoulos, *Empirical Learning through Neural Networks: The Wave-Net Solution*
Bhavik R. Bakshi and George Stephanopoulos, *Reasoning in Time: Modeling, Analysis, and Pattern Recognition of Temporal Process Trends*
Matthew J. Realff, *Intelligence in Numerical Computing: Improving Batch Scheduling Algorithms through Explanation-Based Learning*

Volume 23 (1996)

Jeffrey J. Siirola, *Industrial Applications of Chemical Process Synthesis*
Arthur W. Westerberg and Oliver Wahnschafft, *The Synthesis of Distillation-Based Separation Systems*
Ignacio E. Grossmann, *Mixed-Integer Optimization Techniques for Algorithmic Process Synthesis*
Subash Balakrishna and Lorenz T. Biegler, *Chemical Reactor Network Targeting and Integration: An Optimization Approach*
Steve Walsh and John Perkins, *Operability and Control inn Process Synthesis and Design*

Volume 24 (1998)

Raffaella Ocone and Gianni Astarita, *Kinetics and Thermodynamics in Multicomponent Mixtures*
Arvind Varma, Alexander S. Rogachev, Alexandra S. Mukasyan, and Stephen Hwang, *Combustion Synthesis of Advanced Materials: Principles and Applications*
J. A. M. Kuipers and W. P. Mo, van Swaaij, *Computational Fluid Dynamics Applied to Chemical Reaction Engineering*
Ronald E. Schmitt, Howard Klee, Debora M. Sparks, and Mahesh K. Podar, *Using Relative Risk Analysis to Set Priorities for Pollution Prevention at a Petroleum Refinery*

Volume 25 (1999)

J. F. Davis, M. J. Piovoso, K. A. Hoo, and B. R. Bakshi, *Process Data Analysis and Interpretation*
J. M. Ottino, P. DeRoussel, S., Hansen, and D. V. Khakhar, *Mixing and Dispersion of Viscous Liquids and Powdered Solids*
Peter L. Silverston, Li Chengyue, Yuan Wei-Kang, *Application of Periodic Operation to Sulfur Dioxide Oxidation*

Volume 26 (2001)

J. B. Joshi, N. S. Deshpande, M. Dinkar, and D. V. Phanikumar, *Hydrodynamic Stability of Multiphase Reactors*
Michael Nikolaou, *Model Predictive Controllers: A Critical Synthesis of Theory and Industrial Needs*

Volume 27 (2001)

William R. Moser, Josef Find, Sean C. Emerson, and Ivo M, Krausz, *Engineered Synthesis of Nanostructure Materials and Catalysts*
Bruce C. Gates, *Supported Nanostructured Catalysts: Metal Complexes and Metal Clusters*
Ralph T. Yang, *Nanostructured Absorbents*
Thomas J. Webster, *Nanophase Ceramics: The Future Orthopedic and Dental Implant Material*
Yu-Ming Lin, Mildred S. Dresselhaus, and Jackie Y. Ying, *Fabrication, Structure, and Transport Properties of Nanowires*

Volume 28 (2001)

Qiliang Yan and Juan J. DePablo, *Hyper-Parallel Tempering Monte Carlo and Its Applications*
Pablo G. Debenedetti, Frank H. Stillinger, Thomas M. Truskett, and Catherine P. Lewis, *Theory of Supercooled Liquids and Glasses: Energy Landscape and Statistical Geometry Perspectives*

Michael W. Deem, *A Statistical Mechanical Approach to Combinatorial Chemistry*
Venkat Ganesan and Glenn H. Fredrickson, *Fluctuation Effects in Microemulsion Reaction Media*
David B. Graves and Cameron F. Abrams, *Molecular Dynamics Simulations of Ion–Surface Interactions with Applications to Plasma Processing*
Christian M. Lastoskie and Keith E, Gubbins, *Characterization of Porous Materials Using Molecular Theory and Simulation*
Dimitrios Maroudas, *Modeling of Radical-Surface Interactions in the Plasma-Enhanced Chemical Vapor Deposition of Silicon Thin Films*
Sanat Kumar, M. Antonio Floriano, and Athanassiors Z. Panagiotopoulos, *Nanostructured Formation and Phase Separation in Surfactant Solutions*
Stanley I. Sandler, Amadeu K. Sum, and Shiang-Tai Lin, *Some Chemical Engineering Applications of Quantum Chemical Calculations*
Bernhardt L. Trout, *Car-Parrinello Methods in Chemical Engineering: Their Scope and potential*
R. A. van Santen and X. Rozanska, *Theory of Zeolite Catalysis*
Zhen-Gang Wang, *Morphology, Fluctuation, Metastability and Kinetics in Ordered Block Copolymers*

Volume 29 (2004)

Michael V. Sefton, *The New Biomaterials*
Kristi S. Anseth and Kristyn S. Masters, *Cell–Material Interactions*
Surya K. Mallapragada and Jennifer B. Recknor, *Polymeric Biomaterias for Nerve Regeneration*
Anthony M. Lowman, Thomas D. Dziubla, Petr Bures, and Nicholas A. Peppas, *Structural and Dynamic Response of Neutral and Intelligent Networks in Biomedical Environments*
F. Kurtis Kasper and Antonios G. Mikos, *Biomaterials and Gene Therapy*
Balaji Narasimhan and Matt J. Kipper, *Surface-Erodible Biomaterials for Drug Delivery*

Volume 30 (2005)

Dionisio Vlachos, *A Review of Multiscale Analysis: Examples from System Biology, Materials Engineering, and Other Fluids-Surface Interacting Systems*
Lynn F. Gladden, M.D. Mantle and A.J. Sederman, *Quantifying Physics and Chemistry at Multiple Length- Scales using Magnetic Resonance Techniques*
Juraj Kosek, Frantisek Steěpánek, and Miloš Marek, *Modelling of Transport and Transformation Processes in Porous and Multiphase Bodies*
Vemuri Balakotaiah and Saikat Chakraborty, *Spatially Averaged Multiscale Models for Chemical Reactors*

Volume 31 (2006)

Yang Ge and Liang-Shih Fan, *3-D Direct Numerical Simulation of Gas–Liquid and Gas–Liquid–Solid Flow Systems Using the Level-Set and Immersed-Boundary Methods*
M.A. van der Hoef, M. Ye, M. van Sint Annaland, A.T. Andrews IV, S. Sundaresan, and J.A.M. Kuipers, *Multiscale Modeling of Gas-Fluidized Beds*
Harry E.A. Van den Akker, *The Details of Turbulent Mixing Process and their Simulation*
Rodney O. Fox, *CFD Models for Analysis and Design of Chemical Reactors*
Anthony G. Dixon, Michiel Nijemeisland, and E. Hugh Stitt, *Packed Tubular Reactor Modeling and Catalyst Design Using Computational Fluid Dynamics*

Volume 32 (2007)

William H. Green, Jr., *Predictive Kinetics: A New Approach for the 21st Century*
Mario Dente, Giulia Bozzano, Tiziano Faravelli, Alessandro Marongiu, Sauro Pierucci and Eliseo Ranzi, *Kinetic Modelling of Pyrolysis Processes in Gas and Condensed Phase*
Mikhail Sinev, Vladimir Arutyunov and Andrey Romanets, *Kinetic Models of C_1–C_4 Alkane Oxidation as Applied to Processing of Hydrocarbon Gases: Principles, Approaches and Developments*
Pierre Galtier, *Kinetic Methods in Petroleum Process Engineering*

Volume 33 (2007)

Shinichi Matsumoto and Hirofumi Shinjoh, *Dynamic Behavior and Characterization of Automobile Catalysts*
Mehrdad Ahmadinejad, Maya R. Desai, Timothy C. Watling and Andrew P.E. York, *Simulation of Automotive Emission Control Systems*
Anke Güthenke, Daniel Chatterjee, Michel Weibel, Bernd Krutzsch, Petr Kočí, Miloš Marek, Isabella Nova and Enrico Tronconi, *Current Status of Modeling Lean Exhaust Gas Aftertreatment Catalysts*
Athanasios G. Konstandopoulos, Margaritis Kostoglou, Nickolas Vlachos and Evdoxia Kladopoulou, *Advances in the Science and Technology of Diesel Particulate Filter Simulation*

Volume 34 (2008)

C.J. van Duijn, Andro Mikelic', I.S. Pop, and Carole Rosier, *Effective Dispersion Equations for Reactive Flows with Dominant Péclet and Damkohler Numbers*
Mark Z. Lazman and Gregory S. Yablonsky, *Overall Reaction Rate Equation of Single-Route Complex Catalytic Reaction in Terms of Hypergeometric Series*
A.N. Gorban and O. Radulescu, *Dynamic and Static Limitation in Multiscale Reaction Networks, Revisited*
Liqiu Wang, Mingtian Xu, and Xiaohao Wei, *Multiscale Theorems*

Volume 35 (2009)

Rudy J. Koopmans and Anton P.J. Middelberg, *Engineering Materials from the Bottom Up – Overview*
Robert P.W. Davies, Amalia Aggeli, Neville Boden, Tom C.B. McLeish, Irena A. Nyrkova, and Alexander N. Semenov, *Mechanisms and Principles of 1 D Self-Assembly of Peptides into β-Sheet Tapes*
Paul van der Schoot, *Nucleation and Co-Operativity in Supramolecular Polymers*
Michael J. McPherson, Kier James, Stuart Kyle, Stephen Parsons, and Jessica Riley, *Recombinant Production of Self-Assembling Peptides*
Boxun Leng, Lei Huang, and Zhengzhong Shao, *Inspiration from Natural Silks and Their Proteins*
Sally L. Gras, *Surface- and Solution-Based Assembly of Amyloid Fibrils for Biomedical and Nanotechnology Applications*
Conan J. Fee, *Hybrid Systems Engineering: Polymer-Peptide Conjugates*

Volume 36 (2009)

Vincenzo Augugliaro, Sedat Yurdakal, Vittorio Loddo, Giovanni Palmisano, and Leonardo Palmisano, *Determination of Photoadsorption Capacity of Polychrystalline TiO_2 Catalyst in Irradiated Slurry*

Marta I. Litter, *Treatment of Chromium, Mercury, Lead, Uranium, and Arsenic in Water by Heterogeneous Photocatalysis*

Aaron Ortiz-Gomez, Benito Serrano-Rosales, Jesus Moreira-del-Rio, and Hugo de-Lasa, *Mineralization of Phenol in an Improved Photocatalytic Process Assisted with Ferric Ions: Reaction Network and Kinetic Modeling*

R.M. Navarro, F. del Valle, J.A. Villoria de la Mano, M.C. Alvarez-Galva'n, and J.L.G. Fierro, *Photocatalytic Water Splitting Under Visible Light: Concept and Catalysts Development*

Ajay K. Ray, *Photocatalytic Reactor Configurations for Water Purification: Experimentation and Modeling*

Camilo A. Arancibia-Bulnes, Antonio E. Jiménez, and Claudio A. Estrada, *Development and Modeling of Solar Photocatalytic Reactors*

Orlando M. Alfano and Alberto E. Cassano, *Scaling-Up of Photoreactors: Applications to Advanced Oxidation Processes*

Yaron Paz, *Photocatalytic Treatment of Air: From Basic Aspects to Reactors*

Volume 37 (2009)

S. Roberto Gonzalez A., Yuichi Murai, and Yasushi Takeda, *Ultrasound-Based Gas–Liquid Interface Detection in Gas–Liquid Two-Phase Flows*

Z. Zhang, J. D. Stenson, and C. R. Thomas, *Micromanipulation in Mechanical Characterisation of Single Particles*

Feng-Chen Li and Koichi Hishida, *Particle Image Velocimetry Techniques and Its Applications in Multiphase Systems*

J. P. K. Seville, A. Ingram, X. Fan, and D. J. Parker, *Positron Emission Imaging in Chemical Engineering*

Fei Wang, Qussai Marashdeh, Liang-Shih Fan, and Richard A. Williams, *Electrical Capacitance, Electrical Resistance, and Positron Emission Tomography Techniques and Their Applications in Multi-Phase Flow Systems*

Alfred Leipertz and Roland Sommer, *Time-Resolved Laser-Induced Incandescence*

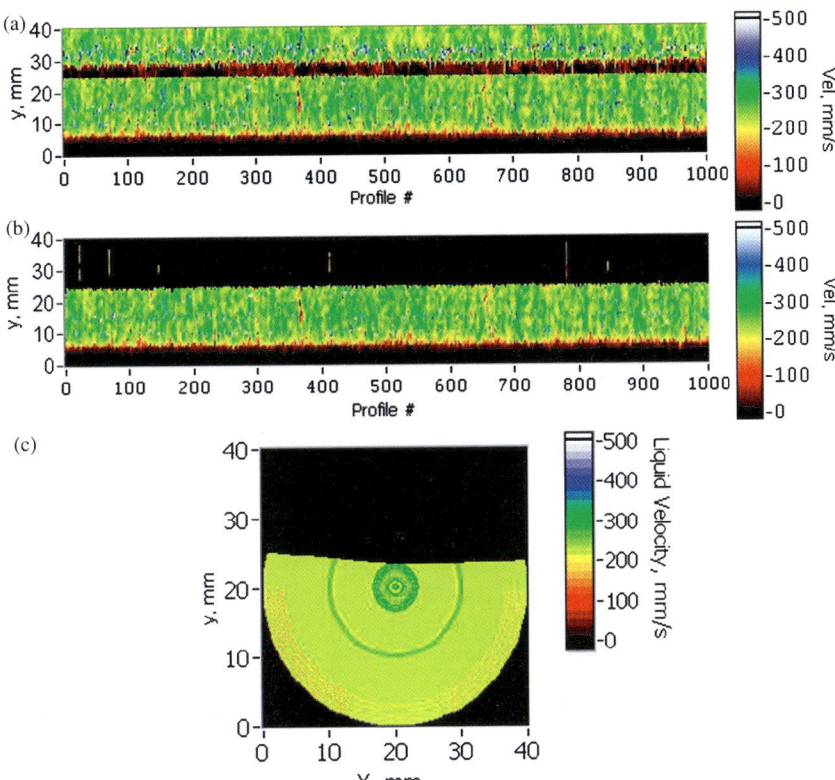

Plate 1 $Q_a = 0.6 \, \text{m}^3/\text{h}$, $\alpha = 0\%$: (a) raw velocity values, (b) liquid velocity values after gas–liquid interface is detected, and (c) flow map, velocity profile # 50 (for Black and White version, see page 21).

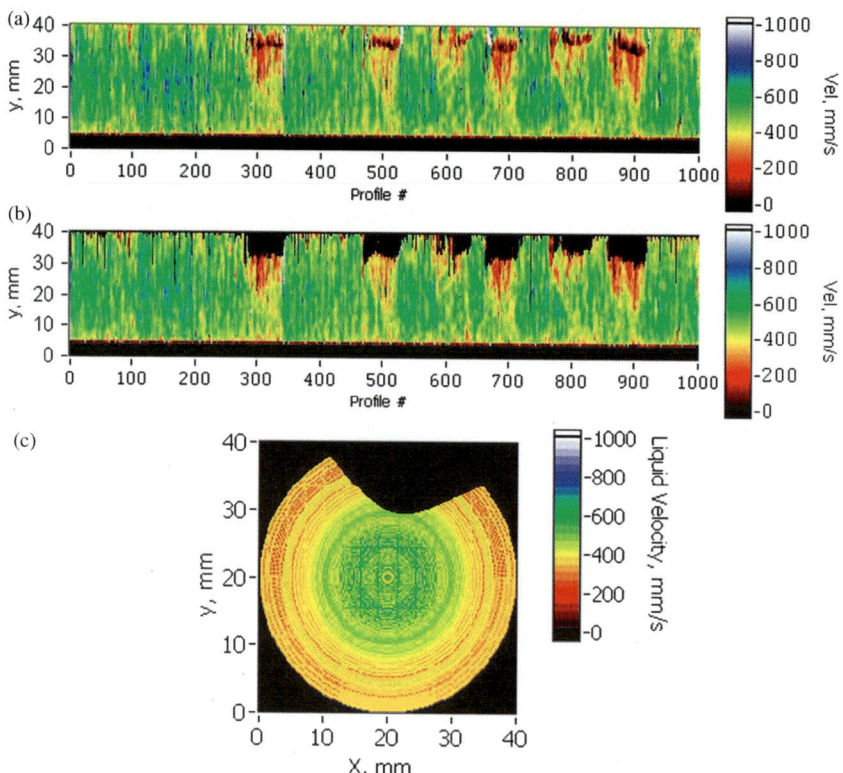

Plate 2 $Q_a = 1.8 \, m^3/h$, $\alpha = 10\%$: (a) raw velocity values, (b) liquid velocity values after gas—liquid interface is detected, and (c) flow map, velocity profile # 495 (for Black and White version, see page 22).

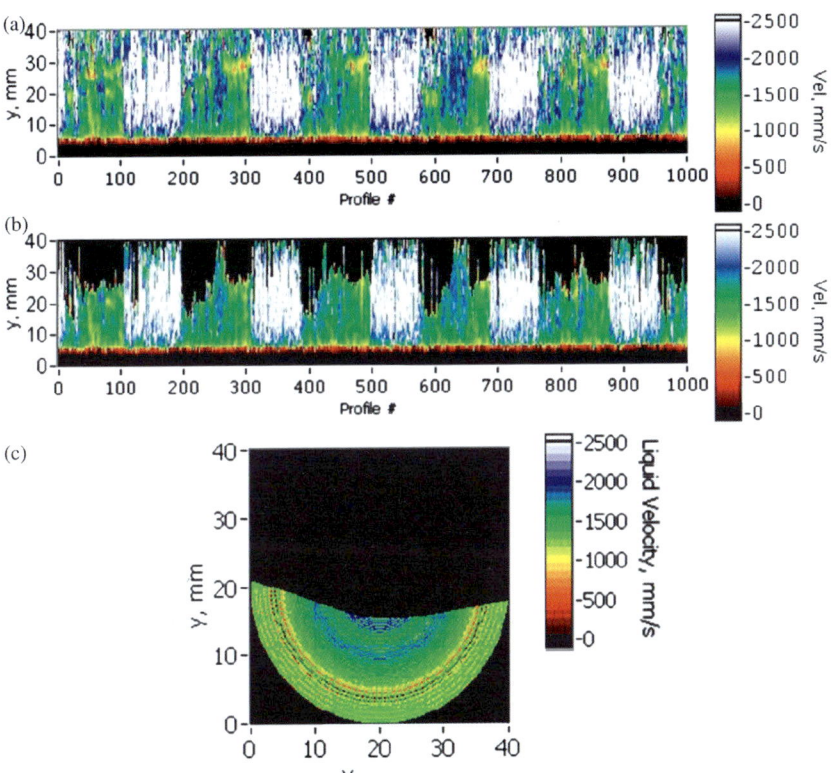

Plate 3 $Q_a = 6.2\,m^3/h$, $\alpha = 30\%$: (a) raw velocity values, (b) liquid velocity values after gas–liquid interface is detected, and (c) flow map, velocity profile # 221 (for Black and White version, see page 23).

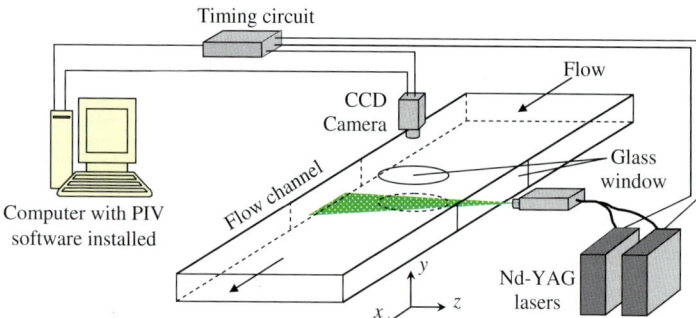

Plate 4 Diagram of a generalized 2D-PIV setup showing all major components: flow channel with the particle seeded fluid flow, laser sheet pulses illuminating one plane in the fluid, a CCD camera imaging the particles in the laser-illuminated sheet in the area of interest, a computer with PIV software installed, a timing circuit communicating with the camera and computer and generating pulses to control the double-pulsed laser. The PIV software setups and controls the major components, and analyses the images to derive a vector representation of flow field (for Black and White version, see page 91).

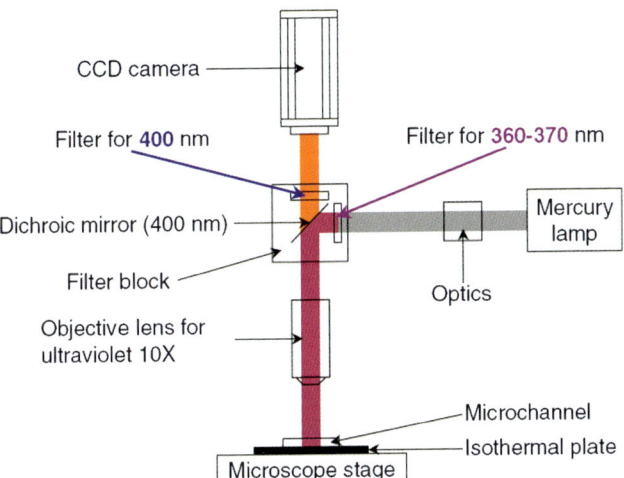

Plate 5 Schematic diagram of a typical μ-PIV (Sato et al., 2003). The light source can be mercury lamp or double-pulsed ND:YAG laser (for Black and White version, see page 104).

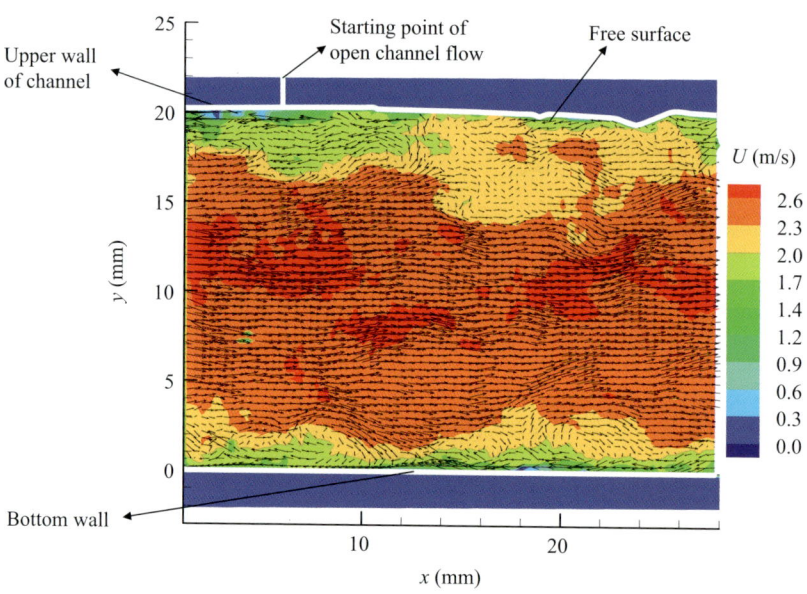

Plate 6 An example of the processed PIV frame, including turbulent velocity field and the position of the wavy free surface (Li et al., 2005c) (for Black and White version, see page 124).

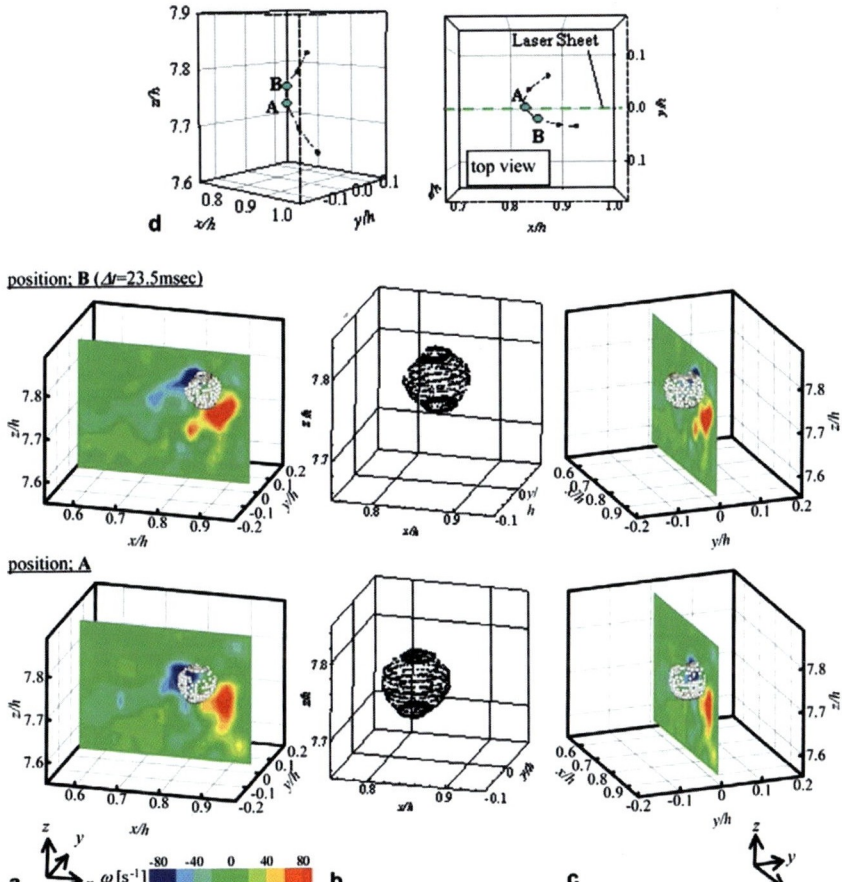

Plate 7 An example of the approximated 3D bubble shape and corresponding flow structure estimated from the measurement using PIV/LIF combining with double-SIT system: (a) characteristic vorticity structure around the bubble (bubble moves in the y–z plane); (b) reconstructed 3D bubble shape; (c) relation between bubble location and measured plane for PIV; and (d) 3D bubble trajectory (Fujiwara et al., 2004a) (for Black and White version, see page 134).

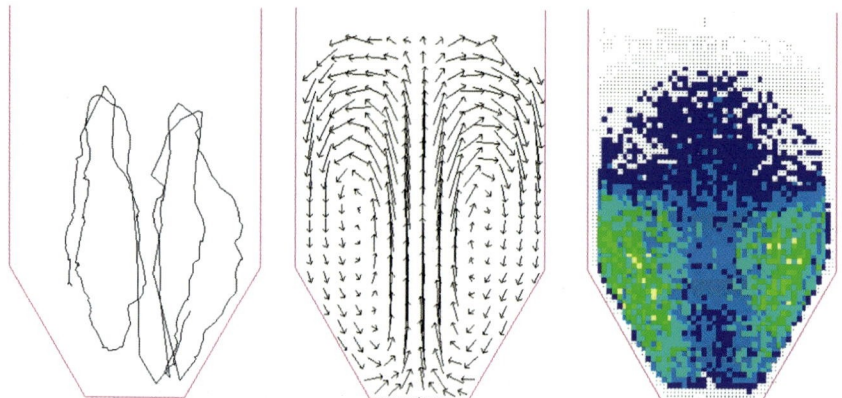

Palte 8 Typical PEPT output for a spouted bed (Left to Right: single trajectory; time-averaged velocity vectors; time-averaged "occupancy", showing denser annular region and leaner spout region) (for Black and White version, see page 153).

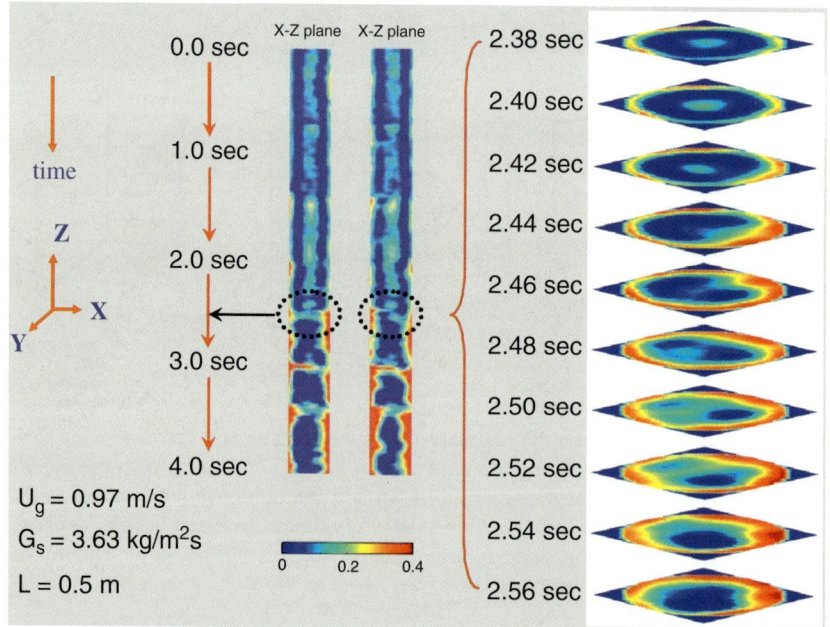

Plate 9 Quasi-3D flow structures for the choking transition at lower part of a CFB riser (Du et al., 2004a) (for Black and White version, see page 189).

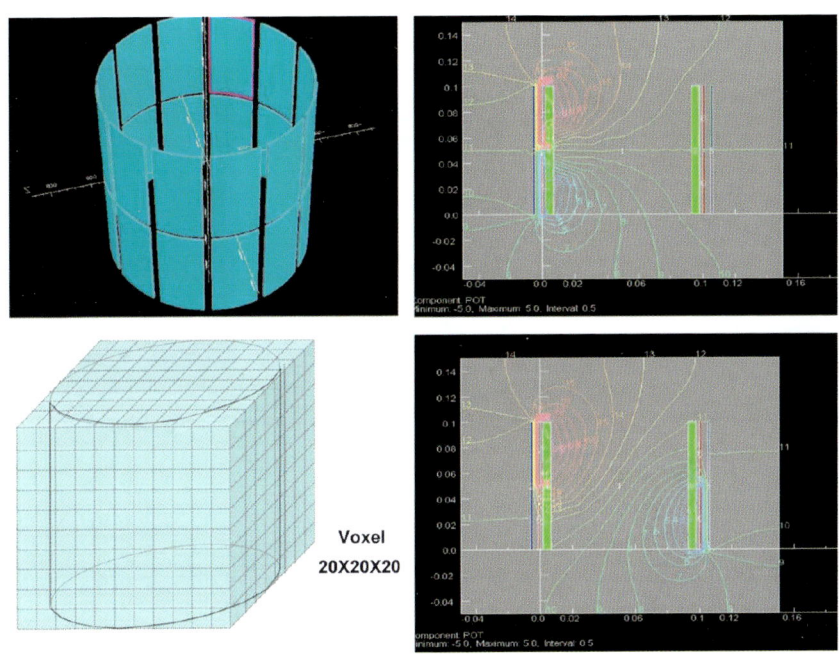

Plate 10 Three-dimensional ECVT sensor configuration and electrical potential distribution (Warsito and Fan, 2005) (for Black and White version, see page 193).

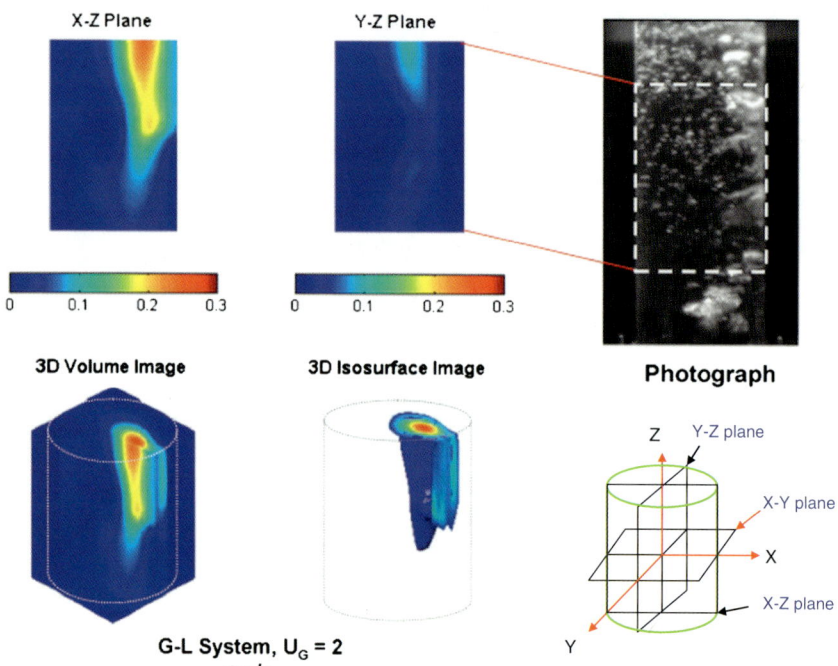

Plate 11 Three-dimensional ECVT images of gas–liquid flow in a bubble column (Warsito and Fan, 2005) (for Black and White version, see page 194).

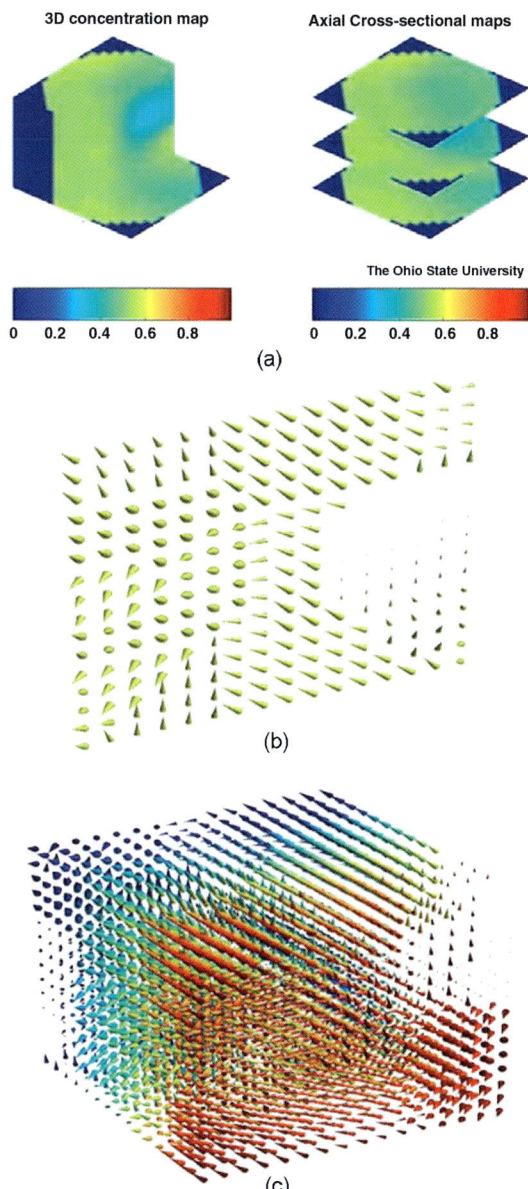

Plate 12 (a) Three-dimensional view of 3D solid concentration with a horizontal gas jet in the fluidized bed; (b) 3D voxel-volume-averaged solid phase velocity vector map in the Y-Z plane of the fluidized bed; (c) 3D voxel-volume-averaged solid phase vector map (Wang et al., 2008) (for Black and White version, see page 195).

(a)

(b)

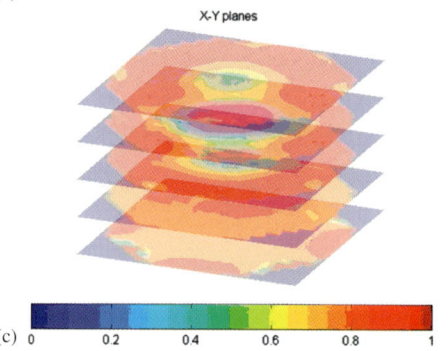

(c)

Plate 13 (a) Configuration of the 32-channel ECVT sensor; (b) fluidization system mounted with the 32-channel ECVT acquisition system; (c) liquid holdup image obtained by the 32-channel ECVT system (Wang et al., 2008) (for Black and White version, see page 197).

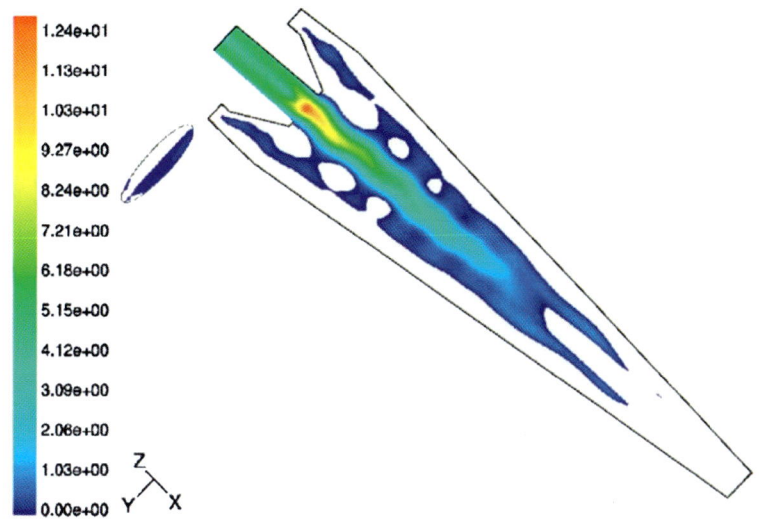

Plate 14 Axial velocity contours in hydrocyclone (Cullivan et al., 2004) (for Black and White version, see page 206).

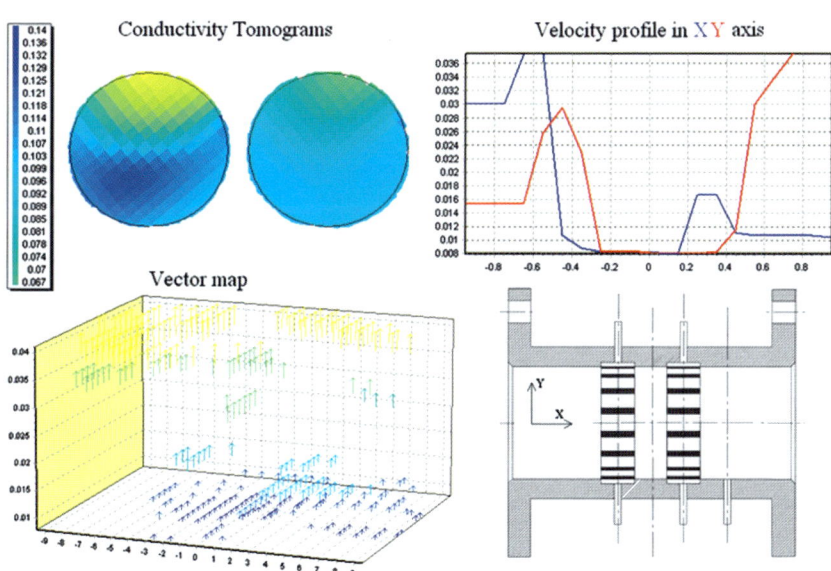

Plate 15 Tomographic images of local oil fraction distributions in oscillatory baffled reactor (top left), with estimated velocity profiles (top right) from cross-correlation signals between adjacent electrode rings (bottom right) and vector map (bottom left) (Vilar et al., 2008; Vilar, 2008) (for Black and White version, see page 208).

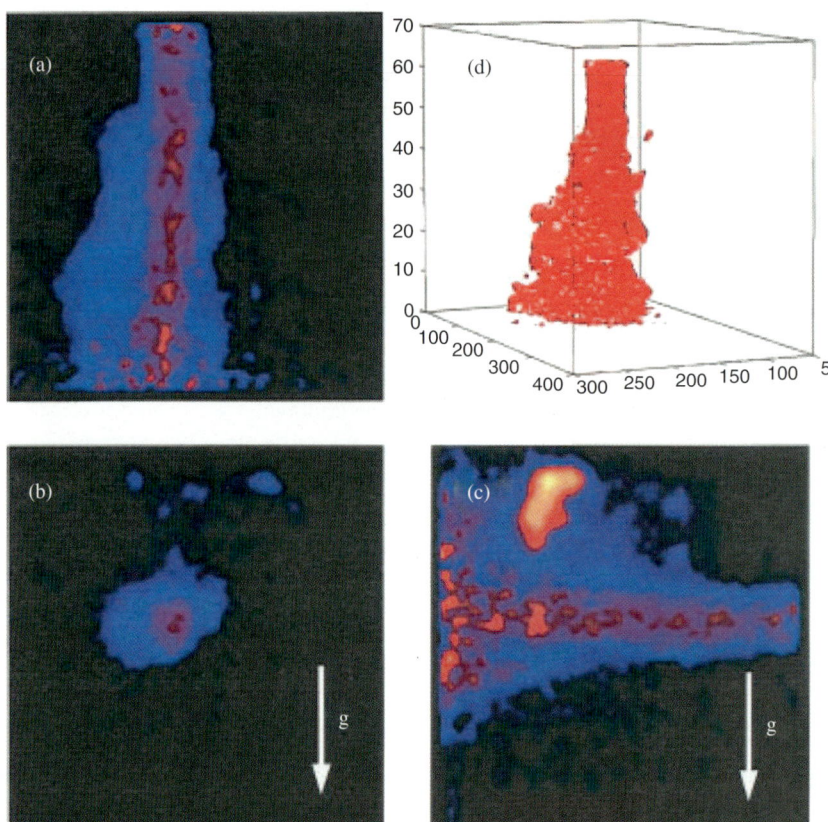

Plate 16 Flow of a fiber suspension through a sudden expansion with an upstream velocity of 0.5 m/s (Heath et al., 2007) (for Black and White version, see page 214).

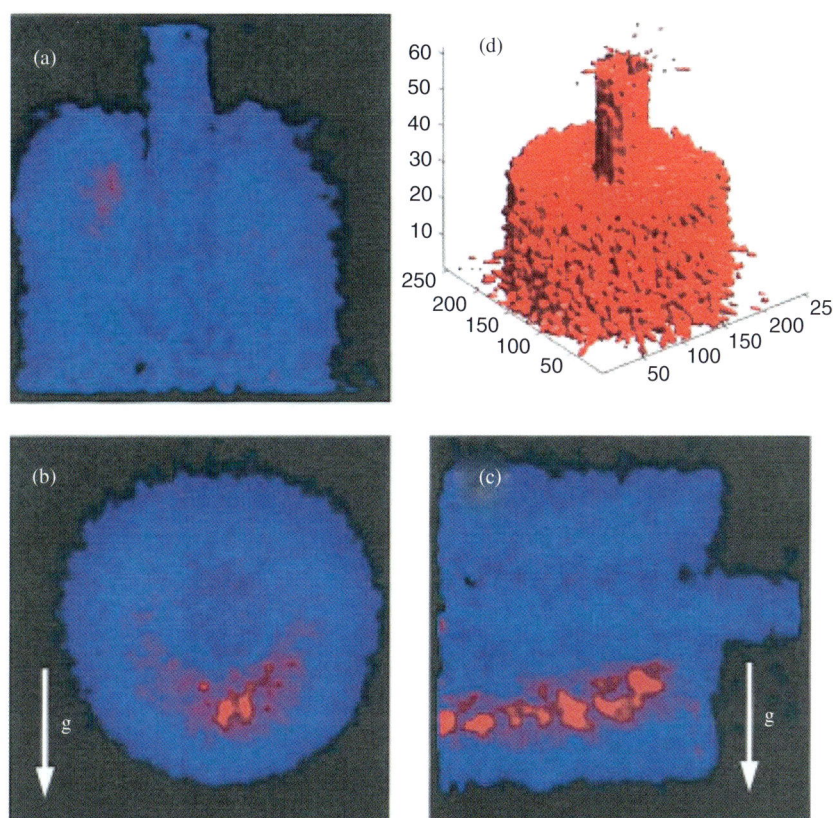

Plate 17 Flow of a fiber suspension through a sudden expansion with an upstream velocity of 0.7 m/s (Heath et al., 2007) (for Black and White version, see page 215).

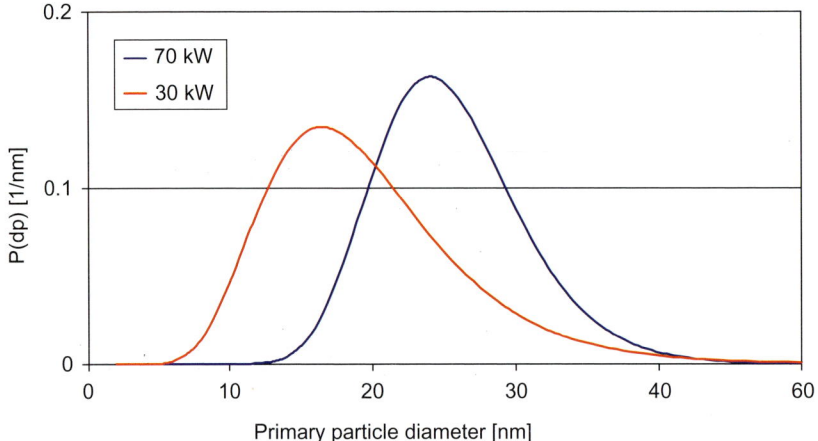

Plate 18 Primary particle size distributions for different plasma power. (Sommer et al., 2004) (for Black and White version, see page 241)